Topics in Geobiology

Volume 54

The Topics in Geobiology series covers the broad discipline of geobiology that is devoted to documenting life history of the Earth. A critical theme inherent in addressing this issue and one that is at the heart of the series is the interplay between the history of life and the changing environment. The series aims for high quality, scholarly volumes of original research as well as broad reviews.

Geobiology remains a vibrant as well as a rapidly advancing and dynamic field. Given this field's multidiscipline nature, it treats a broad spectrum of geologic, biologic, and geochemical themes all focused on documenting and understanding the fossil record and what it reveals about the evolutionary history of life. The Topics in Geobiology series was initiated to delve into how these numerous facets have influenced and controlled life on Earth.

Recent volumes have showcased specific taxonomic groups, major themes in the discipline, as well as approaches to improving our understanding of how life has evolved.

Taxonomic volumes focus on the biology and paleobiology of organisms – their ecology and mode of life – and, in addition, the fossil record – their phylogeny and evolutionary patterns – as well as their distribution in time and space.

Theme-based volumes, such as predator-prey relationships, biomineralization, paleobiogeography, and approaches to high-resolution stratigraphy, cover specific topics and how important elements are manifested in a wide range of organisms and how those dynamics have changed through the evolutionary history of life.

Comments or suggestions for future volumes are welcomed.

Neil H. Landman Department of Paleontology American Museum of Natural History New York, USA E-mail: landman@amnh.org

Peter J. Harries Department of Marine, Earth and Atmospheric Sciences North Carolina State University Raleigh, USA E-mail: pjharrie@ncsu.edu

Ashu Khosla • Omkar Verma • Sachin Kania
Spencer Lucas

Microbiota from the Late Cretaceous-Early Palaeocene Boundary Transition in the Deccan Intertrappean Beds of Central India

Systematics and Palaeoecological, Palaeoenvironmental and Palaeobiogeographical Implications

Springer

Ashu Khosla
Department of Geology
Panjab University
Chandigarh, India

Sachin Kania
Department of Geology
Panjab University
Chandigarh, India

Omkar Verma
Discipline of Geology, School of Sciences
Indira Gandhi National Open University
New Delhi, India

Spencer Lucas
New Mexico Museum of Natural History
and Science
Albuquerque, NM, USA

ISSN 0275-0120
Topics in Geobiology
ISBN 978-3-031-28854-8   ISBN 978-3-031-28855-5 (eBook)
https://doi.org/10.1007/978-3-031-28855-5

This Springer imprint is published by the registered company Springer Nature Switzerland AG
The registered company address is: Gewerbestrasse 11, 6330 Cham, Switzerland

Preface

Micropalaeontological studies of microbiota, including charophytes, ostracods, foraminiferans, fishes, and other organisms at the Cretaceous-Palaeogene boundary, have been of global importance during the last decade, and the present study from Central India is a new addition to this significant area of research at the Cretaceous-Palaeogene boundary. The microbiota of the intertrappean beds in the Chhindwara District of Madhya Pradesh, Central India, are described in a detailed manner for the first time in this book. This study focuses on the microbiota of the central Narmada River region during the Late Cretaceous-Early Palaeocene transition. Recently, the intertrappean beds of the Eastern Deccan Volcanic Province (one of the Deccan Volcanic Province's subprovinces) have received significant attention, resulting in the addition of some significant biotic assemblages to the province's existing record from the Chhindwara area. The recent discoveries and the known fossil record of the Late Cretaceous-Early Palaeocene biota of India led us to document precisely the palaeoecologic and palaeobiogeographic implications together with potential environmental changes of the contemporary biota caused by the lava flows of the Deccan Volcanic Province. In contrast to the Western Deccan Volcanic Province, recent biotic reports from the intertrappean beds of the Chhindwara region (Eastern Deccan Volcanic Province, Central India) strongly suggest that these beds have a huge potential in terms of fossil content. These reports also suggest that these beds contain new and distinct biotic remains. Intriguing implications can be drawn for defining the age limits of the basaltic flows from the record of diverse accumulations of freshwater charophyte, brackish to freshwater ostracod, planktic foraminiferal, and fish assemblages from the intertrappean beds of Jhilmili and adjacent areas of Early Danian (P1a) age (e.g., Keller et al. 2009a, b, c; Sharma and Khosla 2009; Khosla, 2015; Kania et al. 2022; Khosla et al. 2022). Concerns about the sedimentary environments of these intertrappean beds have also been raised by the presence of non-marine taxa – algae, molluscs and vertebrates – in the nearby Singpur and Mohgaon Kalan localities of the Chhindwara region, where brackish water ostracods are also found (Kar and Srinivasan 1998; Khosla and Nagori 2007; Keller et al. 2009a, b, c, 2010; Khosla 2015).

The new discoveries (presented in this book) shed light on various palaeobiogeographic models proposed for the northward drifting Indian plate and help us understand the palaeoecology and palaeoenvironment of the biota. The aim of this study is to provide comprehensive information on the microbiota found in the Chhindwara intertrappean beds along the central Narmada River region of Madhya Pradesh. Despite the fact that ostracods were first described more than 30 years ago, there has never been a thorough analysis or systematic research on the microbiota from the intertrappean beds from the Late Cretaceous to Early Palaeocene period (e.g., Sahni and Khosla 1994; Khosla and Sahni 2000; Whatley and Bajpai 2000a, b; Whatley and Bajpai 2002a, b, c; Whatley et al. 2003a, b; Sharma and Khosla 2009; Khosla 2015; Kania et al. 2022; Khosla et al. 2004, 2009, 2022). Regarding the taxonomic affinities of several Indian ostracods, there is a great deal of uncertainty. This is particularly true of species that were recently thought to be endemic to the Indian subcontinent (e.g., Whatley et al. 2003a, b; Whatley and Bajpai 2006; Whatley 2012; Khosla 2014, 2015; Kapur and Khosla 2019; Khosla and Lucas 2020a, b, c, d, e) but had previously been linked to Chinese and Mongolian taxa (e.g., Mohabey et al. 1993; Sahni and Khosla 1994; Khosla and Sahni 2000). To fill in the gaps, the authors of this book studied the microbiota from the intertrappean beds of the Chhindwara District, Madhya Pradesh. Research on charophytes and foraminiferans can assist in reconstructing biostratigraphy, palaeobiogeography, and global geological events across the Cretaceous-Palaeogene boundary (e.g., Martín-Closas and Serra-Kiel 1986; Keller et al. 2008, 2009a, b; Khosla 2015; Villalba-Breva et al. 2012; Chassagne-Manoukian et al. 2013; Vicente et al. 2015, 2016a, b, 2019; Li et al. 2016, 2019; Tian et al. 2021; Kania et al. 2022; Khosla et al. 2022).

In this book, a thorough assessment of the Deccan volcanic intertrappean beds of Central India, which have produced a wide range of biotic assemblages, is made. By describing the Cretaceous-Palaeogene boundary transition microbiota of Central India, this is the first book of its kind. Taxonomy, including some new species, and fundamental palaeoecological ideas are covered in this overview. The main microfossil groups, as well as their morphology and biostratigraphy, are discussed. Complex palaeoenvironmental and palaeoecological settings are indicated by the biota of charophytes, ostracods, foraminiferans, and fishes. Freshwater charophytes, ostracods and marine microfossils – in particular, brackish water ostracods and planktic foraminiferans – are discussed in greater detail in this coverage than the other groups because they are crucial to the assessment of biostratigraphy, palaeoecology and palaeobiogeography.

The book is organised as follows:

Chapter 1: A general introduction to the intertrappean beds containing microbiota is discussed, followed by outlining the objectives, rationale of the problem and importance of this study. The geographic location and area of study are situated in Chhindwara District in Madhya Pradesh. This chapter presents an analysis of palaeoecology, palaeoenvironments and palaeobiogeographical implications of microbiota-bearing localities and their systematic studies recorded in the Cretaceous-Palaeogene freshwater and brackish marine water succession from the Chhindwara District, Madhya Pradesh (Central India), integrated with biostratigraphical studies.

The study findings show that the sedimentary, palaeosol, floral and faunal context varied during the Late Cretaceous-Early Palaeogene, suggesting a notable change from an arid freshwater to a brackish marine regime that may have been caused by the K-Pg extinctions.

Chapter 2: This chapter provides a review of data collected across intertrappean beds concerning taxonomical records, as well as sedimentological and biostratigraphical analyses of Upper Cretaceous and Early Palaeocene intertrappean successions of peninsular India. The sedimentary basin of Chhindwara, which was deposited on a crystalline basement of Archaean rocks and Gondwanas, contains a predominately freshwater and shallow marine sequence dating from the Upper Cretaceous to the Early Palaeocene. The area is renowned for having one of the best developed Late Cretaceous (Maastrichtian)-Early Palaeocene sequences in the Indian subcontinent, as well as for having a rich charophyte flora, ostracods, planktic foraminiferans and fishes. Four localities in the Chhindwara region, namely Jhilmili, Ghat Parasia, Government well and Shriwas (=Shiraj) well, have produced microbiota from the intertrappean beds. The end-Cretaceous extinction event's effects on subaerial and shallow marine environments are evaluated by the authors critically. As a result, there may have been a less extensive mass extinction of terrestrial flora and fauna during the Upper Cretaceous-Early Palaeogene period than there was in the marine realm.

Chapter 3: This chapter provides detailed information on the geology and stratigraphy of microbiota-bearing intertrappean beds of the Chhindwara district, Central India. In the Chhindwara area, the sections contain the Archaeans (Precambrian age), Gondwanas (Permain to Early Cretaceous), intertrappean beds (Upper Cretaceous-Early Palaeocene) and Deccan traps (Upper Cretaceous-Early Palaeocene). The four stratigraphic sections from the Chhindwara region that were studied have produced a variety of microbiota – charophytes, ostracods, foraminiferans and fishes – that are taxonomically diverse. The intertrappean beds are sandwiched between the volcanic flows of the Deccan Volcanic Province.

Chapter 4: This chapter deals with a detailed description of the microbiotic assemblages, their current status and, lastly, the taxonomic description of charophytes, ostracods, foraminiferans and fishes. It is concerned with the first detailed description of charophytic studies of 10 species (including three new species, i.e., *Chara chhindwaraensis* sp. nov., *? Grambastichara*, *Microchara shivarudrappai* sp. nov., *Platychara* sp. and *Platychara closasi* sp. nov.) belonging to 6 genera, and 34 ostracod species including 3 new species – *Buntonia whittakerensis* sp. nov., *Limnocythere martensi* sp. nov. and *Zonocypris penchi* sp. nov. – 9 foraminiferan species and 2 fish taxa from the uppermost Cretaceous-Early Palaeocene intertrappean beds of Central India.

Chapter 5: This chapter provides a detailed account of palaeoenvironmental and palaeoecological implications of biota from the intertrappean beds in Chhindwara District to document Upper Cretaceous-Early Palaeocene environments of Central India. These intertrappean beds likely formed in a climate that was generally terrestrial, semi-humid to arid, followed by a shallow, alkaline, freshwater/lacustrine environment over a shallow marine environment during low tide intervals.

Biostratigraphically, the recovered taxa point to the studied Jhilmili intertrappean beds being Late Cretaceous to Early Palaeocene in age. The intertrappean beds at Ghat Parasia, Shiraj (=Shriwas) well and Government well can be dated to the Late Cretaceous, particularly to the Maastrichtian period.

Chapter 6: The fish, foraminifera and ostracod fauna and charophyte flora from the Chhindwara intertrappeans exhibit a complex pattern of endemic and Laurasian forms in their palaeobiogeographical affinities. The fossil Charophyta in the current collection only includes six genera, but it exhibits strong similarities to other Late Cretaceous charophytes from China, Mongolia, Europe, America and Africa. Because of this, the *Microchara* and *Platychara* from Chhindwara have affinities with other taxa in Laurasia despite appearing to be endemic to peninsular India. The charophyte assemblage from the Chhindwara area reflects the fact that the Indian subcontinent underwent exchanges of species from various parts of the world during the Cretaceous-Palaeogene time interval, particularly Laurasian species. As a result, associations between the charophytes of Jhilmili and those from Europe, North Africa and North-South America are found. These associations suggest that charophytes were dispersed among these continental masses along some migration routes. A troodontid dinosaur, some pollen, alligatorid crocodiles, anguimorph lizards, pelobatid and discoglossid frogs and other previously known biota from peninsular India also exhibit affinities of Laurasian origin (Khosla and Sahni 2003; Goswami et al. 2013; Khosla and Verma 2015; Verma 2015; Kapur and Khosla 2016, 2019; Verma et al. 2016, 2017; Khosla 2021). The Kohistan-Dras volcanic arcs are thought to have served as a land route for the Laurasian biota to disperse to India (i.e., a northern sweepstakes form of dispersion). Ostracods from the intertrappean beds of Chhindwara area are incredibly endemic. While some species, including *Eucypris* and *Cyclocypris*, left India during the Late Cretaceous and spread to other parts of the planet, others, including *Gomphocythere*, *Paracypretta* and *Cypridopsis*, are wholly endemic to India (Whatley and Bajpai 2006; Khosla 2015, 2021). By following Amirante Ridge, the Seychelles block and the Providence bank during the latest Cretaceous, *Buntonia* disseminated through shallow marine waters from Africa to India via Madagascar. This is shown by the presence of *Buntonia* in Maastrichtian deposits of Madagascar and K-Pg boundary-sediments of the Jhilmili section (India). *Buntonia's* presence in India also raises the possibility of a different dispersal pathway from Africa to India via the Oman-Kohistan-Dras Island Arc, possibly close to or at the K-Pg transition. When compared at the affinities level, the Chhindwara ostracod assemblage shows significant similarities to other intertrappean assemblages from Mamoni (District Kota), Rajasthan; Yanagundi and Chandarki (District Gulbarga), Karnataka; Phulsagar (District Mandla), Madhya Pradesh; and Anjar, Kora and Lakshmipur (District Kachchh), Gujarat. A planktic foraminiferal assemblage found at Jhilmili suggests that the Central Indian region was intermittently well connected to marine environments, and that marine incursions took place inland during high tides or eustatic sea level changes. The shallow marine dispersal of *Igdabatis* from Africa to India or the other way around would have been enabled by the existence of the Oman-Kohistan-Dras Island Arc near or at the Cretaceous-Palaeogene transition between India and Africa.

Lepisosteus's presence in India may be the result of a dispersal event that occurred from the south or north during the Late Cretaceous. The genus *Lepisosteus* may have attained a worldwide distribution during the Cretaceous.

Chandigarh, India — Ashu Khosla
Delhi, India — Omkar Verma
Chandigarh, India — Sachin Kania
Albuquerque, NM, USA — Spencer Lucas

References

Chassagne-Manoukian M, Haddoumi H, Cappetta H, Charriere A, Feist M, Tabuce R, Vianey-Liaud M (2013) Dating the "red beds" of Eastern Moroccan high plateaus, evidence from late cretaceous charophytes and dinosaur eggshells. Geobios 46(5):371–379

Goswami A, Prasad GVR, Verma O, Flynn JJ, Benson RBJ (2013) A troodontid dinosaur from the latest Cretaceous of India. Nat Commun 4:1–5. https://doi.org/10.1038/ncomms2716

Kania S, Khosla A, Verma O, Prashanth M (2022) Charophyte assemblage in the Cretaceous–Palaeogene boundary of the Deccan intertrappeans from Chhindwara District (Madhya Pradesh), Central India. Himal Geol 43(2):383–396

Kapur VV, Khosla A (2016) Late Cretaceous terrestrial biota from India with special reference to vertebrates and their implications for biogeographic connections. In: Khosla A, Lucas SG (eds) Cretaceous period: biotic diversity and biogeography, vol 71. New Mexico Museum of Natural History & Science, Albuquerque, pp 161–172

Kapur VV, Khosla A (2019) Faunal elements from the Deccan volcano-sedimentary sequences of India: a reappraisal of biostratigraphic, palaeoecologic, and palaeobiogeographic aspects. Geol J 54(5):2797–2828

Kar RK, Srinivasan S (1998) Late Cretaceous palynofossils from the Deccan intertrappean beds of Mohgaon-Kalan, Chhindwara District, Madhya Pradesh. Geophytology 27:7–22

Keller G, Adatte T, Gardin S, Bartolini A, Bajpai S (2008) Main Deccan volcanism phase ends near the K-T boundary: evidence from the Krishna-Godavari Basin, SE India. Earth Planet Sci Lett 268:293–311

Keller G, Khosla SC, Sharma R, Khosla A, Bajpai S, Adatte T (2009a) Early Danian planktic foraminifera from Cretaceous-Tertiary intertrappean beds at Jhilmili, Chhindwara District, Madhya Pradesh, India. J Foraminifer Res 39(1):40–55

Keller G, Adatte T, Bajpai S, Mohabey DM, Widdowson M, Khosla A, Sharma R, Khosla SC, Gertsch B, Fleitmann D, Sahni A (2009b) K-T transition in Deccan Traps of central India marks major marine seaway across India. Earth Planet Sci Lett282:10–23

Keller G, Sahni A, Bajpai S (2009c) Deccan volcanism, the KT mass extinction and dinosaurs. J Biosci 34:709–728

Keller G, Adatte T, Pardo A, Bajpai S, Khosla A, Samant B (2010) Cretaceous extinctions: evidence overlooked. Science 328(5981):974–975

Khosla A (2014) Upper Cretaceous (Maastrichtian) charophyte gyrogonites from the Lameta Formation of Jabalpur, Central India: palaeobiogeographic and palaeoecological implications. Acta Geol Pol 64(3):311–323

Khosla A (2015) Palaeoenvironmental, palaeoecological and palaeobiogeographical implications of mixed fresh water and brackish marine assemblages from the Cretaceous–Palaeogene Deccan intertrappean beds at Jhilmili, Chhindwara District, Central India. Rev Mex de Cien Geol 32(2):344–357

Khosla A (2021) Paleobiogeographical inferences of Indian Late Cretaceous vertebrates with special reference to dinosaurs. Hist Biol 33:1431–1442

Khosla A, Lucas SG (2020a) Introduction of Indian Late Cretaceous dinosaur eggs and eggshells of Peninsular India. In: Khosla A, Lucas SG (eds) Late Cretaceous dinosaur eggs and eggshells of Peninsular India: oospecies diversity and taphonomical, palaeoenvironmental, biostratigraphical and palaeobiogeographical inferences, Topics in Geobiology, vol 51. Springer Nature, Cham, pp 1–30

Khosla A, Lucas SG (2020b) Historical background of Late Cretaceous dinosaur studies and associated biota in India. In: Khosla A, Lucas SG (eds) Late Cretaceous dinosaur eggs and eggshells of Peninsular India: oospecies diversity and taphonomical, palaeoenvironmental, biostratigraphical and palaeobiogeographical inferences, Topics in Geobiology, vol 51. Springer Nature, Cham, pp 31–56

Khosla A, Lucas SG (2020c) Geology and stratigraphy of dinosaur eggs and eggshell-bearing infra and intertrappean Beds of Peninsular India. In: Khosla A, Lucas SG (eds) Late Cretaceous dinosaur eggs and eggshells of Peninsular India: oospecies diversity and taphonomical, palaeo environmental,biostratigraphical and palaeobiogeographical inferences, Topics in Geobiology, vol 51. Springer Nature, Cham, pp 57–115

Khosla A, Lucas SG (2020d) Indian Late Cretaceous dinosaur nesting sites and their systematic studies. In: Khosla A, Lucas SG (eds) Late Cretaceous dinosaur eggs and eggshells of Peninsular India: oospecies diversity and taphonomical, palaeoenvironmental, biostratigraphical and palaeobiogeographical inferences, Topics in Geobiology, vol 51. Springer Nature, Cham, pp 117–205

Khosla A, Lucas SG (2020e) Discussion: oospecies diversity, biomineralization aspects, taphonomical, biostratigraphical, palaeoenvironmental, palaeoecological and palaeobiogeographical inferences of the dinosaur-bearing Lameta Formation of Peninsular India. In: Khosla A, Lucas SG (eds) Late Cretaceous dinosaur eggs and eggshells of Peninsular India: Oospecies diversity and taphonomical, palaeoenvironmental, biostratigraphical and palaeobiogeographical inferences, Topics in Geobiology, vol 51. Springer Nature, Cham, pp 207–271

Khosla A, Prasad GVR, Verma O, Jain AK, Sahni A (2004) Discovery of a micromammal-yielding Deccan intertrappean site near Kisalpuri, Dindori District, Madhya Pradesh. Curr Sci 87(3):380–383

Khosla A, Sertich JJW, Prasad GVR, Verma O (2009) Dyrosaurid remains from the intertrappean beds of India and the Late Cretaceous distribution of Dyrosauridae. J Vertebr Paleontol 29(4):1321–1326

Khosla SC, Nagori ML (2007) Ostracoda from the Intertrappean beds of Mohgaon-Haveli, Chhindwara District, Madhya Pradesh. J Geol Soc India 69:209–221

Khosla A, Sahni A (2000) Late Cretaceous (Maastrichtian) ostracodes from the Lameta Formation, Jabalpur cantonment area, Madhya Pradesh, India. J Palaeontol Soc India 45:57–78

Khosla A, Sahni A (2003) Biodiversity during the Deccan volcanic eruptive episode. J Asia Earth Sci 21(8):895–908

Khosla A, Verma O (2015) Paleobiota from the Deccan volcano-sedimentary sequences of India: paleoenvironments, age and paleobiogeographic implications. Hist Biol 27(7):898–914

Khosla A, Kania S, Lucas SG, Verma O (2022) Charophytes from the Cretaceous–Palaeogene transition in the Jhilmili intertrappean beds of Central India. Geol J: 57:4412–4438 https://doi.org/10.1002/gj.4528

Li S, Wang Q, Zhang H, Lu H, Martín-Closas C (2016) Charophytes from the Cretaceous–Paleogene transition in the Pingyi Basin (Eastern China) and their Eurasian correlation. Cretac Res 59:179–200

Li S, Wang QF, Zhang HC, Wan XQ, Martín-Closas C (2019) Charophytes from the Cretaceous–Paleocene boundary in the Songliao Basin (north-eastern China): a Chinese biozonation and its calibration to the Geomagnetic Polarity Time Scale. Pap Palaeontol 5:47–81

Martin-Closas, Serra-Kiel J (1986) Two examples of evolution controlled by large scale abiotic processes: Eocene nummulitids of the south-Pyrenean basin and Cretaceous Charophyta of western Europe. In: Walliser O (Ed), Global Bioevents, pp. 375–380

Mohabey DM, Udhoji SG, Verma KK (1993) Palaeontological and sedimentological observations on non-marine Lameta Formation (Upper Cretaceous) of Maharashtra, India: their palaeontological and palaeoenvironmental significance. Palaeogeogr Palaeoclimatol Palaeoecol 105:83–94
Sahni A, Khosla A (1994) A Maastrichtian ostracode assemblage (Lameta Formation) from Jabalpur Cantonment, Madhya Pradesh, India. Curr Sci 67(6):456–460
Sharma R, Khosla A (2009) Early Palaeocene ostracoda from the Cretaceous-Tertiary (K-T) Deccan intertrappean sequence at Jhilmili, district Chhindwara. Central India J Palaeontol Soc India 54(2):197–208
Tian JX, Chen J, Lu NH, Chen C, Wang H, Li T, Li S, Wang QF, Zhang HC, Wan XQ (2021) Discovery of charophyte flora across the Cretaceous-Paleocene transition in the Jiaolai basin. Palaeoworld 30(3):538–550
Verma O (2015) Cretaceous vertebrate fauna of the Cauvery Basin, southern India: palaeodiversity and palaeobiogeographic implications. Palaeogeogr Palaeoclimatol Palaeoecol 431:53–67
Verma O, Khosla A, Goin FJ, Kaur J (2016) Historical biogeography of the Late Cretaceous vertebrates of India: Comparison of geophysical and paleontological data. In: Khosla A, Lucas SG (eds) Cretaceous period: Biotic diversity and biogeography, vol 71. New Mexico Museum of Natural History & Science, Albuquerque, pp 317–330
Verma O, Khosla A, Kaur J, Prasanth M (2017) Myliobatid and pycnodont fish from the Late Cretaceous of Central India and their paleobiogeographic implications. Hist Biol 29(2):253–265
Vicente A, Martín-Closas C, Arz JA, Oms O (2015) Maastrichtian-basal Paleocene charophyte biozonation and its calibration to the Global Polarity Time Scale in the southern Pyrenees (Catalonia, Spain). Cretac Res 52:268–285
Vicente A, Expósito M, Sanjuan J, Martín-Closas C (2016a) Small sized charophyte gyrogonites in the Maastrichtian of Coll de Nargó, Eastern Pyrenees: An adaptation to temporary floodplain ponds. Cretac Res 57:443–456
Vicente A, Villalba-Breva S, Ferràndez-Cañadell C, Martín-Closas C (2016b) Charophyte biostratigraphy in the Fontllonga type section. Geol Acta 14:349–362
Vicente A, Csiki-Sava Z, Martín-Closas C (2019) European charophyte evolution across the Cretaceous–Paleogene boundary. Palaeogeog Palaeoclimat Palaeoecol 533:109–244
Villalba-Breva S, Martín-Closas C, Marmi J, Gomez B, Fernandez-Marron MT (2012) Peat-forming plants in the Maastrichtian coals of the Eastern Pyrenees. Geol Acta 10:189–207
Whatley RC (2012) The 'Out of India' hypothesis: further supporting evidence from the extensive endemism of Maastrichtian non-marine Ostracoda from the Deccan volcanic region of peninsular India. Rev Paléobiol 11:229–248
Whatley RC, Bajpai S (2000a) A new fauna of Late Cretaceous non-marine Ostracoda from the Deccan intertrappean beds of Lakshmipur, Kachchh (Kutch District), Gujarat, western India. Rev Esp Micropaleontol 32(3):385–409
Whatley RC, Bajpai S (2000b) Further nonmarine Ostracoda from the Late Cretaceous intertrappean deposits of the Anjar region, Kachchh, Gujarat, India. Rev Micropaleontol 43(1):173–178
Whatley RC, Bajpai S (2006) Extensive endemism among the Maastrichtian nonmarine Ostracoda of India with implications for palaeobiogeography and "out of India" dispersal. Rev Esp de Micropaleontol 38(2–3):229–244
Whatley RC, Bajpai S, Srinivasan S (2002a) Upper Cretaceous nonmarine Ostracoda from intertrappean horizons in Gulbarga district, Karnataka state, South India. Rev Esp Micropaleontol 34(2):163–186
Whatley RC, Bajpai S, Srinivasan S (2002b) Upper Cretaceous intertrappean nonmarine Ostracoda from Mohgaonkala (Mohgaon-Kalan), Chhindwara District, Madhya Pradesh state, Central India. J Micropaleontol 21:105–114
Whatley RC, Bajpai S, Whittaker J (2003a) Freshwater ostracoda from the Upper Cretaceous intertrappean beds at Mamoni (Kota district), southeastern Rajasthan, India. Rev Esp de Micropaleontol 35(1):75–86
Whatley RC, Bajpai S, Whittaker JE (2003b) Indian intertrappean ostracoda in the collections of the Natural History Museum, London. Cretac Res 24(1):73–88

Acknowledgements

We are thankful to the Department of Geology, Panjab University, Chandigarh for providing necessary facilities to carry out the research work. The biotic study was carried out using a Leica microscope M 125C and M 205C at the Department of Geology, Micropalaeontology and PURSE Laboratory, Chandigarh, and School of Sciences, Indira Gandhi National Open University, New Delhi. We are thankful to the Central Instrumentation Laboratory of Panjab University, Chandigarh, Indian Institute of Technology, Roorkee, Inter-University Accelerator Centre, New Delhi and Wadia Institute of Himalayan Geology, India, for taking Scanning Electron Microscope photography. The unwavering support provided by Ashu Khosla's wife Prof. Amita, son Jayesh, and mother Prof. Rajinder Kaur Khosla during the composition of this work is also acknowledged. Ashu Khosla acknowledges financial support by the Department of Science and Technology, New Delhi (Grant No. SR/S4/ES–382/2008) and DST PURSE project (grant no. 753/Dean Research dated 29.09.2010 Panjab University, Chandigarh). Omkar Verma is also thankful to his beloved wife Nisha and two children (Vinayak and Shreya) for moral support and DST (Grant No. SR/FTP/ES–33/2008) for financial funding. Sachin Kania is grateful to family members for their immense support, patience and encouragement. Spencer G. Lucas is appreciative of the financial support from the New Mexico Museum of Natural History Foundation (USA).

Contents

About the Authors

Ashu Khosla received an M.Sc. in 1991 and a Ph.D. in Geology from Panjab University in January, 1997, and later undertook a postdoctorate from Montpellier University, France, in 1997–1998, on Indian and European dinosaur eggs and their palaeobiogeographical implications. Presently he is a Professor and Director/Head in the Department of Geology, Panjab University, Chandigarh. His research specialisations are in Palaeontology (Micropaleontology, Vertebrate Palaeontology, Palaeobotany, Palaeobiogeography, Palaeoecology, Palaeoenvironments, Stratigraphy and Biostratigraphy). He has teaching experience of more than 25 years at the undergraduate and post-graduate levels in the following areas: Palaeontology and Stratigraphy. His work has been acknowledged worldwide by palaeontologists and palaeobiogeographers, as it covers diverse issues such as evolution, diversity and biogeography of vertebrates and microbiota associated with the Cretaceous fragmentation and drift of the Indian plate. He has handled many international collaborative programmes and collaborated with many organisations, i.e., Laboratoire de Paléontologie, ISEM, cc 064, Université Montpellier II, Place Eugène Bataillon, 34095 Montpellier, France; Museum of Paleontology, The University of Michigan, Ann Arbor, USA; Geosciences, Princeton University, Princeton NJ 08544, USA; Geology Discipline Group, School of Sciences, Indira Gandhi National Open University, New Delhi; Department of Geology, Lucknow University, Lucknow; Department of Geological Sciences and Museum of Natural History, University of Colorado, UCB 265, Boulder, CO 80309, USA; Instituto de Investigaciones en Biodiversidad y Medioambiente (CONICET-INIBIOMA), Quintral 1250, 8400 San Carlos de Bariloche, Río Negro, Argentina and New Mexico Museum of Natural History and Science, Albuquerque, New Mexico, USA. He has to his credit several exciting fossil discoveries from the Late Cretaceous of India. Prof. Khosla's perseverance and commitment led to the first classification of Indian dinosaur eggs and their comparison with eggs from Europe and South America, the discovery of the Cretaceous-Palaeogene boundary (central India), discoveries of the biota from dinosaur coprolites and discoveries of Cenomanian-Turonian sauropod and Maastrichtian theropod dinosaur skeletal material, exotic mammals, ostracods, charophytes and planktic foraminiferans from Upper Cretaceous to Early Palaeocene rocks. He has published several research papers in peer-reviewed national/international journals, apart from a few in press, in high impact factor journals, including two papers in *Science* and other journals, for instance, *Earth and Planetary Science Letters*, *Global and Planetary Change*, *Palaeogeography, Palaeoclimatology, Palaeoecology*, *Journal of Asian Earth Sciences*, *Journal of Vertebrate Palaeontology*, *Geological Journal, Cretaceous Research*, *Historical Biology*, *Acta Geologica Polonica,* etc. He has already successfully completed six research projects funded by the Department of Science and Technology (Government of India), New Delhi. He has published two important books. First on the global Cretaceous (*Cretaceous Period: Biotic Diversity and Biogeography*). The volume was published in the *New Mexico Museum of Natural History and Science Bulletin* in 2016 and the second on the Indian Late Cretaceous dinosaur eggs of peninsular India in October, 2020 (Springer Nature, Switzerland).

Omkar Verma is an Assistant Professor of Geology at the Indira Gandhi National Open University, New Delhi, India. He received his M.Sc. and Ph.D. degrees in Geology from the University of Jammu, Jammu and Kashmir, India. He is the recipient of a Senior Research Fellowship and Research Associateship of the Council of Scientific and Industrial Research, New Delhi. He also completed a research project funded by the Department of Science and Technology, Govt. of India, New Delhi. He led several expeditions to the Cretaceous (145 to 66 million years ago) deposits of western, central, southern and southeastern India in search of small vertebrates that had lived in the shadow of dinosaurs. He has published more than 47 research articles in peer-reviewed national and international journals. He is the life member of the Geological Society of India, Indian Science Congress Association, Palaeontological Society of India, Indian Geological Congress and Gondwana Geological Society. His research focuses on palaeobiodiversity, biotic evolution, palaeobiogeography, palaeoecology and palaeoclimate with reference to the northward drift of the Indian plate.

Sachin Kania did his masters at Kurukshetra University, Haryana, India. Presently, he has submitted a Ph.D. in the Department of Geology, Panjab University, Chandigarh, India. His research interests include micropalaeontology, the Cretaceous-Palaeogene boundary, palaeoecology and palaeobiogeography.

Spencer G. Lucas is a stratigrapher and palaeontologist who has been Curator of Geology and Palaeontology at the New Mexico Museum of Natural History and Science (Albuquerque, New Mexico, USA) since 1988. He received his B.A. degree from the University of New Mexico (1976) and M.S. (1979) and Ph.D. (1984) degrees from Yale University. His research has focused on biostratigraphic problems of the late Palaeozoic, Mesozoic and early Cenozoic. He is a palaeontologist with a specialisation in the study of vertebrate fossils and continental deposits, particularly in New Mexico. He has extensive field experience in the western United States as well as northern Mexico, Costa Rica, Jamaica, Kazakstan, Nicaragua, Soviet Georgia and the People's Republic of China. Lucas has published more than

1000 scientific articles, co-edited 14 books and is the author of 3 dinosaur books. He has been on the editorial boards of *Ichnos*, *Geological Society of America Bulletin*, *New Mexico Geology*, *New Mexico Geological Society Guidebook*, *Journal of Palaeogeography* and *Revista Geologica de America Central*, among others. In 1991, he founded the *New Mexico Museum of Natural History and Science Bulletin* and is its Chief Editor. He has conducted extensive research on Cretaceous rocks and fossils in North America and Asia and has 27 years of museum experience and 5 years of teaching experience at the university level. Lucas has raised approximately $1,000,000,000 in grants and contracts. He has been awarded Clay Minerals Society, Pioneer in Clay Science Lecture Award in 2007; Elected Honorary Member, New Mexico Geological Society in 1994; 1989 *Coelophysis* Society Research Award, New Mexico Museum of Natural History and Science in 1989 and Philip M. Orville Prize for outstanding graduate student research in geology, Yale University in 1982. He is an honorary member of the New Mexico Geological Society and life member of the New Mexico Academy of Science.

Chapter 1
Introduction to Indian Late Cretaceous-Early Palaeocene Microbiota from the Deccan Intertrappean Beds of the Chhindwara District, Madhya Pradesh, India

1.1 Introduction

The Cretaceous Period is the longest and last period of the Mesozoic Era, spanning about 79 million years from 145 to 66 million years ago (Ma). It represents an important phase in Earth's history because numerous tectonic and biotic events occurred during this period (Khosla and Lucas 2016, 2021). It was during the Cretaceous that the already fragmented supercontinent, Pangaea, fully separated into the Laurasia and Gondwana continents and that were further separated into smaller continents (Krause et al. 2019; Langer et al. 2019; Khosla 2021). These continents moved in different directions under the influence of mantle convection currents and, thus, created large-scale geographic isolation, which led to the emergence of distinct biotic realms both in the Laurasian and the Gondwanan continents. Apart from this, flowering plants (angiosperms) first appeared in the Early Cretaceous, and the largest land ruling reptiles – dinosaurs – suddenly disappeared at the end of the Cretaceous. During the last three decades, diverse Cretaceous-Palaeocene biotic remains comprising charophytes, ostracods, foraminiferans, molluscs, fishes, amphibians, reptiles (including marine forms such as ichthyosaurs, plesiosaurs and mosasaurs), birds, small mammals and flora (including angiosperms) have been reported from the Cretaceous and Palaeocene deposits across the globe (Khosla and Lucas 2016). These reports provided many new useful insights for understanding the diversity, phylogeny, extinction patterns, palaeoecology, palaeoenvironment, palaeobiology and palaeobiogeography of many Cretaceous-Palaeocene ecosystems (e.g. Kielan-Jaworowska et al. 2004; Pereda-Suberbiola 2009; Csiki-Sava et al. 2015; Khosla and Lucas 2016, 2020a; Khosla et al. 2022).

India was once a part of the Gondwana continents, although it began to split from Africa during the Middle Jurassic (ca. 165–150 Ma), but major rifting, drifting and faunal dynamic events occurred on the landmass during Cretaceous-Palaeocene

A. Khosla et al., *Microbiota from the Late Cretaceous-Early Palaeocene Boundary Transition in the Deccan Intertrappean Beds of Central India*,
Topics in Geobiology 54, https://doi.org/10.1007/978-3-031-28855-5_1

time (Storey et al. 1995; Coffin and Rabinowitz 1987; Khosla and Verma 2015; Verma and Khosla 2018; Verma et al. 2016, 2022; Khosla 2021). Therefore, the Cretaceous-Palaeocene is considered as a special time for the Indian plate as it was wholly separated from Gondwana landmasses and undertook its northward journey across the Tethys Sea towards the Equator (Chatterjee et al. 2017; Verma et al. 2016; Khosla and Lucas 2020a). Furthermore, its northward movement was associated with the shifting latitude and climate belts over about 60° of latitude, from 30° south to 30° north, during the Cretaceous-Palaeocene interval before it finally docked with Asia in the Eocene (e.g. Barron et al. 1995; Kapur and Khosla 2016; Chatterjee et al. 2017; Khosla et al. 2022).

An important tectonic event associated with the Indian plate is the eruption of the Deccan Volcanic Province or Deccan Traps at the Cretaceous-Palaeogene boundary. While crossing over the Reunion hotspot at the close of the Cretaceous Period, the Indian plate witnessed extensive volcanic lava flows that gave rise to the Deccan Volcanic Province in peninsular India. It has been proposed that the enormous amounts of the Deccan volcanic eruptions resulted in the gassing out of tremendous amount of volcanic gases, particularly carbon dioxide and sulphur dioxide, into the atmosphere, which not only altered the global climate system by imposing global warming, acidic rains and increasing biotic stress but also resulted in the biotic mass extinctions at the Cretaceous-Palaeogene boundary (Courtillot et al. 1986, 1988; Jay and Widdowson 2008; Keller et al. 2011a, b, 2012; Renne et al. 2013; Khosla and Bajpai 2021).

Geophysical data and plate tectonic models show that after separating from Gondwana, the Indian plate remained as an isolated entity for more than 45 million years during the Cretaceous-Palaeogene. In addition to this, the Indian plate also experienced environmental stress at the terminal Cretaceous due to the volcanic eruptions. Assuming the northward drift of the Indian plate occurred during a long and extended period of geographic isolation as per the geophysical data, there were changing climatic belts through the drifting of the plate during the Cretaceous-Palaeocene. Subsequently, the Indian plate faced environmental changes triggered by the periodic eruptions of the Deccan basaltic lava flows, and these factors would have provided appropriate environmental conditions and time to produce an endemic biota, as is the case of the Cenozoic biota of some southern continents, particularly Africa, Australia and Madagascar (e.g. Rana and Wilson 2003; Whatley 2012; Khosla and Verma 2015; Verma et al. 2016; Boyer et al. 2010). Therefore, it has long been expected that the Cretaceous-Palaeocene deposits of India may contain a highly endemic biota, which would provide crucial information on various biotic aspects of palaeobiodiversity, palaeoecology, palaeoenvironment, endemism, dispersal or vicariance of the drifting Indian plate (Hocutt 1987; Whatley 2012; Khosla 2014, 2015; Khosla et al. 2015; Verma et al. 2016).

1.2 Rationale of the Problem

As currently understood, the global Late Cretaceous-Early Palaeocene fossil record is relatively well known from the northern landmasses, i.e. North America, Europe, China, Japan and Central Asia. As a consequence, a reliable picture of the faunal diversity and palaeoecology, palaeoenvironment and palaeobiogeography of major biotic elements such as ostracods, charophytes, fishes, reptiles, amphibians and mammals has emerged (e.g. Kielan-Jaworowska et al. 2004; Pereda-Suberbiola 2009; Csiki-Sava et al. 2015 and reference therein). However, the knowledge of the Late Cretaceous-Early Palaeocene biota of the southern continents is far from adequate because the fossil record is relatively poor. Interestingly, the recent significant biotic discoveries from the southern continents include a tribosphenic mammal, *Brasilestes stardusti* from the Late Cretaceous of Brazil (Castro et al. 2018); a bothremydid turtle, *Kinkonychelys rogersi*; a notosuchian crocodile, *Simosuchus clarki*; gondwanatherian mammals, *Vintana sertichi* and *Adalatherium hui* from the Late Cretaceous of Madagascar (Buckley et al. 2000; Gaffney et al. 2009; Krause et al. 2014, 2020) and *Galulatherium jenkinsi* from the Cretaceous of Tanzania (O'Connor et al. 2019); an australosphenid mammal, *Ausktribosphenos nyktos* from the Early Cretaceous of Australia (Rich et al. 1997); abelisaurid dinosaur, *Rugops primus* from the Aptian-Albian and Cenomanian of Africa (Sereno et al. 2004); and ostracods from the Cretaceous of South America (Piovesan et al. 2013). These have promulgated a new debate on the biogeographic origins and distribution patterns of these biotic elements.

Among the southern continents, the Cretaceous-Palaeocene fossil record of India is relatively poor (Khosla 2015; Khosla and Verma 2015; Kapur and Khosla 2019; Kania et al. 2022). It is important to note that India has numerous Cretaceous-Palaeocene outcrops, which occur both in the peninsular and Himalayan regions of India, but the fossils of charophytes, ostracods, foraminiferans and vertebrates are recorded from the Upper Cretaceous deposits and only a few from Lower Palaeocene sediments of the peninsular region of the country (e.g. Keller et al. 2008, 2009a; Khosla 2014, 2015; Khosla and Verma 2015; Verma, 2015; Verma et al. 2016; Kapur and Khosla 2019; Khosla and Lucas 2020a; Khosla 2021; Khosla et al. 2021, 2022). It is a noteworthy observation that the fossil record of the above-mentioned biotic elements in India is yet to be documented from most of the Lower Cretaceous sedimentary units. From the Upper Cretaceous and Lower Palaeocene sedimentary successions of India, biotic remains from the sediments associated with the Deccan Volcanic Province have received special attention (e.g. Khosla and Sahni 2003; Khosla 2014, 2015; Khosla and Verma 2015; Khosla and Lucas 2020a; Prasad et al. 2021; Khosla 2021; Kania et al. 2022; Khosla et al. 2022). Numerous studies have shown that the Deccan volcano-sedimentary sequences are very significant because they contain important biotic remains that can help us understand biodiversity patterns, including the impact of volcanic eruptions on the environment and its contemporary biota during the Deccan volcanic episode in the context of northward drift of the Indian plate (Khosla and Sahni 2003; Khosla and Verma 2015; Verma et al.

2016; Khosla and Lucas 2020a). In addition, these fossil remains are also very useful for evaluating the role of the Indian subcontinent as a geographic entity for the origin, evolution and dispersal of many biotic forms during its northward drift as an isolated landmass and the demarcation of the Cretaceous-Palaeogene boundary in peninsular India (Keller et al. 2008, 2009b; Khosla 2015; Khosla et al. 2022).

The Deccan Traps (also termed as Deccan Volcanic Province) of peninsular India comprise one of the world's most extensive mainland basalt fields, encompassing roughly 500,000 km^2 of western, central and southern India, as well as southernmost Pakistan (Wadia 1919; Verma and Khosla 2019; Khosla and Lucas 2020a). Recently, there has been a sudden resurgence of interest in various aspects of this volcanic field, including its age and duration. One of the primary goals of this expanded investigation has been the suggestion that Deccan volcanism may have played a significant role in the Late Cretaceous mass extinctions (e.g. Courtillot et al. 1986, 1988; Duncan and Pyle 1988; Chenet et al. 2007; Jay and Widdowson 2008; Sharma and Khosla 2009; Keller et al. 2009a, b, c, 2011a, b, 2012, 2020; Malarkodi et al. 2010; Renne et al. 2013; Fernández and Khosla 2015; Font et al. 2015; Khosla 2015, 2021; Khosla and Verma, 2015; Schoene et al. 2015; Fantasia et al. 2016; Kapur and Khosla 2016, 2019; Khosla et al. 2016; Verma et al. 2016, 2017; Kundal et al. 2018; Kapur et al. 2019; Verma and Khosla 2018, 2019; Kale et al. 2020a, b; Khosla and Lucas 2020a, b, c, d, e, 2021; Khosla et al. 2021; Khosla and Bajpai 2021; Kania et al. 2022; Khosla et al. 2022; Wilson et al. 2022). The Deccan Traps are made up of multiple layers of solidified flood basalt that are over 2000 m thick (Singh 1981; Courtillot et al. 1986, 1988; Tandon et al. 1995; Khosla and Sahni 2003).

Over the last three decades, our understanding of the Deccan Traps and infra- and intertrappean sedimentary beds associated with this volcanic activity has greatly improved (e.g. Courtillot et al. 1986, 1988; Srinivasan 1991; Chenet et al. 2007, 2008; Khosla and Nagori 2007a, b; Keller et al. 2008, 2009a, b, c, 2010a, b; Malarkodi et al. 2010; Gertsch et al. 2011; Keller et al. 2011a, b, 2012; Bajpai et al. 2013; Fernández and Khosla 2015; Khosla and Verma 2015; Fantasia et al. 2016; Kapur et al. 2019; Khosla 2021; Verma and Khosla 2019; Khosla and Lucas 2020a, b, c, d, e; Khosla and Bajpai 2021; Khosla et al. 2022). Different researchers have questioned the age and total duration of the Deccan volcanic activity.

Previously, the total duration of volcanic eruptions was estimated to be between 3 and 5 million years or even 7 and 8 million years (Sheth et al. 2001). However, it is now thought that the time span of Deccan volcanic eruptions was less than 1 million years within magnetic polarity C29R (Courtillot et al. 1986; Duncan and Pyle 1988). Thus, data from different studies such as palaeomagnetic, geochronologic, radiometric, biostratigraphic, chemostratigraphic and sedimentologic studies show that 90% of the 3500-m-thick Deccan volcanic lava pile erupted in less than 1 million years, during magnetic polarity C29R (Khosla and Lucas 2020a). And, it appears that each eruptive event lasted for a very short time, i.e. less than a decade (e.g. Chenet et al. 2008, 2009; Keller et al. 2008, 2009a, b; Keller et al. 2010a, b; Gertsch et al. 2011; Courtillot and Fluteau 2014; Font et al. 2015; Khosla 2015; Khosla and Verma 2015; Schoene et al. 2015; Kapur et al. 2019; Eddy et al. 2020;

Khosla and Lucas 2020a). Some workers, notably Keller et al. (2009a, b, 2011a, b, 2012), Khosla (2015) and Khosla and Lucas (2020a), have concluded that the Deccan volcanism is directly related to the Cretaceous-Palaeogene mass extinction, particularly of dinosaurs.

There have been three major Deccan volcanic events documented to date, with two of them occurring in the Krishna-Godavari basin (Keller et al. 2008). And, the Deccan Traps erupted in three stages in the Western Ghats, according to Chenet et al. (2007, 2008), Keller et al. (2011a, b) and Fantasia et al. (2016).

At 67.5 Ma, phase I began with a smaller eruption near the base of C30N (Late Maastrichtian). This was followed by a 2-year period of silence (Chenet et al. 2007; Khosla and Lucas 2020a; Punekar et al. 2014). The Deccan phase II accounted for 80% of volcanic activity. During this phase, the total volume of lava extruded was about 106 km^3 (Font et al. 2015) or perhaps more than 1.1 million km^3 of basalt (Fantasia et al. 2016). Various workers likewise recommended that the Cretaceous-Palaeogene mass extinctions happened during phase II and went on for an extremely brief time period, principally in planktic foraminiferal zones CF2-CF1, traversing the last 120–160 k.y. of the Late Maastrichtian palaeomagnetic chron C29R (Punekar et al. 2014; Khosla and Lucas 2020a). Punekar et al. (2014) proposed an immediate connection between phase II volcanic emissions and mass extinctions, asserting that 50% of planktic foraminiferans disappeared before the first mega flows, and more than 50% vanished after the first mega flows. The phase III (~64.5 Ma) started in the Early Danian (C29N) and included 14% of the volcanic movement and caused fewer extinctions (Chenet et al. 2007; Jay and Widdowson 2008; Keller et al. 2009b, c, 2011a, b; Khosla and Lucas 2020a, b).

A major portion of the Deccan Volcanic Province lies south of the Narmada River and, in the western, central and south eastern parts of peninsular India, is known as the Main or Western Deccan Volcanic Province (Vaidyanadhan and Ramakrishnan 2010). An isolated lava pile having a thickness of about 900 m, located on the eastern part of the Main Deccan Volcanic Province in Madhya Pradesh, covers areas around Chhindwara, Seoni and Jabalpur and is known as the Eastern Deccan Volcanic Province or Mandla Lobe (Shrivastava and Pattanayak 2002; Kania et al. 2022). The lava flows of these Deccan Volcanic Provinces are separated by intervening thin sedimentary layers and volcanic ash beds. It may be noted that beds of volcanic ash associated with the lava flows are a very common feature of the traps.

The sedimentary beds sandwiched between two successive lava flows are known as intertrappean beds, whereas the sedimentary beds that lie just below the first or the oldest lava flow are called infratrappean beds or Lameta Formation. The intertrappean beds are 1–6 m thick and are exposed at Mamoni (Kota District), Rajasthan; Dayapur, Kora, Anjar and Lakshmipur (Kachchh District), Gujarat; Chandarki, Gurmatkal and Yanagundi (Gulbarga District), Karnataka; Khandala, Aastha and Nagpur in Maharashtra; Rangapur and Naskal (Rangareddy District), Telangana; Barela, Ranipur and Padwar (Jabalpur District), Jhilmili; and Mohgaon Kalan (Chhindwara District) and Kisalpuri (Dindori District), Madhya Pradesh (Khosla and Verma 2015; Khosla and Lucas 2020a). Lithologically, marls, silty clays, claystones, mudstones, channel sandstones, siltstones, shales, limestones, conglomerates, calcretised palaeosols and silicified cherts make up the infra- and intertrappean beds (e.g. Khosla and Sahni 2003; Khosla and Verma 2015; Kapur et al. 2019; Khosla 2021; Khosla et al. 2021).

The infra- and intertrappean beds have been extensively studied over the last three decades, leading to a better understanding of their fossil biotic content (e.g. Khosla 1994, 2001; Khosla and Sahni 1995, 2000, 2003; Khosla et al. 2004, 2009; Whatley and Bajpai 2005, 2006; Prasad et al. 2007a, b, 2010; Sharma et al. 2008; Keller et al. 2009a, b, c; Bajpai et al. 2013; Khosla and Verma 2015; Fernández and Khosla 2015; Khosla et al. 2015, 2016; Verma et al. 2016, 2017; Kapur and Khosla 2016, 2019; Kapur et al. 2019; Khosla 2021; Khosla et al. 2021; Khosla and Bajpai 2021; Prasad et al. 2021; Kania et al. 2022; Khosla et al. 2022; Wilson et al. 2022) and the palaeobiogeographic relationships of the Indian subcontinent during its northward passage (Loyal et al. 1996, 1998; Prasad et al. 2007a, b, 2010; Verma et al. 2012, 2016, 2017; Kapur and Khosla 2016, 2019; Khosla et al. 2016; Chatterjee et al. 2017; Kapur et al. 2019; Khosla 2021; Khosla et al. 2021, 2022; Khosla and Bajpai, 2021; Prasad et al. 2021; Kania et al. 2022). These sedimentary units are exposed in many places along the southern, south-eastern, eastern, north-eastern and north-western margins of the Deccan Volcanic Provinces in peninsular India and yield abundant fossil biotic remains such as vertebrates, including fishes, anurans, crocodiles, turtles, dinosaurs, lizards, snakes and mammals, together with microfossils consisting of palynomorphs, charophytes, ostracods, foraminiferans, molluscs and macroplant fossils (e.g. Khosla and Sahni 2003; Whatley and Bajpai 2005, 2006; Whatley 2012; Khosla 2014, 2015, 2021; Khosla and Verma 2015; Kapur and Khosla 2016, 2019; Kapur et al. 2019; Verma and Khosla 2019; Khosla and Lucas 2020a, b, c, d, e; Khosla 2021; Kania et al. 2022; Khosla et al. 2022).

Now, the majority of intertrappean beds surrounding the main Deccan basaltic province have been equated with the classic chemostratigraphic units of the Western Ghats sections' Ambenali and Poladpur formations (Widdowson et al. 2000; Khosla 2015; Verma and Khosla 2019; Khosla and Lucas 2020a). These beds have yielded ostracod assemblages as well as diagnostic dinosaur remains, for example, eggshells (e.g. Sahni and Khosla 1994a, b; Whatley and Bajpai 2000a, b; Whatley et al. 2002a, b; Vianey-Liaud et al. 2003; Whatley and Bajpai 2005, 2006; Khosla et al. 2005; Khosla and Nagori 2007a, b; Fernández and Khosla 2015; Khosla 2017; Kapur et al. 2019; Khosla 2021) and a palynofloral assemblage consisting of *Aquilapollenites-Gabonisporites-Ariadnaesporites* (Kar and Srinivasan 1998; Thakre et al. 2017). The intertrappean beds of Rajahmundry in Andhra Pradesh have been assigned an Early Palaeocene age based on planktic and benthic foraminiferans, brackish water ostracods and pollen (Keller et al. 2008, 2009a, b, 2011a, b; Malarkodi et al. 2010; Kapur and Khosla 2019; Khosla and Lucas 2020a). The Oil and Natural Gas Commission's exploration of the intertrappean beds at Narsapur has also yielded the *Aquilapollenites* palynofloral assemblage (Kar et al. 1998; Kar and Srinivasan 1998) and planktic foraminiferal assemblages of Maastrichtian age (Govindan 1981; Keller et al. 2008, 2011a, b).

From the palaeobiogeographic point of view, numerous biogeographic models, comprising endemism, vicariance, dispersal (sweepstakes, filters and corridors),

"Noah's Ark," "Docked Noah's Ark," "Viking funeral ship" and "land spans," have been advanced for the northward drifting Indian plate in order to understand its Cretaceous-Palaeogene biogeography (Prasad and Sahni 1999, 2009; Khosla and Verma 2015; Verma et al. 2016, 2017; Khosla and Lucas 2020e). The palaeobiotic assemblages recovered from the sedimentary sequences associated with the Western Deccan Volcanic Province show an anomalous composition of the biota, consisting of both Gondwanan and Laurasian forms and some endemic, ancient lineages as well, which clearly indicate a complex Late Cretaceous-Early Palaeocene biogeography of India (e.g. Khosla et al. 2004, 2009; Prasad et al. 2007a, b, 2010; Goswami et al. 2011, 2013; Verma et al. 2016; Wilson et al. 2022). Moreover, the existence of both southern and northern biotic dispersals/connections has also been advanced in order to explain the presence of mixed biota on the plate, but it has also been stressed that more discoveries are required to enhance understanding of the Late Cretaceous biogeography of the Indian plate (Verma 2015; Khosla and Verma 2015; Verma et al. 2016; Kapur and Khosla 2016, 2019; Khosla 2021).

Recently, the intertrappean beds of the Eastern Deccan Volcanic Province have received considerable attention, which resulted in the addition of some significant biotic assemblages from the intertrappean beds exposed in the Dindori-Chhindwara area of the province. The biotic assemblages include foraminiferans, ostracods, fishes, frogs, lizards, turtles, crocodiles and mammals (e.g. Khosla et al. 2004, 2009; Keller et al. 2009a, b; Prasad et al. 2007a, b, 2010; Sharma and Khosla 2009; Verma et al. 2012, 2016; Rage et al. 2020). In spite of the recent discoveries, the known fossil record of the Late Cretaceous-Early Palaeocene biota of India is not sufficient and, hence, does not permit speculation on the possible impact of environmental changes triggered by the Deccan volcanic lava flows on the contemporary biota and to precisely document their palaeoecological, palaeoenvironmental and palaeobiogeographical implications.

The recent biotic reports from the intertrappean beds exposed in the Chhindwara region of the Eastern Deccan Volcanic Province clearly indicate that these beds have a vast potential in terms of fossil content, which could yield new and dissimilar biotic remains when compared to the Western Deccan Volcanic Province. The record of a planktic foraminiferal assemblage from the intertrappean beds of Jhilmili area of Early Danian (P1a) age (Keller et al. 2009a, b, c; Sharma and Khosla 2009) and lying just north of Chhindwara town and in the heart of peninsular India has intriguing implications for restricting the age limits of basaltic flows. The occurrence of non-marine taxa, for example, algae, molluscs and vertebrates in conjunction with brackish water ostracods in the nearby Singpur and Mohgaon-Kalan localities (Kar and Srinivasan 1998; Khosla and Nagori 2007a, b; Keller et al. 2009a, b, c, 2010a, 2011a, b; Samant and Mohabey 2009; Sharma and Khosla 2009; Khosla 2015), has also raised concerns about the sedimentary environments of these intertrappean beds. The new finds may prove useful for a better understanding of the palaeoecology and palaeoenvironment of the biota and may throw light on various biogeographic models proposed for the northward drifting Indian plate. The present research work was formulated against this background with the certain objectives.

1.3 Objectives

The analysis of the microbiotic assemblages of the intertrappean beds of the Eastern Deccan Volcanic Province at District Chhindwara, Madhya Pradesh, is documented in this book. The microbiotas of the central Narmada River region, such as charophytes, ostracods, planktic foraminiferans and fishes, have received special attention in this study.

Investigations were conducted with the following objectives:

1. To document the diversity of the Upper Cretaceous-Lower Palaeocene biota of the Chhindwara area
2. To reconstruct the palaeoecology and palaeoenvironments of the Eastern Deccan Volcanic Province
3. To infer the palaeobiogeographical implications of the fossil biota in the context of the drifting Indian plate

1.4 Location of the Study Area

To meet the above listed objectives, four stratigraphic sections of intertrappean beds associated with the Deccan Volcanic Province, Madhya Pradesh, Central India, were investigated (Fig. 1.1). These sections are Jhilmili (22° 02′ 44″:79° 09′ 22″), Ghat Parasia (22° 03′ 53.74″: 79° 02′ 45.57″), Government well (22° 01′ 06.61″: 79° 11′ 10.69″) and Shriwas (=Shiraj) well (22° 01′ 17.8″: 79° 11′ 06.3″), all exposed in the Chhindwara District, Madhya Pradesh, Central India (Fig. 1.2).

1.4.1 Jhilmili Section

The Jhilmili outcrop is exposed on a small hill and situated along the right bank of the Pench River near the village Jhilmili on the left side of the Chhindwara-Seoni road (Fig. 1.2). It is an approximately 14-m-thick succession that consists of claystones, palaeosols, siltstones, marlstones and limestones and has been divided into six lithological units, including two units of lava flows. Unit 1 represents the lowest weathered lava flows of the Deccan basalts. The overlying unit 2 is 6 m thick and consists of coarse-grained sandstone, purple siltstone, red-clayey siltstone and red claystone. Unit 3 is 60 cm thick and consists of yellow and pink claystone and yellow clayey limestone (Khosla et al. 2022).

Palaeontologically, unit 3 is very important because it has yielded diverse assemblages of freshwater and brackish water ostracods, charophytes, planktic foraminiferans and marine benthic calcareous chlorophytes, which together mark the Cretaceous-Palaeogene transition (Keller et al. 2009a, b; Khosla 2015; Kundal et al. 2018; Kania et al. 2022; Khosla et al. 2022). Freshwater, lacustrine to brackish

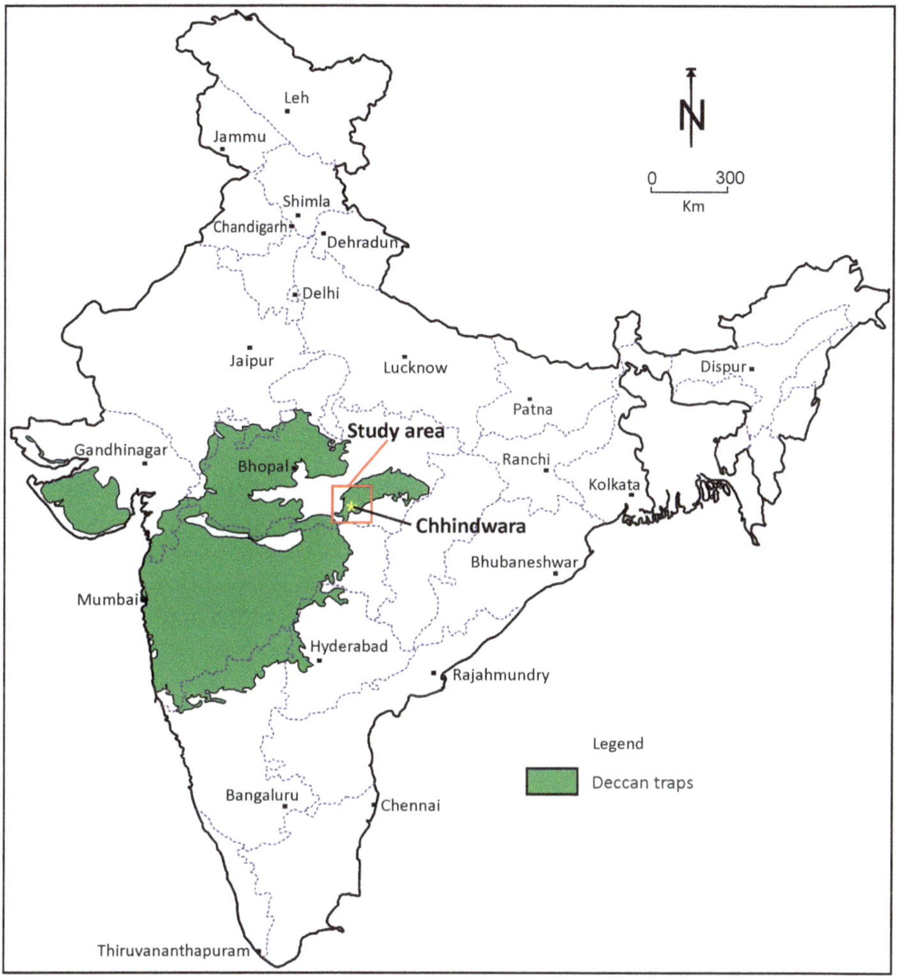

Fig. 1.1 Map of India showing the Deccan Volcanic Province. The study area is marked by a square (in red colour)

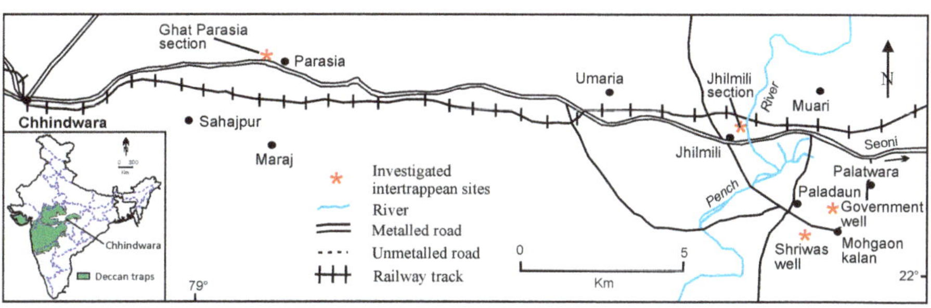

Fig. 1.2 Map showing investigated sites in the Chhindwara District, Madhya Pradesh. Inset is a map of India showing the Deccan Volcanic Province

marine environments have been inferred for this unit based on the presence of microfossils (e.g. Keller et al. 2009a, b; Khosla 2015; Kundal et al. 2018; Kania et al. 2022; Khosla et al. 2022). The charophytes, ostracods and foraminiferans described in the present study were recovered from this unit. Unit 4 consists of red clay and siltstone. Unit 5 encompasses green-grey siltstone with rare layers of fine sand. The topmost unit 6 consists of weathered flows of the Deccan basalt.

After processing and sorting of the samples at laboratory, a rich assemblage consisting of charophytes (*Platychara perlata* Peck and Reker 1947, *Platychara raoi* Bhatia and Mannikeri 1976, *Platychara sahnii* Bhatia and Mannikeri 1976, *Platychara compressa* Peck and Reker 1948, *Platychara* sp., *Peckichara* cf. *varians* Grambast 1957, *Nemegtichara* cf. *grambasti* Bhatia et al. 1990b, ?*Grambastichara* sp., *Microchara shivarudrappai* sp. nov. and *Chara chhindwaraensis* sp. nov.), ostracods (*Buntonia whittakerensis* sp. nov., *Neocyprideis raoi* Jain 1978, *Limnocythere deccanensis* Khosla et al. 2005, *Limnocythere martensi* sp. nov., *Frambocythere tumiensis anjarensis* Bhandari and Colin 1999, *Gomphocythere strangulata* Jones 1860, *Gomphocythere paucisulcatus* Whatley et al. 2002b, *Gomphocythere dasyderma* Whatley et al. 2002a, *Paracypretta subglobosa* Sowerby 1840, *Paracypretta jonesi* Bhatia and Rana 1984, *Paracypretta verruculosa* Whatley et al. 2002a, *Strandesia jhilmiliensis* Khosla et al. 2011a, *Stenocypris cylindrica* Sowerby in Malcolmson, 1840, *Periosocypris megistus* Whatley et al. 2012, *Zonocypris spirula* Whatley and Bajpai 2000a, *Zonocypris viriensis* Khosla and Nagori 2005, *Zonocypris penchi* sp. nov., *Cypridopsis astralos* Whatley et al. 2002a, *Cypridopsis hyperectyphos* Whatley and Bajpai 2000a, *Eucypris pelasgicos* Whatley and Bajpai 2000a, *Cyclocypris amphibolos* Whatley et al. 2002a, *Cypria cyrtonidion* Whatley and Bajpai 2000a and *Talicypridea pavnaensis* Khosla et al. 2005), foraminiferans (*Subbotina triloculinoides* Plummer 1926, *Globanomalina compressa* Plummer 1926, *Woodringina hornerstownensis* Olsson 1960, *Woodringina claytonensis* Loeblich and Tappan 1957b, *Hedbergella holmdelensis* Olsson 1964, *Guembelitria cretacea* Cushman 1933, *Parasubbotina pseudobulloides* Plummer 1926, *Globigerinelloides aspera* Ehrenberg 1854, *Globigerina* (*Eoglobigerina*) *pentagona* Morozova 1961) and fishes (*Lepisosteus indicus* Woodward, 1908) was recovered from unit 3 of the Jhilmili section.

1.4.2 Government Well

The government well is located approximately 0.4 km northeast of Mohgaon-Kalan village and approximately 34 km east of Chhindwara town in Madhya Pradesh, Central India (Fig. 1.2). This well section is about 100 cm thick and is located between two lava flows. In ascending order, it is composed of carbonaceous shale, black laminated carbonaceous shale and green clay with an intercalation of calcareous shale. Ostracods (*Frambocythere tumiensis lakshmiae* Whatley and Bajpai 2000a, *Gomphocythere strangulata* Jones 1860, *Gomphocythere paucisulcatus*

Whatley et al. 2002b, *Gomphocythere* sp. 1, *Cypridopsis elachistos* Whatley et al. 2002b, ?*Eucypris verruculosa* Whatley et al. 2002a and *Cyprois rostellum* Whatley and Bajpai 2000a), gastropods (*Lymnaea* sp.) and fishes (*Lepisosteus indicus* Woodward 1908 and Osteoglossidae genus and species indeterminate) were recovered from the carbonaceous shale.

1.4.3 Shriwas (=Shiraj) Well

The Shriwas well is situated around 0.5 km southwest of Mohgaon-Kalan village and nearly 34 km east of Chhindwara town, Madhya Pradesh, Central India (Fig. 1.2). The intertrappean beds of this section are around 1 m thick and sandwiched between two basaltic lava flows. Lithologically, the Shriwas well section consists of green shale, silicified green shale, lignitic shale and carbonaceous shale. Based on unidentified egg shells of avian dinosaurs together with ostracods and palynoflora, an Upper Maastrichtian age has been assigned to these intertrappeans (Srinivasan 1996; Kumaran et al. 1997; Kar and Srinivasan 1998; Thakre et al. 2017). Ostracods (*Frambocythere tumiensis lakshmiae* Whatley and Bajpai 2000a, *Gomphocythere strangulata* Jones 1860, *Gomphocythere paucisulcatus* Whatley et al. 2002b, *Zonocypris labyrinthicos* Whatley et al. 2002b, *Zonocypris gujaratensis* Bhandari and Colin 1999, *Cypridopsis elachistos* Whatley et al. 2002b and *Cyprois rostellum* Whatley and Bajpai 2000a) and fishes (*Igdabatis indicus* Prasad and Cappetta 1993 and *Lepisosteus indicus* Woodward, 1908) have been recovered from the lignitic shale, and ostracods (*Zonocypris labyrinthicos* Whatley et al. 2002b and *Zonocypris gujaratensis* Bhandari and Colin, 1999) were recovered from carbonaceous shale.

1.4.4 Ghat Parasia

The Ghat Parasia intertrappean site is located approximately 7 km east of Chhindwara and is exposed on both sides of the Chhindwara-Seoni road (Fig. 1.2). The thickness of the section is 170 cm. It consists of reddish chert, fossiliferous clayey limestone, greenish chertified clay, fossiliferous black chert and hard clayey limestone. Reddish chert, fossiliferous clayey limestone, greenish chertified clay, fossiliferous black chert and hard clayey limestone at this site have all yielded ostracods, gastropods, charophytes, fish teeth and scales.

The recovered biota is represented by *Platychara closasi* sp. nov. (charophyte), *Limnocythere deccanensis* Khosla et al. 2005, *Frambocythere tumiensis anjarensis* Bhandari and Colin 1999, *Gomphocythere paucisulcatus* Whatley et al. 2002b, *Periosocypris megistus* Whatley et al. 2012, *Candona* sp., *Eucypris* sp. 1, *Cyclocypris amphibolos* Whatley et al. 2002a, *Cyprois rostellum* Whatley and Bajpai 2000a,

Cyprois sp. and *Darwinula* sp. (ostracods), and *Igdabatis indicus* Prasad and Cappetta 1993, *Lepisosteus indicus* Woodward 1908 and Osteoglossidae genus and species indeterminate (fishes). The presence of the above fauna in the Ghat Parasia intertrappean beds supports a Late Cretaceous (Maastrichtian) age assignment.

1.5 Methodology

The following methodology and approach were employed in the present work:

1. As a first step towards achieving the objectives of the proposed research work, an extensive published and relevant literature survey was carried out.
2. Based on the literature survey, the Upper Cretaceous-Lower Palaeocene intertrappean beds of the Eastern Deccan Volcanic Province exposed in and around the Chhindwara District, Madhya Pradesh, Central India, were selected for field investigation. The micropalaeontological field investigations were carried out at four selected intertrappean sections namely, Jhilmili, Government well, Shriwas (=Shiraj) well and Ghat Parasia. These field investigations were conducted in October 2018, April 2019 and January 2020. For the systematic sampling of these intertrappean sites, about 1000 kg of samples were collected for the recovery of microfossils.
3. The lateral and vertical extent of the fossiliferous sites was delineated, and lithologs were constructed. Bulk sampling of sediments was made from the Jhilmili site in the field. The collected samples were brought into the laboratory for processing and study. In the laboratory, screen-washing techniques were used to recover microfossils from the collected sediments.
4. Mechanical and chemical techniques were used for the extraction of microfossils from hard and consolidated samples such as claystone, shale, siltstone, mudstone and limestone, which were brought to laboratory from various sites of the study area. The samples were disintegrated into smaller fragments and later kept in a bucket after being weighed. Subsequently, according to their weight, they were treated with 5% acetic acid, which allowed them to disaggregate into a slurry that was sieved using sieves of different mesh sizes such as 1.0, 0.80, 0.65, 0.35 and 0.2 mm. The residue so obtained was dried and sieved by using different sets of ASTMs.
5. The screen-washed residue obtained after applying the above-said microfossil recovery techniques was sorted under the binocular stereoscopic zoom microscope (Leica M 125 and M 205C). The specimens were measured with the help of a binocular microscope reticle.
6. The well-preserved specimens were selected, sputter-coated with gold and photographed using a JEOL JSM-25S SEM at the Department of Geology, Panjab University; JEOL JSM 6400 SEM at the Central Instrumentation Laboratory of Panjab University, Chandigarh; field emission scanning electron microscope (JEOL JSM-7610F) at the Inter-University Accelerator Centre, New Delhi; and

the Carl Zeiss EVO-40 EP SEM at the Wadia Institute of Himalayan Geology, Dehradun, India.

7. The biotic assemblage of the studied horizons is represented by microfossils such as charophytes, ostracods, foraminiferans and fishes.
8. Systematic identifications of the charophytes, ostracods, foraminiferans and fishes were carried out, and their continental and global affinities were evaluated.
9. The biostratigraphic analysis of the recovered biota from the intertrappean beds of the Eastern Deccan Volcanic Province compared with those known from the Western Deccan Volcanic Province was carried out to document age, palaeoecology and palaeoenvironment of the studied horizons.
10. Finally, the recovered fossil data comprising shells or tests, scales and tooth remains were integrated to obtain a comprehensive picture of the taxonomic position, diversity, palaeoecology, palaeoenvironment and continental as well as global affinities of the Late Cretaceous-Early Palaeocene biota of the study area.
11. A detailed description of charophytes, ostracods, foraminiferans and fishes recovered from the Upper Cretaceous-Lower Palaeocene deposits of Chhindwara (Madhya Pradesh), India is presented.

1.6 Significance

The following are the significant scientific outcomes of this research work:

1. The results of the present research work provided a better picture of the Late Cretaceous-Early Palaeocene taxonomic diversity of microfossils such as charophytes, ostracods, foraminiferans and fishes of the Eastern Deccan Volcanic Province.
2. It allowed a precise reconstruction of the palaeoecology and palaeoenvironment of the intertrappean beds of the Chhindwara region and a correlation with the intertrappean biota known from the Western Deccan Volcanic Province.
3. It provided a better picture of the palaeobiogeographic implications of the Upper Cretaceous-Lower Palaeocene biota of the Eastern Deccan Volcanic Province of Central India in the context of the northward drift of the Indian plate.

1.7 Repository of the Fossil Specimens

The microfossils, such as charophytes, ostracods, foraminiferans and fishes, are described in this book. The specimens so described are housed in the Micropalaeontological Laboratory and are stored with Sachin Kania of the Department of Geology, Panjab University, Chandigarh, India. Sachin Kania is responsible for collecting the material from the field, maceration and assigning the numbers to the microfossils. They are prefixed by the following catalogue abbreviations: MPL/SK/JML/, MPL/SK/GW/, MPL/SK/SW/, MPL/SK/GP/, MPL/SK/GP/CV/, MPL/SK/JML/CV/, and MPL/SK/GW/CV/, which are suffixed by the catalogue number. MPL, Micropalaeontological Laboratory; SK, Sachin Kania; JML, Jhilmili; GW, Government well; SW, Shriwas (=Shiraj) well; GP, Ghat Parasia; and CV, Chhindwara vertebrates.

1.8 Book Organisation

This book has six chapters in all. This **Chapter 1** presents a rationale behind the selection of the research problem and gives a concise background of microfossil studies pertaining to charophytes, ostracods, foraminiferans and fishes in peninsular India. It also discusses the objectives of the research, geology of the study area, methodology used to achieve the objectives of this study and repository of described specimens. **Chapter 2** covers the previous work done so far on charophytes, ostracods, foraminiferans and fishes in the Upper Cretaceous and Lower Palaeocene infra- and intertrappean beds of peninsular India. **Chapter 3** deals with general Late Cretaceous geology of India and geology of the studied intertrappean beds of the Chhindwara District, Madhya Pradesh.

Chapter 4 is devoted to the systematic description, comparison, biologic affinities and geological as well as geographical distribution of the recovered taxa. **Chapter 5** describes components of palaeoecology and palaeoenvironment. It gives a detailed account of the palaeoecological and palaeoenvironmental implications of the studied biota of charophytes, ostracods, foraminiferans and fishes. **Chapter 6** discusses the biogeographic significance of the described taxa in the context of the northward drift of the Indian plate close to or at the Cretaceous-Palaeogene boundary.

1.9 Conclusions

1. The Cretaceous-Palaeocene is a special time interval for the Indian plate as the plate was completely separated from Gondwana landmasses and undertook its northward journey across the Tethys Sea towards the Equator up to Asia. This movement was associated with the shifting latitude and climate belts by about 60° of latitude, from 30° south to 30° north, and enormous Deccan volcanic eruptions straddling the Cretaceous-Palaeogene boundary. The Deccan Volcanic Province is divided into four subprovinces, the main Deccan plateau, Malwa plateau, Mandla lobe and Saurashtra plateau. It contains numerous associated fossiliferous sedimentary sequences in the form of infratrappean and intertrappean beds. These beds are a rich storehouse of palaeobiodiversity of biota that lived prior to and during the Deccan volcanic activity.

2. The fossiliferous sedimentary sequences of the main Deccan plateau, Malwa plateau and Saurashtra plateau are well studied as compared with those of the Mandla lobe. However, during the last two decades, sedimentary sequences of the Mandla lobe, Central India, have yielded many significant fossil biotic remains, which are useful to interpreting biodiversity, evolution and biogeography and to demarcate the Cretaceous-Palaeogene boundary within the lava flows of the Mandla lobe.
3. The present research was carried out with the objective of improving the microfossil record of the Mandla lobe of the Deccan Volcanic Province for a better understanding of biodiversity and evaluating their palaeoecological, palaeoenvironmental and palaeobiogeographical significance. To achieve the desired objectives, detailed palaeontological field investigations were carried out in the intertrappean beds exposed in Jhilmili, Ghat Parasia, Government well and Shriwas (=Shiraj) well, Chhindwara District, Madhya Pradesh, Central India.

References

Bajpai S, Holmes J, Bennett C, Mandal N, Khosla A (2013) Palaeoenvironment of Northwestern India during the Late Cretaceous Deccan volcanic episode from trace-element and stable-isotope geochemistry of intertrappean ostracod shells. Glob Planet Chang 107:82–90

Barron EJ, Fawsett PJ, Peterson WH, Pollard D, Thompson SL (1995) "A simulation" of mid-Cretaceous climate. Paleoceanography 10:953–962

Bhandari A, Colin JP (1999) Limnic ostracods of the intertrappean sediments (terminal Maastrichtian-basal Paleocene) of the Anjar region (Kachchh, Gujarat state), India: systematics, paleoecology, and paleobiogeographic affinities. J Micropaleontol 42(1):3–20

Bhatia SB, Mannikeri MS (1976) Some Charophyta from the Deccan intertrappean beds near Nagpur, Central India. Geophytology 6(1):75–81

Bhatia SB, Rana RS (1984) Palaeogeographic implications of the Charophyta and Ostracoda of the intertrappean beds of peninsular India. Mém Soc Geol Fr 147:29–35

Bhatia SB, Srinivasan S, Bajpai S, Jolly A (1990b) Microfossils from the Deccan intertrappean bed at Mamoni, District Kota, Rajasthan: additional taxa and age implication. In: Sahni A, Jolly A (eds) Cretaceous event stratigraphy and the correlation of the Indian nonmarine strata. A Seminar cum Workshop IGCP 216 and 245, Chandigarh, pp 118–119

Boyer DM, Prasad GVR, Krause DW, Godinot M, Goswami A, Verma O, Flynn JJ (2010) New postcrania of Deccanolestes from the Late Cretaceous of India and their bearing on the evolutionary and biogeographic history of euarchontan mammals. Naturwissenschaften 94(7):365–377

Buckley GA, Brochu CA, Krause DW, Pol D (2000) A pug-nosed crocodyliform from the Late Cretaceous of Madagascar. Nature 405:941–944

Castro MC, Goin FJ, Ortiz-Jaureguizar E, Vieytes EC, Tsukui K, Ramezani J, Batezelli A, Marsola JC, Langer MC (2018) A Late Cretaceous mammal from Brazil and the first radioisotopic age for the Bauru Group. R Soc Open Sci 5(5):180482. https://doi.org/10.1098/rsos.180482

Chatterjee S, Scotese CR, Bajpai S (2017) The restless Indian plate and its epic voyage from Gondwana to Asia: its tectonic, paleoclimatic, and paleobiogeographic evolution. Geol Soc Am Spl Paper 529:1–147

Chenet AL, Quidelleur X, Fluteau F, Courtillot V (2007) $^{40}Ar/^{39}Ar$ dating of the main Deccan large igneous province: further evidence of KTB age and short duration. Earth Planet Sci Lett 263:1–15

Chenet AL, Fluteau F, Courtillot V, Gerard M, Subbarao KV (2008) Determination of rapid Deccan eruptions across the Cretaceous-Tertiary boundary using palaeomagnetic secular variation: results from a 1200 m thick section in the Mahabaleshwar. J Geophys Res 113:B04101. https://doi.org/10.1029/2006JB004635

Chenet AL, Courtillot V, Fluteau F, Gerard M, Quidelleur X, Khadri SFR, Subbarao KV, Thordarson T (2009) Determination of rapid Deccan eruptions across the Cretaceous–Tertiary boundary using paleomagnetic secular variation: 2. Constraints from analysis of eight new sections and synthesis for a 3500m thick composite section. J Geophys Res 114:B06103. https://doi.org/10.1029/2008JB005644

Coffin MF, Rabinowitz PD (1987) Reconstruction of Madagascar and Africa: evidence from the Davie Fracture Zone and western Somali Basin. J Geophys Res 92(B9):9385–9406

Courtillot V, Fluteau F (2014) A review of the embedded time scales of flood basalt volcanism with special emphasis on dramatically short magmatic pulses. In: Keller G, Kerr AC (eds) Volcanism, impacts, and mass extinctions: causes and effects, vol 505. Geological Society of America, Boulder, pp 301–317

Courtillot V, Besse J, Vandamme D, Jaeger JJ, Cappetta H (1986) Deccan flood basalts at the Cretaceous/Tertiary boundary. Earth Planet Sci Lett 80:361–374

Courtillot V, Feraud G, Maluski H, Vandamme D, Moreau MG, Besse J (1988) Deccan flood basalts and the Cretaceous-Tertiary boundary. Nature 333(6176):843–846

Csiki-Sava Z, Buffetaut E, Ősi A, Pereda-Suberbiola X, Brusatte SL (2015) Island life in the Cretaceous-faunal composition, biogeography, evolution, and extinction of land-living vertebrates on the Late Cretaceous European archipelago. ZooKeys 469:1–161

Cushman JA (1933) Some new foraminiferal genera. Contrib Cushman Lab Foram Res 8:4–7

Duncan RA, Pyle DG (1988) Rapid eruption of the Deccan flood basalts, Western India. Nature 333:841–843

Eddy MP, Schoene B, Samperton KM, Keller G, Adatte T, Khadri SFR (2020) U-Pb zircon age constraints on the earliest eruptions of the Deccan Large Igneous Province, Malwa Plateau, India. Earth Planet Sci Lett 540:116249. https://doi.org/10.1016/j.epsl.2020.116249

Ehrenberg CG (1854) Mikrogeologie. Das Erden und Felsen schaffende Wirken des unsicht barkleinense selbständige Lebens auf der Erde. Leopold Voss, Leipzig, pp xxviii + 374

Fantasia A, Adatte T, Spangenberg JE, Font E (2016) Palaeoenvironmental changes associated with Deccan volcanism, examples from terrestrial deposits of Central India. Palaeogeogr Palaeoclimatol Palaeoecol 441:165–180

Fernández MS, Khosla A (2015) Parataxonomic review of the Upper Cretaceous dinosaur eggshells belonging to the oofamily Megaloolithidae from India and Argentina. Hist Biol 27(2):158–180

Font E, Ponte J, Adatte T, Fantasia A, Florindo F, Abrajevitch A, Mirão J (2015) Tracing acidification induced by Deccan Phase 2 volcanism. Palaeogeogr Palaeoclimatol Palaeoecol 441:181–197

Gaffney ES, Krause DW, Zalmou IS (2009) *Kinkonychelys*, a new sidenecked turtle (Pelomedusoides: Bothremydidae) from the Late Cretaceous of Madagascar. Am Mus Novit 3662:1–25

Gertsch B, Keller G, Adatte T, Garg R, Prasad V, Berner Z, Fleitmann D (2011) Environmental effects of Deccan volcanism across the Cretaceous–Tertiary transition in Meghalaya, India. Earth Planet Sci Lett 310:272–285

Goswami A, Prasad GVR, Upchurch P, Boyer DM, Seiffert ER, Verma O, Gheerbrant E, Flynn JJ (2011) A radiation of arboreal basal eutherian mammals beginning in the Late Cretaceous of India. Proc Natl Acad Sci USA 108:16333–16338

Goswami A, Prasad GVR, Verma O, Flynn JJ, Benson RBJ (2013) A troodontid dinosaur from the latest Cretaceous of India. Nat Commun 4:1–5. https://doi.org/10.1038/ncomms2716

Govindan A (1981) Foraminifera from the infra- and intertrappean subsurface sediments of Narsapur Well-1 and age of the Deccan Trap flows. In: Khosla SC, Kachhara RP (eds) Proceedings of the IXth Indian Colloquium on Micropaleontology and Stratigraphy, Udaipur, pp 81–93

Grambast L (1957) Ornamentation de la gyrogonite et systematique chez les charophytes fossiles. Rev Gén Bot 64:339–362
Hocutt CH (1987) Evolution of the Indian Ocean and the drift of India: a vicariant event. Hydrobiologia 150:203–223
Jain SP (1978) Ostracodes des "inter-trappean beds" (Eocene Inferieur) de Kateru, Rajahmundry, Côte est de l' Inde. Rev Microplaeontol 21(2):51–58
Jay AE, Widdowson M (2008) Stratigraphy, structure and volcanology of the SE Deccan continental flood basalt province: implications for eruptive extent and volumes. J Geol Soc Lond 167:177–188
Jones T (1860) Notes on the fossil Cypride. Q J Geol Soc London 16:154–189
Kale V, Bodas M, Chatterjee P, Pande K (2020a) Emplacement history and evolution of the Deccan Volcanic Province, India. Episodes 43(1):278–299
Kale V, Dole G, Shandilya P, Pande K (2020b) Stratigraphy and correlations in Deccan Volcanic Province, India: quo vadis? GSA Bull 132(3/4):588–607
Kania S, Khosla A, Verma O, Prashanth M (2022) Charophyte assemblage in the Cretaceous–Palaeogene boundary of the Deccan intertrappeans from Chhindwara District (Madhya Pradesh), Central India. Himal Geol 43(2):383–396
Kapur VV, Khosla A (2016) Late Cretaceous terrestrial biota from India with special reference to vertebrates and their implications for biogeographic connections. In: Khosla A, Lucas SG (eds) Cretaceous period: biotic diversity and biogeography, vol 71. New Mexico Museum of Natural History & Science, Albuquerque, pp 161–172
Kapur VV, Khosla A (2019) Faunal elements from the Deccan volcano-sedimentary sequences of India: a reappraisal of biostratigraphic, palaeoecologic, and palaeobiogeographic aspects. Geol J 54(5):2797–2828
Kapur VV, Khosla A, Tiwari N (2019) Paleoenvironmental and paleobiogeographical implications of the microfossil assemblage from the Late Cretaceous intertrappean beds of the Manawar area, District Dhar, Madhya Pradesh, Central India. Hist Biol 31(9):1145–1160
Kar RK, Srinivasan S (1998) Late Cretaceous palynofossils from the Deccan intertrappean beds of Mohgaon-Kalan, Chhindwara District, Madhya Pradesh. Geophytology 27:7–22
Kar RK, Sahni A, Ambwani K, Singh RS (1998) Palynology of Indian onshore-offshore Maastrichtian sequence in India: implications for correlation and paleogeography. Indian J Petrol Geol 7:39–49
Keller G, Adatte T, Gardin S, Bartolini A, Bajpai S (2008) Main Deccan volcanism phase ends near the K-T boundary: evidence from the Krishna-Godavari Basin, SE India. Earth Planet Sci Lett 268:293–311
Keller G, Khosla SC, Sharma R, Khosla A, Bajpai S, Adatte T (2009a) Early Danian planktic foraminifera from Cretaceous-Tertiary intertrappean beds at Jhilmili, Chhindwara District, Madhya Pradesh, India. J Foram Res 39(1):40–55
Keller G, Adatte T, Bajpai S, Mohabey DM, Widdowson M, Khosla A, Sharma R, Khosla SC, Gertsch B, Fleitmann D, Sahni A (2009b) K-T transition in Deccan Traps of central India marks major marine seaway across India. Earth Planet Sci Lett 282:10–23
Keller G, Sahni A, Bajpai S (2009c) Deccan volcanism, the KT mass extinction and dinosaurs. J Biosci 34:709–728
Keller G, Adatte T, Pardo A, Bajpai S, Khosla A, Samant B (2010a) Cretaceous extinctions: evidence overlooked. Science 328(5981):974–975
Keller G, Adatte T, Pardo A, Bajpai S, Khosla A, Samant B (2010b) Comment on the 'Review' article by Schulte and 40 co-authors: the Chicxulub asteroid impact and mass extinction at the Cretaceous- Paleogene Boundary: Geoscientist Online. The Geological Society of London, London. Retrieved May 5, 2010, https://www.geolsoc.org.uk/gsl/geoscientist/features/keller/page7669.html
Keller G, Bhowmick PK, Upadhyay H, Dave A, Reddy AN, Jaiparkash BC, Adatte T (2011a) Deccan volcanism linked to the Cretaceous-Tertiary boundary mass extinction: new evidence from ONGC wells in the Krishna-Godavari Basin. J Geol Soc India 78:399–428

Keller G, Abramovich S, Adatte T, Berner Z (2011b) Biostratigraphy, age of the Chicxulub impact, and depositional environment of the Brazos River KTB sequences. In: Keller G, Adatte T (eds) The end-Cretaceous mass extinction and the Chicxulub impact in Texas, vol 100. Society for Sedimentary Geology (SEPM) Special Publication, pp 81–122

Keller G, Adatte T, Bhowmick PK, Upadhyay H, Dave A, Reddy AN, Jaiprakash BC (2012) Nature and timing of extinctions in Cretaceous–Tertiary planktic foraminifera preserved in Deccan intertrappean sediments of the Krishna–Godavari Basin, India. Earth Planet Sci Lett 341–344:211–221

Keller G, Mateo P, Monkenbusch J, Thibault N, Punekar J, Spangenberg JE, Abramovich S, Ashckenazi-Polivoda S, Schoene B, Eddy MP, Samperton KM, Khadri SFR, Adatte T (2020) Mercury linked to Deccan traps volcanism, climate change and the end-Cretaceous mass extinction. Glob Planet Chang 194:103312. https://doi.org/10.1016/j.gloplacha.2020.103312

Khosla A (1994) Petrographical studies of Late Cretaceous pedogenic calcretes of the Lameta Formation at Jabalpur and Bagh. Bull India Geol Assoc 27(2):117–128

Khosla A (2001) Diagenetic alterations of Late Cretaceous dinosaur eggshell fragments of India. Gaia 16:45–49

Khosla A (2014) Upper Cretaceous (Maastrichtian) charophyte gyrogonites from the Lameta Formation of Jabalpur, Central India: palaeobiogeographic and palaeoecological implications. Acta Geol Pol 64(3):311–323

Khosla A (2015) Palaeoenvironmental, palaeoecological and palaeobiogeographical implications of mixed fresh water and brackish marine assemblages from the Cretaceous–Palaeogene Deccan intertrappean beds at Jhilmili, Chhindwara District, Central India. Rev Mex de Cien Geol 32(2):344–357

Khosla A (2017) Evolution of dinosaurs with special reference to Indian Mesozoic ones. Wisd Her 8(1–2):281–292

Khosla A (2021) Paleobiogeographical inferences of Indian Late Cretaceous vertebrates with special reference to dinosaurs. Hist Biol 33:1431–1442

Khosla A, Bajpai S (2021) Dinosaur fossil records from India and their palaeobiogeographic implications: an overview. J Palaeosci 70:193–212

Khosla A, Lucas SG (2016) Cretaceous period: biotic diversity and biogeography – an Introduction. In: Khosla A, Lucas SG (eds) Cretaceous period: biotic diversity and biogeography, vol 71. New Mexico Museum of Natural History & Science, Albuquerque, pp 1–4

Khosla A, Lucas SG (2020a) Introduction of Indian Late Cretaceous dinosaur eggs and eggshells of Peninsular India. In: Khosla A, Lucas SG (eds) Late Cretaceous dinosaur eggs and eggshells of Peninsular India: oospecies diversity and taphonomical, palaeoenvironmental, biostratigraphical and palaeobiogeographical inferences, Topics in Geobiology, vol 51. Springer Nature, Cham, pp 1–30

Khosla A, Lucas SG (2020b) Historical background of Late Cretaceous dinosaur studies and associated biota in India. In: Khosla A, Lucas SG (eds) Late Cretaceous dinosaur eggs and eggshells of Peninsular India: oospecies diversity and taphonomical, palaeoenvironmental, biostratigraphical and palaeobiogeographical inferences, Topics in Geobiology, vol 51. Springer Nature, Cham, pp 31–56

Khosla A, Lucas SG (2020c) Geology and stratigraphy of dinosaur eggs and eggshell-bearing infra- and intertrappean Beds of Peninsular India. In: Khosla A, Lucas SG (eds) Late Cretaceous dinosaur eggs and eggshells of Peninsular India: oospecies diversity and taphonomical, palaeoenvironmental, biostratigraphical and palaeobiogeographical inferences, Topics in Geobiology, vol 51. Springer Nature, Cham, pp 57–115

Khosla A, Lucas SG (2020d) Indian Late Cretaceous dinosaur nesting sites and their systematic studies. In: Khosla A, Lucas SG (eds) Late Cretaceous dinosaur eggs and eggshells of Peninsular India: oospecies diversity and taphonomical, palaeoenvironmental, biostratigraphical and palaeobiogeographical inferences, Topics in Geobiology, vol 51. Springer Nature, Cham, pp 117–205

Khosla A, Lucas SG (2020e) Discussion: oospecies diversity, biomineralization aspects, taphonomical, biostratigraphical, palaeoenvironmental, palaeoecological and palaeobiogeographical inferences of the dinosaur-bearing Lameta Formation of Peninsular India. In: Khosla A, Lucas SG (eds) Late Cretaceous dinosaur eggs and eggshells of Peninsular India: Oospecies diversity and taphonomical, palaeoenvironmental, biostratigraphical and palaeobiogeographical inferences, Topics in Geobiology, vol 51. Springer Nature, Cham, pp 207–271

Khosla A, Lucas, SG (2021) End-Cretaceous Extinctions. In: Elias S, Alderton, D (eds) Encyclopedia of Geology, 2nd edition. Earth Systems and Environmental Sciences (Academic Press, Elsevier), pp. 665–678. https://doi.org/10.1016/B978-0-12-409548-9.12473-X

Khosla SC, Nagori ML (2005) A restudy of ostracode fauna from the intertrappean beds of the Anjar, Kachchh District, Gujarat. J Geol Soc India 66(5):573–580

Khosla SC, Nagori ML (2007a) Ostracoda from the Intertrappean beds of Mohgaon-Haveli, Chhindwara District, Madhya Pradesh. J Geol Soc India 69:209–221

Khosla SC, Nagori ML (2007) A revision of the Ostracoda from the intertrappean beds of Takli, Nagpur District, Maharashtra. J Paleontol Soc India 52:1–15

Khosla A, Sahni A (1995) Parataxonomic classification of Late Cretaceous dinosaur eggshells from India. J Palaeontol Soc India 40:87–102

Khosla A, Sahni A (2000) Late Cretaceous (Maastrichtian) ostracodes from the Lameta Formation, Jabalpur cantonment area, Madhya Pradesh, India. J Palaeontol Soc India 45:57–78

Khosla A, Sahni A (2003) Biodiversity during the Deccan volcanic eruptive episode. J Asia Earth Sci 21(8):895–908

Khosla A, Verma O (2015) Paleobiota from the Deccan volcano-sedimentary sequences of India: paleoenvironments, age and paleobiogeographic implications. Hist Biol 27(7):898–914

Khosla A, Prasad GVR, Verma O, Jain AK, Sahni A (2004) Discovery of a micromammal-yielding Deccan intertrappean site near Kisalpuri, Dindori District, Madhya Pradesh. Curr Sci 87(3):380–383

Khosla SC, Nagori ML, Mohabey DM (2005) Effect of Deccan volcanism on non-marine Late Cretaceous ostracode fauna: a case study from Lameta Formation of Dongargaon area (Nand-Dongargaon basin), Chandrapur District, Maharashtra. Gond Geol Magaz 8:133–146

Khosla A, Sertich JJW, Prasad GVR, Verma O (2009) Dyrosaurid remains from the intertrappean beds of India and the Late Cretaceous distribution of Dyrosauridae. J Vertebr Paleontol 29(4):1321–1326

Khosla SC, Nagori ML, Jakhar SR, Rathore AS (2011a) Early Danian lacustrine–brackish water Ostracoda from the Deccan Intertrappean beds near Jhilmili, Chhindwara District, Madhya Pradesh, India. Micropaleontology 57(3):223–245

Khosla A, Chin K, Alimohammadin H, Dutta D (2015) Ostracods, plant tissues, and other inclusions in coprolites from the Late Cretaceous Lameta Formation at Pisdura, India: taphonomical and palaeoecological implications. Palaeogeogr Palaeoclimatol Palaeoecol 418:90–100

Khosla A, Chin K, Verma O, Alimohammadin H, Dutta D (2016) Paleobiogeographical and paleoenvironmental implications of the freshwater Late Cretaceous ostracods, charophytes and distinctive residues from coprolites of Lameta Formation at Pisdura, Chandrapur District (Maharashtra), Central India. In: Khosla A, Lucas SG (eds) Cretaceous period: biotic diversity and biogeography, vol 71. New Mexico Museum of Natural History & Science, Albuquerque, pp 173–184

Khosla A, Fernández MS, Salgado L, Sahni A (2021) Comparative morphotaxonomical affinities, palaeoecological implications and cladistic analyses of ornithoid eggshells (oofamily Laevisoolithidae) from the Late Cretaceous intertrappean beds of Anjar, District Kachchh, Gujarat, India. Hist Biol 33(1):2829–2844

Khosla A, Kania S, Lucas SG, Verma O (2022) Charophytes from the Cretaceous–Palaeogene transition in the Jhilmili intertrappean beds of Central India. Geol J. 57(11):4412–4438. https://doi.org/10.1002/gj.4528

Kielan-Jaworowska Z, Cifelli RL, Luo Z-X (2004) Mammals from the age of dinosaurs: origins, evolution and structure. Columbia University Press, New York, pp 1–630

Krause DW, Hoffmann S, Wible JR, Kirk EC, Schultz JA, von Koenigswald W, Dumont ER et al (2014) First cranial remains of a gondwanatherian mammal reveal remarkable mosaicism. Nature 515(7528):512–517

Krause DW, Hoffmann S, Hu Y, Wible JR, Rougier GW, Kirk EC, Groenke JR, Rogers RR, Rossie JB, Schultz JA, Evans AR, von Koenigswald W, Rahantarisoa LJ (2020) Skeleton of a Cretaceous mammal from Madagascar reflects long-term insularity. Nature 581:421–427

Krause DW, Sertich JW, O'Connor MO, Rogrs KS, Rogers RR (2019) The Mesozoic biogeographic history of Gondwanan terrestrial vertebrates: insight's from Madagascar fossil record. Ann Rev Earth Planet Sci 47:519–553

Kumaran KPN, Bonde SD, Kanikar MD (1997) An *Aquilapollenites* associated palynoflora from Mohgaon-Kalan and its stratigraphic implications for age and stratigraphic correlation of Deccan intertrappean beds. Curr Sci 72:590–592

Kundal P, Humane S, Humane SK, Petkar SP (2018) Discovery of marine benthic chlorophycean algae in Early Danian Deccan intertrappean at Jhilmili, Central India: new insights into existence of marine seaway close to Cretaceous-Paleogene boundary. J Paleontol Soc India 63(2):203–211

Langer MC, Martins NO, Manzig PS, Ferreira GDS, Marsola JCA, Fortes E, Lima R, Lucas Sant'ana CF, Vidal LDS, Lorençato RHDS, et al. (2019) A new desert dwelling dinosaur (theropoda, Noasaurinae) from the Cretaceous of south Brazil. Sci Rep 9:9379

Loeblich AR, Tappan H (1957b) *Woodringina,* a new foraminiferal genus (Heterohelicidae) from the Paleocene of Alabama. J Wash Acad Sci 47:39–40

Loyal RS, Khosla A, Sahni A (1996) Gondwanan dinosaurs of India: affinities and palaeobiogeography. Mem Queens Mus 39(3):627–638

Loyal RS, Mohabey DM, Khosla A, Sahni A (1998) Status and palaeobiology of the Late Cretaceous Indian theropods with description of a new theropod eggshell oogenus and oospecies, *Ellipsoolithus khedaensis*, from the Lameta Formation, District Kheda, Gujarat, western India. Gaia 15:379–387

Malarkodi N, Keller G, Fayazudeen PJ, Mallikarjuna UB (2010) Foraminifera from the early Danian intertrappean beds in Rajahmundry quarries, Andhra Pradesh. J Geol Soc India 75:851–863

Malcolmson JG (1840) On the fossils of the Eastern portion of the Great Basaltic District of India. Trans Geol Soc London S2(5):537–575

Morozova VG (1961) Datsko-Montskie planktonnye foraminifery yuga SSSR (Danian-Montian Planktonic Foraminifera of the Southern USSR). Paleontol Zh 2:8–19

O'Connor PM, Krause DW, Stevens NJ, Groenke JR, MacPhee RDE, Kalthoff DC, Roberts EM (2019) A new mammal from the Turonian–Campanian (Upper Cretaceous) Galula Formation, southwestern Tanzania. Acta Palaeontol Pol 64(1):65–84

Olsson RK (1960) Foraminifera of latest Cretaceous and earliest Tertiary age in the New Jersey coastal plain. J Paleontol 34:1–58

Olsson RK (1964) Late Cretaceous planktonic foraminifera from New Jersey and Delaware. Micropaleontology 10:157–188

Peck RE, Reker CC (1947) Cretaceous and Lower Cenozoic charophyta from Peru. Am Mus Nat Hist Novit 1369:1–6

Peck RE, Reker CC (1948) Eocene Charophyta from North America. J Paleontol 22(1):85–90

Pereda-Suberbiola X (2009) Biogeographical affinities of Late Cretaceous continental tetrapods of Europe: a review. Bull Soc Géol Fr 180(1):57–71

Piovesan EK, Nicolaidis DD, Fauth G, Viviers MC (2013) Ostracodes from the Aptian–Santonian of the Santos, Campos and Espírito Santo basins, Brazil. J S Am Earth Sci 48:240–254

Plummer HJ (1926) Foraminifera of the midway formation in Texas. Univ Tex Bull 2644:1–206

Prasad GVR, Cappetta H (1993) Late Cretaceous selachians from India and the age of the Deccan Traps. Palaeontology 36(1):231–248

Prasad GVR, Sahni A (1999) Were there size constraints on biotic exchanges during the northward drift of the Indian plate? Proc Natl Acad Sci 65A:377–396

Prasad GVR, Sahni A (2009) Late Cretaceous continental vertebrate fossil record from India: Palaeobiogeographical insights. Bull de la Soc Géol de France 180:369–381

Prasad GVR, Verma O, Sahni A, Krause DW, Khosla A, Parmar V (2007a) A new Late Cretaceous Gondwanatherian mammal from Central India. Proc Indian Natl Sci Acad 73(1):17–24

Prasad GVR, Verma O, Sahni A, Parmar V, Khosla A (2007b) A Cretaceous hoofed mammal from India. Science 318:937

Prasad GVR, Verma O, Gheerbrant E, Goswami A, Khosla A, Parmar V, Sahni A (2010) First mammal evidence from the Late Cretaceous of India for biotic dispersal between India and Africa at the K/T transition. C R Palevol 9:63–71

Prasad GVR, Verma O, Sahni A, Khosla A (2021) Cretaceous mammals of India–stratigraphic distribution, diversity and intercontinental affinities. J Palaeosci 70:173–192

Punekar J, Keller G, Khozyem H, Hamming C, Adatte T, Tantawy AA, Spangenberg JE (2014) Late Maastrichtian–early Danian high–stress environments and delayed recovery linked to Deccan volcanism. Cretac Res 49:63–82

Rage JC, Prasad GVR, Verma O, Khosla A, Parmar V (2020) Anuran lissamphibian and squamate reptiles from the Upper Cretaceous (Maastrichtian) Deccan intertrappean sites in Central India, with a review of lissamphibian and squamate diversity in the northward drifting Indian plate. In: Prasad GVR, Patnaik R (eds) Biological consequences of plate tectonics, Vertebrate paleobiology and paleoanthropology. Springer, Cham, pp 99–121

Rana RS, Wilson GP (2003) New Late Cretaceous mammals from the intertrappean beds of Rangapur, India and paleobiogeographic framework. Acta Palaeontol Pol 48:331–348

Renne PR, Deino AL, Hilgen FJ, Kuiper KF, Mark D, Mitchell WS et al (2013) Time scales of critical events around the Cretaceous–Paleogene boundary. Science 339:684–687

Rich TH, Vickers-Rich P, Constantine A, Flannery TF, Kool L, van Klaveren N (1997) A tribosphenic mammal from the Mesozoic of Australia. Science 278(5342):1438–1442

Sahni A, Khosla A (1994a) A Maastrichtian ostracode assemblage (Lameta Formation) from Jabalpur Cantonment, Madhya Pradesh, India. Curr Sci 67:456–460

Sahni A, Khosla A (1994b) Palaeobiological, taphonomical and palaeoenvironmental aspects of Indian Cretaceous sauropod nesting sites. Gaia 10:215–223

Samant B, Mohabey DM (2009) Palynoflora from Deccan volcano-sedimentary sequence (Cretaceous–Palaeogene transition) of central India: implications for spatio-temporal correlation. J Biosci 34:811–823

Schoene B, Samperton KM, Eddy MP, Keller G, Adatte T, Bowring SA et al (2015) U–Pb geochronology of the Deccan Traps and relation to the end-Cretaceous mass extinction. Science 347:182–184

Sereno PC, Wilson JA, Conrad JL (2004) New dinosaurs link southern landmasses in the mid-Cretaceous. Proc R Soc B271:1325–1330

Sharma R, Khosla A (2009) Early Palaeocene Ostracoda from the Cretaceous-Tertiary (K-T) Deccan intertrappean sequence at Jhilmili, District Chhindwara, Central India. J Paleontol Soc India 54(2):197–208

Sharma R, Bajpai S, Singh MP (2008) Freshwater Ostracoda from the Paleocene-age Deccan intertrappean beds of Lalitpur (Uttar Pradesh), India. J Paleontol Soc India 53(2):81–87

Sheth HC, Pande K, Bhutani R (2001) 40Ar/39Ar age of a national geological monument: the Gilbert Hill basalt, Deccan Traps, Bombay. Curr Sci 80:1437–1440

Shrivastava JP, Pattanayak SK (2002) Basalts of the Eastern Deccan Volcanic Province, India. Gond Res 5:649–665

Singh IB (1981) Palaeoenvironment and palaeogeography of Lameta Group sediments (Late Cretaceous) in Jabalpur area, India. J Palaeontol Soc India 26:38–53

Sowerby JdC (1840) Explanations of the plates and wood-cuts. In: Malcolmson JC (ed) On the fossils of the eastern portion of the Great Basaltic District of India, vol 2, no 5. Transactions of the Geological Society, London, pp 511–567

Srinivasan S (1991) Geology and Micropalaeontology of Deccan Trap associated sediments of Northern Karnataka, Peninsular India. Unpublished PhD thesis, Panjab University, pp 1–175

Srinivasan S (1996) Late Cretaceous eggshells from the Deccan volcano-sedimentary sequences of central India. Mem Geol Soc India 37:321–336

Storey M, Mahoney JJ, Saunders AD, Duncan RA, Kelley SP, Coffin MF (1995) Timing of hot spot-related volcanism and the breakup of Madagascar and India. Science 267:852–855

Tandon SK, Sood A, Andrews JE, Dennis PF (1995) Palaeoenvironment of the dinosaur bearing Lameta Beds (Maastrichtian), Narmada Valley, Central India. Palaeogeogr Palaeoclimatol Palaeoecol 117:153–184

Thakre D, Samant B, Mohabey DM, Sangode S, Srivastava P, Kapgate DK, Mahajan R, Upreti N, Manchester SR (2017) A new insight into age and environments of intertrappean beds Mohgaon-Kalan, Chhindwara District, Madhya Pradesh using palynology, megaflora, magnetostratigraphy and clay mineralogy. Curr Sci 112(11):2193–2197

Vaidyanadhan R, Ramakrishnan M (2010) Geology of India, vol II. Geological Society of India, Bangalore

Verma O (2015) Cretaceous vertebrate fauna of the Cauvery Basin, southern India: palaeodiversity and palaeobiogeographic implications. Palaeogeogr Palaeoclimatol Palaeoecol 431:53–67

Verma O, Khosla A (2018) Application of internet technology in assembling literature for palaeontological research. Iran J Sci Technol Trans A Sci 4:1715–1723

Verma O, Khosla A (2019) Developments in the stratigraphy of the Deccan Volcanic Province, peninsular India. Compt Rendus Geosci 351:461–476

Verma O, Prasad GVR, Khosla A, Parmar V (2012) Late Cretaceous gondwanatherian mammals of India: distribution, interrelationships and biogeographic implications. J Paleontol Soc India 57:95–104

Verma O, Khosla A, Goin FJ, Kaur J (2016) Historical biogeography of the Late Cretaceous vertebrates of India: comparison of geophysical and paleontological data. In: Khosla A, Lucas SG (eds) Cretaceous Period: biotic diversity and biogeography, vol 71. New Mexico Museum of Natural History and Science, Albuquerque, pp 317–330

Verma O, Khosla A, Kaur J, Prasanth M (2017) Myliobatid and pycnodont fish from the Late Cretaceous of central India and their paleobiogeographic implications. Hist Biol 29(2):253–265

Verma O, Prashanth M, Greco R, Khosla A, Singh K (2022) Geological education scenario in India and role of open educational resources in the light of COVID-19 pandemic. Earth Sci Res J 26(2):239–254

Vianey-Liaud M, Khosla A, Geraldine G (2003) Relationships between European and Indian dinosaur eggs and eggshells of the oofamily Megaloolithidae. J Vertebr Paleontol 23(3):575–585

Wadia DN (1919) Geology of India. Tata McGraw-Hill Publishing Co., New Delhi, pp 1–508

Whatley RC (2012) The 'Out of India' hypothesis: further supporting evidence from the extensive endemism of Maastrichtian non-marine Ostracoda from the Deccan volcanic region of peninsular India. Rev Paléobiol 11:229–248

Whatley RC, Bajpai S (2000a) A new fauna of Late Cretaceous non-marine Ostracoda from the Deccan intertrappean beds of Lakshmipur, Kachchh (Kutch District), Gujarat, western India. Rev Esp Micropaleontol 32(3):385–409

Whatley RC, Bajpai S (2000b) Further nonmarine Ostracoda from the Late Cretaceous intertrappean deposits of the Anjar region, Kachchh, Gujarat, India. Rev Micropaleontol 43(1):173–178

Whatley RC, Bajpai S (2005) Some aspects of the paleoecology and distribution of non-marine Ostracoda from Upper Cretaceous intertrappean deposits and the Lameta Formation of peninsular India. J Paleontol Soc India 50(2):61–76

Whatley RC, Bajpai S (2006) Extensive endemism among the Maastrichtian nonmarine Ostracoda of India with implications for palaeobiogeography and "Out of India" dispersal. Rev Esp Micropaleontol 38(2–3):229–244

Whatley RC, Bajpai S, Srinivasan S (2002a) Upper Cretaceous nonmarine Ostracoda from intertrappean horizons in Gulbarga district, Karnataka state, South India. Rev Esp Micropaleontol 34(2):163–186

Whatley RC, Bajpai S, Srinivasan S (2002b) Upper Cretaceous intertrappean nonmarine Ostracoda from Mohgaonkala (Mohgaon-Kalan), Chhindwara District, Madhya Pradesh state, Central India. J Micropaleontol 21:105–114

Whatley RC, Khosla SC, Rathore AS (2012) *Periosocypris megistus* n. gen. and n. sp.: a new gigantic non-marine cyprid ostracod from the Maastrichtian Lameta Formation of India. J Palaeontol Soc India 57(2):113–117

Widdowson M, Pringle MS, Fernandez OA (2000) A post K–T boundary (Early Palaeocene) age for Deccan-type feeder dykes, Goa, India. J Pet 41(7):1177–1194

Wilson GP, Renne PR, Samant B, Mohabey DM, Dhobale A, Tholt AJ, Tobin TS, Widdowson M, Anantharaman S, Dassarma DS, Wilson JA (2022) New mammals from the Naskal intertrappean site and the age of India's earliest eutherians. Palaeogeogr Palaeoclimatol Palaeoecol 591(3):110857. https://doi.org/10.1016/j.palaeo.2022.110857

Woodward AS (1908) On some fish remains from the Lameta Beds at Dongargaon, Central Province. Mem Geol Sur India Palaeontol Indica NS 3:1–6

Chapter 2
Historical Background of Late Cretaceous-Early Palaeocene Microbiotic Assemblages from the Sediments Associated with Deccan Volcanic Province, peninsular India

2.1 Introduction

In peninsular India, Upper Cretaceous deposits are better preserved than Lower Palaeocene deposits (Naqvi 2005; Valdiya 2016). Indeed, Upper Cretaceous deposits – the Bagh and Lameta formations of the Narmada basin, Central India; the Kallamedu Formation of the Cauvery basin, South India; and the sedimentary sequences (i.e. infra- and intertrappean beds) associated with the Deccan Volcanic Province in peninsular India – have yielded a rich microfossil assemblage consisting of charophytes, ostracods, gastropods, foraminiferans and vertebrates (e.g. Khosla and Sahni 2003; Khosla et al. 2004, 2009, 2015, 2011a, b, 2022; Prasad et al. 2007a, b, 2010, 2013, 2015; Keller et al. 2009a, b; Prasad 2012; Whatley and Bajpai, 2002a, b; Whatley 2012; Verma et al. 2012, 2016, 2017; Bajpai et al. 2013; Khosla and Verma 2015; Verma 2015; Thakre et al. 2017; Verma and Khosla 2018, 2019; Kapur and Khosla 2019; Khosla and Lucas 2020a, b, c, d, e; Joyce and Bandypadhyay 2020; Rage et al. 2020; Khosla 2021; Verma et al. 2022). Based on foraminiferans, the Cretaceous-Palaeogene boundary was recently delineated for the first time in a sedimentary sequence associated with the Deccan Volcanic Province in Central India (Keller et al. 2009a, b; Sharma and Khosla 2009; Khosla 2015; Khosla and Lucas 2021; Kania et al. 2022; Khosla et al. 2022). Further, this discovery clearly indicates the presence of Palaeocene sediments in peninsular India. This chapter reviews previous work on the charophytes, ostracods, foraminiferans and fishes known from the rocks of Upper Cretaceous-Lower Palaeocene deposits of peninsular India.

A. Khosla et al., *Microbiota from the Late Cretaceous-Early Palaeocene Boundary Transition in the Deccan Intertrappean Beds of Central India*, Topics in Geobiology 54, https://doi.org/10.1007/978-3-031-28855-5_2

2.2 Charophytes

Charophytes are known from both the Lameta Formation (infratrappeans) and the intertrappean beds of peninsular India. Mohabey et al. (1993) documented the presence of two genera of charophytes, *Microchara* sp. and *Platychara perlata*, without describing and illustrating them, from the Lameta Formation of the Nand and Dongargaon areas (Chandrapur District), Maharashtra. Khosla et al. (2016) recently illustrated and described one charophyte taxon (*Microchara* sp.) recovered from coprolites in the Lameta Formation of Pisdura, Maharashtra. Shome and Chandel (2013) described four genera of charophytes from the Papro Formation (infratrappean), Lalitpur District, Uttar Pradesh, as well as gastropods, fishes and a few fragmentary and isolated amphibian vertebrae. The charophytes: *Stephanochara*, *Chara*, *Nemegtichara* and *Microchara* were discovered in 1- to 2-m-thick fossiliferous chert. The Papro Formation has been linked to the Upper Cretaceous infratrappean beds of Pisdura, Maharashtra. Shome and Chandel (2013) hypothesised that the infratrappean beds formed in freshwater to lacustrine environments.

Subsequently, Khosla (2014) reported three charophyte species – *Platychara sahnii*, *Nemegtichara grambasti* and *Microchara* sp. – from two localities, Bara Simla and the Chui hills, in the Lameta Formation at Jabalpur, Madhya Pradesh. These charophytes were found in the Bara Simla hill's green marl unit and the Chui hill's siltstone unit. Khosla (2014) conducted a detailed systematic study of charophytes (gyrogonites) from the Lameta Formation for the first time. The recovered charophytes were found to be associated with infratrappean assemblages from the Nand, Dongargaon and Dhammi-Pavna sections of Maharashtra, as well as some intertrappean beds of Kora (Gujarat), Rangapur (Telangana) and Gurmatkal (Karnataka). Khosla (2014) also compared *Platychara*, *Microchara* and *Nemegtichara* to those found in Upper Cretaceous deposits in Europe, Asia, America, Africa, China and Mongolia. The charophyte assemblage Khosla documented indicates that the Lameta sediments were deposited on a floodplain in lacustrine and palustrine conditions.

In the intertrappean beds, Bhatia and Mannikeri (1976) reported charophytes from the intertrappean beds near Nagpur (Maharashtra). They described *Platychara perlata*, *P. raoi*, *P. sahnii*, *Peckichara varians*, *Microchara vestita* and *M. sausari*. Subsequently, Bhatia and Rana (1984) studied the same species and reconstructed their palaeobiogeography. Bhatia (1992) presented a detailed account of the palaeobiogeography of Late Cretaceous charophytes of India. In the meantime, Chanda et al. (1989) recovered charophytes from the intertrappean beds exposed at Kateru area, Krishna-Godavari basin, Andhra Pradesh. They reported *Platychara perlata*, *P. compressa*, *P. rajahmundrica*, *P. sahnii*, *Peckichara varians*, *Peckisphaera clavata*, *Microchara vestita*, *M. tunicta*, *Microchara* sp. and *Harrisichara leptocera*.

Bhatia et al. (1990a) investigated the charophytes from the Mamoni intertrappean beds of Rajasthan. These researchers described *Nemegtichara grambasti* and *Pseudoharrisichara baytikshanensis* from these intertrappean beds. Following that, Bhatia et al. (1990b) reported two species of charophytes, *Platychara perlata* and

Nemegtichara grambasti, from the mammal-rich Rangapur intertrappean beds near Hyderabad, Telangana. Srinivasan et al. (1992, 1994) recovered a diverse charophyte assemblage, including eight genera and ten species, from the Gurmatkal intertrappean beds, Karnataka and Kora (District Kachchh), Gujarat. These species were *Harrisichara muricata*, *Stephanochara* cf. *S. levis*, *Platychara perlata*, *P. compressa*, *P. sahnii*, *Peckichara varians*, *Nemegtichara grambasti*, *Grambastichara* sp., *Microchara* sp. and *Chara* sp. The majority of these taxa were recovered for the first time from India and found in close association with ostracods, favouring their placement close to the Cretaceous-Palaeogene transition. Recently, Kapur et al. (2019) described *Platychara* cf. *perlata* from intertrappean beds exposed near Manawar (Dhar District), Madhya Pradesh. These workers suggested a freshwater palustrine/lacustrine depositional environment for the Manawar intertrappean beds.

2.3 Ostracods

Ostracods are a group of invertebrates that belong to the Arthropoda class Crustacea. They span a wide geological range, from Cambrian to recent. They are tiny organisms with shells composed of two valves or carapaces. Their dimensions range from 0.1 to 32 mm (Mahalakshmi and Hussain 2015). Because of their diversity, small size, mode of preservation and widespread distribution, they are an effective tool for the bio-zonation of marine and non-marine environments. They are also useful in the reconstruction of palaeoenvironment, palaeoecology and palaeobiogeography (e.g. Bhatia 1984; Seeling et al. 2004; Whatley and Bajpai 2006; Puckett 2012; Andreu et al. 2012; Whatley 2012; Khosla et al. 2013; Khosla and Verma 2015; Khosla 2015; Rathore et al. 2017; Kapur et al. 2019). Ostracods have been well studied from various Cretaceous rocks of India by many workers (e.g. Bhaita 1984; Mathur and Verma 1988; Bhatia and Rana 1984; Singh and Powal 1989; Bhatia et al. 1990a, b, 1996; Mohabey and Udhoji 1990; Srinivasan 1991; Mohabey et al. 1993; Mallikarjuna and Nagaraja 1996; Mohabey 1996; Sahni and Khosla 1994; Udhoji and Mohabey 1996; Singh 1997; Bhandari and Colin 1999; Khosla and Sahni 2000; Whatley and Bajpai 2000a, b, 2005; Bajpai and Whatley 2001; Whatley et al. 2002a, b, 2003a, b, 2012; Khosla and Nagori 2007a, b; Andreu et al. 2007; Sharma et al. 2008; Khosla et al. 2009, 2010; Sharma and Khosla 2009; Khosla et al. 2011a, b, 2013, 2015, 2016; Bajpai et al. 2013; Khosla 2015; Rathore et al. 2017; Rathore 2018; Chaudhary et al. 2017, 2019; Chaudhary and Nagori 2019; Kapur et al. 2019; Kshetrimayum et al. 2021).

The Bagh Formation is exposed in Central India along the Narmada valley for about 350 km, from Barwaha (Madhya Pradesh) in the east to Rajpipla (Gujarat) in the west. This formation was named after the Bagh caves in Madhya Pradesh's Dhar District. In ascending order, the Bagh Formation is divided into distinct units, the Nimar Sandstone, Nodular Limestone and Coralline Limestone (Khosla et al. 2003; Naqvi 2005; Valdiya 2016). Ammonites, brachiopods, foraminiferans, lamellibranchs, ostracods and other biotic remains are abundant in these beds. These beds have also yielded sharks, dinosaur bones, eggs and eggshells (Khosla et al. 2003; Verma 2015; Khosla and Lucas 2020b). The biotic assemblages as a whole suggest a terrestrial to marine environment (epi-continental sea) and a Cenomanian to Coniacian age for the Bagh Formation (Naqvi 2005; Valdiya 2016).

Jain (1961) conducted a pioneering work on ostracods in the Bagh Formation, reporting an assemblage of 22 species from the Coralline Limestone unit exposed at the Thuati village, Dhar District, Madhya Pradesh. Following this, Roy-Chowdhury and Sastri (1962) identified four taxa in the Deola-Chirakhan marl of the Nodular Limestone of the Man River Section in the lower Narmada valley. Guha and Ghosh (1970) discovered 17 ostracod species of the genus *Cytherelloidea* in the Deola-Chirakhan marl unit of the Bagh Formation. Though the ostracods were reported, the authors did not provide any micrographs or sketches of them. Later, Jain (1975) recovered 32 taxa from the Coralline Limestone exposed at the Thuati village, Dhar District, Madhya Pradesh. However, he described only five species: *Cytherelloidea khoslai*, *C. oudiapurensis*, *C. raoi*, *C. subgranulosa* and *C. thuatiensis*.

Chaudhary et al. (2017) described eight new ostracod taxa, including *Cytherelloidea awaldaensis* and *C. rosebaidaensis*. The other six are *Cytherelloidea oertelii*, *C. oudiapurensis*, *C. raoi*, *C. subgranulosa*, *C. thuatiensis* and *Cytherelloidea* sp., which were found in the creamy-marl and Coralline Limestone exposed in the Nala and river sections of Rosebadia near the Ratitalai village, Jeerabad (near Hanumanpura and Awalda villages), Dhar District, Madhya Pradesh. According to these authors, the genus *Cytherelloidea* is a cosmopolitan ostracod genus found worldwide in the Cretaceous shallow marine periphery basins of the eastern side of the Gondwanan continents such as South America, Africa, the Middle East, Madagascar, Australia and India. The genus *Cytherelloidea* was discovered in the Neocomian of South Africa and later in the Algoa basin of South Africa (Dingle 1969; Brenne and Oerti 1976); Coniacian-Santonian deposits of the Essaouria basin (Atlantic Atlas, Morocco); Albian-Turonian sediments of the Antsiranana region, northern Madagascar (Babinot et al. 2009); and the Upper Albian of Brazil (Piovesan et al. 2013). *Cytherelloidea* has been found in the Jaisalmer basin, Rajasthan (Singh 1997; Andreu et al. 2007), as well as in Upper Cretaceous rocks of the Ariyalur basin (Sastry et al. 1972).

Chaudhary and Nagori (2019) described six taxa of the genus *Rostrocytherdia*, including two new forms, *R. decurtata* and *R. divergens*, as well as two previously reported forms, *R. baghensis* and *R. jaisalmerensis*, from Rajasthan's Jaisalmer basin. *R. baghensis*, for example, has a close relationship with *R. cerasmoderma* from Argentina (Ballent and Whatley 2007). They also discovered a single indeterminate form of ostracod in the Nodular Limestone and siliceous limestone exposed in Rojabedia, Ratitalai and Hanumanpura, Bagh Town (Dhar District), Madhya Pradesh.

Chaudhary et al. (2019) described 47 ostracod taxa from the Bagh Formation cropping out in the various areas of Madhya Pradesh and Gujarat. Out of 47, 10 new species were described – *Ovocytheropteron baghensis*, *Amicytheridea bilthanaensis*, *Paijenborchella jeerabadensis*, *Cytheropteron ratitalaiensis*, *C. hanumanpuraensis*, *Eocytheropteron bilthanaensis*, *Curfsina coarctata*, *C. hanumanpuraensis*,

Nigeroloxoconcha baghensis and *N. diluta*. Other recovered taxa were *Cytherella* sp., *C. okadai*, *Cytherelloidea awaldaensis*, *C. oertlii*, *C. oudiapurensis*, *C. raoi*, *C. rosebaidaensis*, *C. subgranulosa*, *C. thuatiensis*, *Cytherelloidea* sp., *Bythocypris mohani*, *Bairdoppilata* sp., *Paracypris acutocaudata*, *P. jaini*, *Rostrocytheridea baghensis*, *Rostrocytheridea* sp., cf. *Rostrocytheridea cerasmoderma*, *R. decurtata*, *R. divergens*, *R. jaisalmerensis*, *Parakrithe oudiapurensis*, *?Majungaella batei*, *?M. rajendrai*, *Amphicytherura* aff. *A. yakhiniensis*, *Bythoceratina tewarii*, *Neocytherideis reymenti*, *Veeniacythereis raoi*, *Cytheropteron* sp., *Haughtonileberis derooi*, *H. thuatiensis*, *Makatinell abilthanaensis*, *M. punctata*, *Sapucariella jodhpurensis* and *Sapucariella* sp. cf. *S. subtriangulata* from the creamy-marl, nodular limestone and siliceous limestones exposed at 12 localities/villages – Rojabediya, Ratitalai, Hanumanpura, Jeerabad, Awalda, Bilthana, Kakanpur, Bhundmariya, Moti Chikli and Badkeshwar Mahadev temple, near Bagh town and on the Bagh-Kukshi road, Madhya Pradesh. Later, Chaudhary et al. (2017) described and illustrated three ostracod species belonging to the genus *Makatinella* from the siliceous limestone and marl exposed at Bilthana (Rajpipla), Gujarat and Ratitalai village, Dhar District, Madhya Pradesh. These three species were *M. bilthanaensis*, *M. punctata* and *M. thuatiensis*.

Chaudhary and Nagori (2019) investigated the following genera: *Cytherelloidea*, *Rostrocytheridea*, *Makatinella*, *Sapucariella*, *Majungaella*, *Nigeroloxoconcha*, *Bythoceratina* and *Amphicytherura* from the Bagh Formation, Narmada basin. They reconstructed the palaeobiogeographic association of these ostracods with other Gondwanan landmasses during the Cenomanian-Turonian-Coniacian periods.

Overlying the Bagh Formation is the Lameta Formation (also known as infratrappean beds), which is a very prolific stratigraphic unit for fossils of dinosaur skeletons, eggs, eggshells and coprolites (Khosla and Lucas 2020b, c, d). The term Lameta Formation was coined by Medlicott (1860) after the Lameta Ghat, where the type section of the Lameta Formation is well exposed on the upstream side of the Narmada River, Jabalpur, Madhya Pradesh. Matley (1921) divided the Lameta Formation into five distinct stratigraphic units (in ascending order), the Green Sandstone, Lower Limestone, Mottled Nodular Bed, Upper Limestone and Upper Sandstone. The Lameta Formation overlies either Precambrian basement or the Jabalpur Formation of the Gondwana Supergroup and is overlain by the Deccan Traps (Matley 1921). The Upper Cretaceous Lameta Formation occurs in at least six basins: (1) Jabalpur, (2) Sagar, (3) Balasinor-Jhabua, (4) Ambikapur-Amarkantak, (5) Nand-Dongargaon and (6) Salbardi-Belkher. It covers parts of Madhya Pradesh, Gujarat and Maharashtra states in peninsular India (Mohabey 1996; Khosla et al. 2011b; Srivastava and Mankar 2015; Verma et al. 2016; Khosla and Lucas 2020a). The Lameta Formation of these basins has yielded a diverse assemblage of ostracods.

Brookfield and Sahni (1987) conducted the first work on the ostracods of the Lameta Formation, recovering four genera – *Candona*, *Paracypretta*, *Cyprois* and *Metacypris* – from the variegated shale band of the Bara Simla hill without providing any illustrations or descriptions. Later, Sahni and Khosla (1994) collected nine ostracod species – *Mongolocypris* cf. *M. gigantea*, *Paracandona jabalpurensis*, *Cypridea* (*Pseudocypridina*) sp., *Cyclocypris transitoria*, *Mongolianella palmosa*,

Candona altanulaensis, *Cypridopsis bugintsavicus*, *Eucypris* sp. and *Altanicypris* sp. – from the green marl and variegated shale bands of Bara Simla hill and the grey siltstone of the Chui hill sections of the Lameta Formation, Jabalpur, Madhya Pradesh.

Subsequently, Khosla and Sahni (2000) described a diverse assemblage of ostracods from the Bara Simla hill green marl band and variegated shale band, as well as the grey siltstone of the Chui hill sections, including 10 genera and 15 species. The following ostracods were discovered: *Mongolianella palmosa*, *Altanicypris bhatiai*, *Eucypris* sp., *Mongolianella* sp., *M. bajshintsavica*, *Mongolianella* cf. *khamarieniensis*, *M. ?gigantea*, *Candona altanulaensis*, *Cypridopsis bugintsavicus*, *Candona* (*Candona*) cf. *C.* (C) *hubeiensis*, *Paracandona jabalpurensis*, *Cypridea* (*Pseudocypridina*) sp., *Cyclocypris transitoria*, *Darwinula* sp. 1 and *Darwinula* sp. 2. These ostracods were thought to be closely related to Chinese and Mongolian forms. *Altanicypris*, *Candona*, *Cypridopsis*, *Darwinula* and *Mongolianella* are some distinct ostracod genera of the Lameta Formation.

Khosla et al. (2005) reported 17 ostracod species, including 5 new forms, from the Lameta Formation of the Pavna region, Chandrapur District, Maharashtra. These authors recovered ostracod taxa from the marlite and gypsiferous clays at Pavna village. The reported species were the following: *Limnocythere deccanensis*, *Cypridopsis dongargaonensis*, *C. sahnii*, *Mongolianella ashui*, *Cypridea pavnaensis*, *Frambocythere tumiensis anjarensis*, *Gomphocythere falsicarinata*, *G. paucisulcatus*, *G. strangulata*, *Paracypretta anjarensis* (revised as *P. jonesi*), *Zonocypris spirula*, *Cypridopsis hyperectyphos*, *Mongolocypris* sp., *Eucypris amphibolos* and *Cypria cyrtonidion.*

Meanwhile, Whatley and Bajpai (2005) and Khosla et al. (2011b) revisited India's Late Cretaceous ostracod assemblages and revised species names previously reported by Khosla and Sahni (2000). *Altanicypris bhatiai* was assigned to *Paracypretta jonesi, Mongolianella palmosa to M. cylindrica* and *Cypridopsis bugintsavicus to C. wynnei*; *Cypridopsis* sp. became *Cypridopsis* sp. 2 and *Candona altanulaensis to C. amosi*; *Candona (Candona)* cf. *C. (C.) hubeiensis* became *Eucandona kakamorpha*; *Cyclocypris transitoria converted to Cyclocypris sahnii*; *Darwinula* sp. 1 converted to *Darwinula torpedo*; *Darwinula* sp. 2 was reassigned to *Cypridopsis legitima*; and gen. et sp. A placed in *Eucypris* sp. cf. *E. catantion.*

Khosla et al. (2010) recorded 17 ostracod species, including two new forms (*Zonocypris pseudospirula* and *Ilyocypris pisduraensis*), from the grey limestone and white silty sandstone of the Lameta Formation, Pisdura, Chandrapur District, Maharashtra. The other recorded species were the following: *Paracypretta jonesi*, *Stenocypris cylindrica*, *Cypridopsis hyperectyphos*, *C. mohgaonensis*, *Eucypris pelasgicos*, *Cyclocypris amphibolos*, *Cypria cyrtonidion*, *Candona amosi*, *Eucandona kakamorpha*, *Haplocythereidea* sp. and *Ilyocypris* sp. They divided the ostracod genera into two groups based on palaeoecology: *Limnocythere*, *Frambocythere*, *Gomphocythere*, *Candona* and *Eucandona* are non-swimming, endobenthonic or epibenthonic walkers/crawlers, whereas *Zonocypris*, *Paracypretta*, *Eucypris*, *Cypria*, *Cyclocypris*, *Cypridopsis* and *Stenocypris* are active swimmers. *Limnocythere* could be endobenthonic (McKenzie 1971). The two genera

Frambocythere and *Gomphocythere* are unmistakable epibenthonic crawlers and walkers. *Candona* adapted to a life of walking, crawling and burrowing. Based on ostracod taxa, the grey siltstone and white silty sandstone of the Lameta Formation of Pisdura were interpreted to have been deposited in a pool or lake, while red clays accumulated in reducing environments.

Khosla et al. (2011b) conducted a comparative study of ostracods from the two Lameta Formation-bearing basins: Jabalpur (Madhya Pradesh) and Nand-Dongargaon (Maharashtra). They found 41 ostracod species, including 5 new forms – *Paracypretta indica*, *Cypridopsis ashui*, *C. huenei*, *?Candona chuiensis* and *Cypridea* (*Pseudocypridina*) *jabalpurensis*. The other reported species were the following: *Limnocythere bajpai*, *L. deccanensis*, *Frambocythere tumiensis anjarensis*, *Gomphocythere paucisulcatus*, *Scabriculocypris* sp., *?Altanicypris* sp., *Paracypretta jonesi*, *P. verruculosa*, *Paracypretta* sp., *G. strangulata*, *?Mongolocypris* sp. cf. *M. gigantea*, *Stenocypris cylindrica*, *Zonocypris gujaratensis*, *Z. pseudospirula*, *Z. spirula*, *Cypridopsis astralos*, *C. dongargaonensis*, *C. hyperectyphos*, *C. mohgaonensis*, *C. sahnii*, *C. wynnei*, *Candona amosi*, *Eucypris intervolcanus*, *E. pelasgicos*, *Eucypris* sp., *Cyclocypris amphibolos*, *C. sahnii*, *Cypria cyrtonidion*, *Eucandona kakamorpha*, *Paracandona jabalpurensis*, *Neuquenocypris* sp., *N. pisduraensis*, *Talicypridea pavnaensis*, *Wolburgiopsis* sp. and *Cyprois rostellum*. These ostracod taxa were found in the grey siltstone, green marl and variegated shale bands of the Chui, Bara Simla and Chhota Simla hills, respectively, of the Jabalpur basin, as well as the red clay and limestone bands of the Pisdura and Dongargaon hill sections. The recovered ostracod assemblages demonstrate a strong endemism to the Indian subcontinent.

Whatley et al. (2012) described the species *Periosocypris megistus* from the Bara Simla hill and Chui hill recovered from the green marl, variegated shale band and grey siltstone units of the Lameta Formation of Jabalpur area. They placed *Mongolocypris* cf. *M. gigantea* in *Periosocypris megistus*.

Later, Khosla et al. (2013) investigated the palaeoecology of the ostracods described by Khosla et al. (2011b). They proposed various palaeoenvironments (e.g. shallow marine/estuarine and freshwater) for the Lameta Formation of the Jabalpur area. They contended that the ostracod assemblages share zoogeographic affinities with forms known from Africa and South America. *Altanicypris*, *Neuquenocypris*, *Paracypretta*, *Stenocypris* and *Wolburgiopsis* are the distinctive ostracod genera recovered from these continents.

Khosla et al. (2015) recovered seven ostracod taxa from the red silty clays of Pisdura, Chandrapur District, Maharashtra, by macerating coprolites. The following taxa had been discovered from the coprolites: *?Mongolianella* sp., *Cypridea* (*Pseudocypridina*) sp., *Cypridopsis* sp., *Eucypris* sp., *Gomphocythere* sp., *G. paucisulcatus* and *Paracypretta* sp. It was proposed that ostracod-bearing units were deposited in permanent or temporary freshwater lakes or ponds based on the assemblages. Following this, Khosla et al. (2016) illustrated and described seven ostracod taxa recovered from the Lameta Formation of Pisdura, Maharashtra. *Gomphocythere*, *Paracypretta*, *Mongolianella* and *Cypridopsis* are considered strongly endemic to India (Whatley and Bajpai 2006), and *Cyclocypris* and *Eucypris* dispersed "out of

India" to other parts of the world during the latest Cretaceous (Khosla 2015). The endemic ostracods suggest that the Indian plate was isolated during its northward journey in the Late Cretaceous.

A diverse assemblage of ostracods was also reported from the intertrappean beds of the Deccan Traps. The intertrappean beds are sandwiched between two successive lava flows of the Deccan Traps and were deposited during non-volcanic activity by the disruption of drainage systems, resulting in the formation of small and large water bodies in the forms of ponds and lakes (e.g. Khosla and Verma 2015). These beds are usually found on the Deccan Traps' southern, south-eastern, eastern, north-eastern and north-western borders. These intertrappeans beds are mostly made up of limestone, chert, shale, clay, claystone, marl, mudstone, siltstone and occasionally sandstone and grit. These beds were deposited in small patches that extended only a few metres vertically and horizontally. The intertrappean beds are the home to a diverse range of biotas, including ostracods, gastropods, charophytes and megaflora and some palynomorphs, fishes, frogs, turtles, dinosaurs, snakes, mammals, etc. (Khosla and Sahni 2003; Goswami et al. 2011, 2013; Bajpai et al. 2013; Khosla and Verma 2015; Verma and Khosla 2019; Khosla and Lucas 2020a; Rage et al., 2020; Prasad et al. 2021). These beds have produced significant biotic elements, such as foraminiferans, which can be used to demarcate the Cretaceous-Palaeogene boundary in the lava flows of the Deccan Traps in peninsular India. Furthermore, during the Cretaceous-Palaeogene transition, foraminiferans favoured marine incursions in the central part of India (Keller et al. 2009a, b; Khosla 2015).

Sowerby (in Malcolmson 1840), Carter (1852) and Jones (1860) carried out the earliest work on ostracods of the intertrappean beds. Sowerby (in Malcolmson 1840) accidentally discovered carapaces of ostracods and provided the first systematic description of two ostracod species, *Cypris subglobosa* and *C. cylindrica*, from the Sichel hills (= Nirmal range), Telangana. Later, Carter (1852) collected three ostracod species from the shale unit of intertrappean beds exposed in Bombay, including *Cypris cylindrica*, *C. semimarginata* and *Cypris* sp. Subsequently, Jones (1860) described five ostracod species – *Cypris cylindrica*, *C. hislopi*, *C. hunteri*, *C. strangulata* and *C. subglobosa* – from the intertrappean beds of Nagpur, Maharashtra.

Sastri (1961) collected four ostracod species from intertrappean beds exposed at Kuntamuru village near Rajahmundry. The described species are the following: *Cytherella* sp., *Cytheropteron* sp., ?*Eucythere* sp. and *Loxoconcha* sp. Subsequently, Sastri (1963) reported five more species from the same locality. These five species are *Bairdia subdeltoidea*, *Cythere (?Xestoleberis) ranikotiana (= Cythere ranikotiana)*, *Cythereis bowerbanki*, *Cythereis* cf. *mersondaviesi* and *Cytherelloidea* sp.

Bhalla (1965) collected 16 ostracod species from the intertrappean beds located in Panangadi, Andhra Pradesh. The collected species are *Cytherella* sp., *Cytherella laticostata*, *Brachycythere* sp., *Bythocypris* sp., *Costa* sp., *Hermanites* sp. A, *Hermanites* sp. B, *Krithe bartonensis*, *Leguminocythereis* sp., *Quadracythere* sp., ?*Schizocythere* sp., *Schizocythere forestensis* and *Xestolrberis subglobosa*.

Guha (1970) studied five ostracod species, including one unidentified form, from the intertrappean beds of peninsular India. These species are *Cythereis* cf.

tamulicus, *Hermanites* cf. *pondicheriensis*, *Paracypris* sp., *Protobuntonia* sp., *Xestoleberis* sp. and *Ovocytheridea*. Jain (1978) examined and illustrated 12 ostracod species, including two new forms, from the Kateru intertrappeans, Rajahmundry, Andhra Pradesh. The studied species are the following: *Ovocytheridea raoi*, *Protobuntonia hartmanni*, *Bairdia* sp., ?*Bythocypris* sp., *Cytherella* sp. cf. *munsteri*, *Cytherelloidea* sp. cf. *C. keiji*, ?*Cytheridea* sp., *Hermanites* sp. cf. *Hermanites cracens*, *Imnocythere* sp., *Loxoconcha* sp., *Quadracythere ubauadrata* and ?*Xestoleberis* sp. Bhatia and Rana (1984) illustrated and described five ostracod species – *Paracypretta jonesi*, *Mongolianella hislopi*, *Cyprois* sp., *Candoniella* sp. and *Metacypris strangulata* – from the intertrappean beds situated near Nagpur, Maharashtra.

Later, Mathur and Verma (1988) illustrated and described six ostracod species from the intertrappean beds exposed at Narli, Jhalawar District, Rajasthan. The recovered species were the following: *Cyprois* sp. (= *Cypria cyrtonidion*), *Moenocypris hunteri* (= *Cypris*), *Mongolianella* sp. (= *Mongolianella cylindrica*), *Pseudoeucypris* sp. and *Paracypretta* sp. Bhatia et al. (1990a) discovered six ostracod species from a 1-m-thick cherty marl and chert overlying the volcanic rocks near the village Mamoni, Kota District, Rajasthan. The described species are the following: *Mongolianella palmosa*, *M. cylindrica*, *Mongolianella* sp., *Altanicypris szczechura*, *Paracypretta subglobosa*, *Frambocythere tumiensis anjarensis*, *Gomphocythere strangulata* and ?*Mongolianella cylindrica*. Bhandari and Colin (1999) examined 11 ostracod species from the intertrappean beds exposed in the Anjar (Kachchh District), Gujarat. The studied species are the following: *Frambocythere tumiensis anjarensis*, *Zonocypris gujaratensis*, *Gomphocythere* sp.1, *Limnocythere* sp. 1, *Typhlocypris* cf. *areecta*, *Candona* cf. *sinensis*, ?*Paracandona* sp.1, *Mongolianella* sp. 1, *Mongolianella* sp. 2, *Potamocypris* sp. 1 and *Valdonniella* sp. 1.

It should be noted that during 2000 to 2012, the late Robin C. Whatley made significant contributions to understanding the various aspects of ostracods from the intertrappean beds. In addition, Whatley and his colleagues revised the names of numerous ostracod species known from various localities in peninsular India. Whatley and Bajpai (2000a) discovered 17 ostracod species in the intertrappean bed at Lakshmipur (Kachchh District), Gujarat. The described species are the following: *Altanicypris deccanesis*, *Centrocypris megalapos*, *Cypria cyrtonidion*, *Cypridopsis hyperectyphos*, *C. wynnei*, *C. legitima*, *Eucypris pelasgicos*, *E. intervolcanus*, *Zonocypris spirula*, *Frambocythere tumiensis lakshmiae*, *Gomphocythere gomphatatos*, *Gomphocythere* sp., *Limnocythere falsicarinata*, *Gomphocythere strangulata*, *Stenocypris cylindrica*, *Paracandona firmamentum*, *Pseudocypris ectopos*, *Cyprois rostellum* and *Cyprois* sp. Subsequently, Whatley and Bajpai (2000b) revised *Paracypretta bhatiai* and *Limnocypridea ecphymatos* that were previously reported from the Anjar intertrappean, Gujarat (Bhandari and Colin 1999). Following that, Whatley and Bajpai (2000c) investigated the combined data from the intertrappean beds of Lakshmipur, Anjar and Kora in the Kachchh District of Gujarat to better understand India's palaeozoogeographical position during its northward movement. According to the authors, the ostracods recovered from the aforementioned areas are endemic to the Indian subcontinent.

Bajpai and Whatley (2001) studied 21 ostracod taxa from the intertrappean locality exposed at Kora. The studied species are the following: *Eucypris catantion*, *E. pelasgicos*, *E. intervolcanus*, *Mongolianella cylindrica*, *Limnocypridea ecphymatos*, *Talicypridea* sp., *Cyprois rostellum*, *C. polygonum*, *Zonocypris spirula*, *Cyclocypris amphibolos*, *Paracypretta elizabethae*, *Paracypretta* sp., *Cypridopsis* sp. 1, *Cypridopsis* sp. 2, *Cypridopsis palaichthonos*, *Sarsicypridopsis* sp., *Cetacella* sp., *Cypria* sp., *Cypria cyrtonidion*, *C. intertrappeana* and *Candona amosi.*

Whatley et al. (2002a) discovered nine new ostracod species from intertrappean sections of 3.5-m-thick yellowish-white cherty marl and weathered chert at Chandarki and 2-m-thick claystone and marls at Yanagundi, Gulbarga District, Karnataka. The reported ostracod species are the following: *Darwinula torpedo*, *Frambocythere tumiensis anjarensis*, *Gomphocythere strangulata*, *G. akalypton*, *G. dasyderma*, *Paracypretta bhatiai*, *Zonocypris spirula*, *Cypridopsis wynnei*, *C. hyperectyphos*, *C. astralos*, *Mongolianella cylindrica*, *Eucypris intervolcanus*, *E. verruculosa*, *Candona amosi*, *C. mysorephaseolus*, *Eucandona kakamorpha*, *Cypridopsis amphibolos*, *Cypria cyrtonidion* and *Paracandona* sp. The ostracod assemblage indicates that they were strongly endemic to the Indian subcontinent during its northward voyages towards Asia. Following that, Whatley et al. (2002b) collected nine ostracod species, including three new forms, from the Mohgaon-Kalan intertrappean beds' carbonaceous shale unit in the Chhindwara District, Madhya Pradesh. The described ostracod assemblages included *Gomphocythere paucisulcatus*, *Cypridopsis elachistos*, *Zonocypris labyrinthicos*, *Frambocythere tumiensis lakshmiae*, *Cypria cyrtonidion*, *Mongolianella* sp., *Paracypretta bhatiai* and *Eucypris pelasgicos.*

Whatley et al. (2002c) described eight ostracod species, including two new forms, from several intertrappean localities in the peninsular India. The listed species were the following: *Altanicypris deccanensis*, *Cypridopsis legitima*, *Cyclocypris amphibolos*, *Paracypretta elizabethae*, *Paracypretta jonesi*, *Zonocypris spirula*, *Limnocyther*e sp. A and *Gomphocythere* sp. A.

Whatley et al. (2003a) found eight ostracod species, including two new ones, in a 1-m-thick cherty marl and chert overlying a Deccan volcanic sequence near Mamoni (Kota District), Rajasthan. The described species are the following: *Frambocythere tumiensis anjarensis*, *Gomphocythere dasyderma*, *Paracypretta subglobosa*, *Cypridopsis hyperectyphos*, *Mongolianella cylindrica*, *M. subarcuata*, *Eucypris catantion* and *Cyclocypris amphibolos.* Subsequently, Whatley et al. (2003b) described and illustrated the eight ostracod species recovered from the intertrappean localities of peninsular India housed in the Natural History Museum, London. The reported species are the following: *Cypria cylindrica* (=*Mongolianella cylindrica*), *C. hislopi* (=*M. hislopi*), *C. hunteri* (=?*Moenocypris hunteri*), *Limnocythere* sp., *Cypridopsis palaichthonos*, *C. wynnei*, *C. alphospilotos* sp. and ?*Eucypris verruculosa*. After that, Whatley et al. (2003c) described and illustrated *Paracypretta subglobosa*, *P. elizabethae* sp., *P. jonesi*, *Cypris semimarginata*, *Cypris cylindrical*, *Cypris hislopi*, *Cypris strangulata* and *Cypris* sp. from the shale unit of the Bombay intertrappean beds of Maharashtra.

Whatley and Bajpai (2005) investigated eight areas in which 24 localities (located in the Lameta Formation and intertrappean beds) containing 93 valid ostracod species were examined in order to reconstruct their palaeoecology. The non-marine ostracods were divided into two groups based on these locations: swimmers included *Darwinulacea*, *Ilyocypris*, *Eucypridinae*, *Cypris*, *Paracypretta*, *Atlanicypris*, *Cypridopsis*, *Sarscypridopsis* and *Cyclocypris*, whereas non-swimmers included *Limnocythere*, *Gomphocythere*, *Frambocythere*, Candonidae and Cyprideinae. These workers deduced that the genus *Limnocythere* inhabits the temporary pools and that a large number of members of this family required permanent water bodies. *Darwinula* lives in permanent bodies of water, primarily lakes and ponds, but also in streams. *Cypris* inhabits temporary pools. It is frequently dispersed by the wind and can sometimes be found in the upper atmosphere. It can also reproduce itself (Whatley 1992). *Eucypris* lives in temporary pools that dry out during the summer. *Cypridopsis* is found in permanent ponds and lakes and is rarely found in streams or rivers.

Khosla and Nagori (2007a) reported 12 ostracod species, including 2 new forms, from the 1.5-m-thick green fossiliferous clay found in an excavation in Mohgaon-Haveli, Chhindwara District, Madhya Pradesh. They described *Limnocythere bajpaii*, *Cypridopsis mohgaonensis*, *Darwinula torpedo*, *Limnocythere deccanensis*, *Frambocythere tumiensis anjarensis*, *Paracypretta jonesi*, *Zonocypris spirula*, *Z. gujaratensis*, *Paracandona firmamentum*, *Cyclocypris amphibolos*, *Cypria cyrtonidion*, *Cyprois rostellum*, *Cypridopsis* sp., *Eucypris* sp. A, *Eucypris* sp. B and *Eucypris* sp. C. Subsequently, Khosla and Nagori (2007b) collected 24 species of ostracod, including 1 new form, from a 2.5-m-thick white sandy marl at the intertrappean locality exposed in Gitti Khadan (the stone quarry), Takli, Nagpur, Maharashtra. The described species are the following: *Gomphocythere whatleyi*, *Darwinula torpedo*, *Limnocythere deccanensis*, *Frambocythere tumiensis anjarensis*, *Gomphocythere falsicarinata*, *Paracypretta jonesi*, *Zonocypris gujaratensis*, *Z. labyrinthicos*, *Z. spirula*, *Cypridopsis elachistos*, *C. hyperectyphos*, *Mongolianella cylindrica*, *M. subarcuata*, *Eucypris intervolcanus*, *E. pelasgicos*, ?*E. verruculosa*, *Candona amosi*, *Cyclocypris amphibolos*, *Cypria cyrtonidion*, *Cyprois rostellum*, *Eucypris* sp., *Mongolianella* sp. and ?*Talicypridea* sp.

Sharma et al. (2008) recovered 14 ostracod species from a 0.50- to 1-m-thick grey and black chert unit of the Palaeocene intertrappean locality at Lalitpur, Uttar Pradesh. The described species are the following: *Gomphocythere akalypton*, *G. paucisulcatus*, *Eucypris intervolcanus*, *E. catantion*, *Eucypris* sp., *Cypria cyrtonidion*, *Cypridopsis hyperectyphos*, *Cypridopsis* sp., *Mongolianella subarcuata*, *M. cylindrica*, *Frambocythere tumiensis*, *Paracypretta elizabethae*, *Zonocypris spirula* and *Cyprois rostellum*.

Khosla et al. (2009b) only illustrated and examined a previously reported ostracod species (*Cypris cylindrica*) from the Sichel hills, Telangana. Following that, the same species was also discovered in the Lakshmipur intertrappean beds of Gujarat. Finally, Khosla and his co-workers reassigned *Cypris cylindrica* as *Stenocypris cylindrica*. Other species were the following: *Candona amosi*, *Cypria cyrtonidion*, *Cypridopsis hyperectyphos*, *Cypridopsis legitima*, *Eucypris intervolcanus*, *E.*

pelasgicos, *Gomphocythere falsicarinata*, *G. paucisulcatus*, *Frambocythere tumiensis anjarensis*, *Zonocypris spirula* and *Z. viriensis*.

Sharma and Khosla et al. (2009) collected 12 genera and 17 species from a 0.60-m-thick yellow and pink clay unit in the Upper Cretaceous-Lower Palaeocene Jhilmili intertrappean section of the Chhindwara District, Madhya Pradesh. The recovered species are the following: *Gomphocythere strangulata*, *Limnocythere* sp., *L. deccanensis*, *L. falsicarinata*, *Frambocythere tumiensis anjarensis*, *Neocyprideis raoi*, *Darwinula torpedo*, *Paracypretta* sp., *P. subglobosa*, *P. jonesi*, *Cypridopsis hyperectyphos*, *Zonocypris spirula*, *Z. viriensis*, *Mongolianella cylindrica*, *Cypria cyrtonidion*, *Limnocypridea ecphymatos* and *Paracandona firmamentum*.

Khosla et al. (2011a, 2015) described 20 ostracod species, including 1 new and 2 brackish water ostracods – *Neocyprideis raoi* and *Buntonia* sp. – and remaining freshwater ostracods were the following: *Limnocythere deccanensis*, *Frambocythere tumiensis anjarensis*, *Gomphocythere paucisulcatus*, *G. strangulata*, *Centrocypris megalopos*, *Heterocypris* sp., *Paracypretta jonesi*, *P. verruculosa*, *Stenocypris cylindrica*, *Strandesia jhilmiliensis*, *Zonocypris spirula*, *Z. viriensis*, *Zonocypris* sp., *Cypridopsis hyperectyphos*, *Eucypris pelasgicos*, *Cyclocypris amphibolos*, *Cypria cyrtonidion* and *Darwinula torpedo*, from 60-cm-thick pink clays and a laminated claystone unit of the Jhilmili intertrappean beds. Furthermore, Khosla et al. (2015) also undertook biostratigraphic, palaeoecologic and palaeoenvironmental investigations of these beds based on the recovered ostracod assemblage.

Bajpai et al. (2013) investigated trace element (Mg/Ca and Sr/Ca) and stable isotope (^{18}O and ^{13}C) data of the ostracods from the Lakshmipur, Kora, Rajahmundry and Asifabad intertrappean beds in order to reconstruct Late Cretaceous climate and palaeoenvironments. They discovered that the Lakshmipur site has ^{18}O and ^{13}C values ranging from −0.5 to −15.99% and – 2.56 to −17.99%, respectively, indicating no systematic changes in either oxygen or carbon. The fact that lake water has a higher ^{18}O value than meteoritic water suggests that the Lakshmipur section was close to the palaeocoastline. The ^{18}C value indicates that the lake functioned as a closed hydrological system with abundant evaporative cooling in a semi-arid to sub-humid climate. The combined result of these ratios suggested additional evidence for the palaeolake and that it was a closed hydrological system.

Rathore et al. (2017) reported 12 ostracod species belonging to 9 genera from olive-green pebbly mudstone (0.45 m) and pale-yellow shaly marl (0.30 m) units of a 0.75-m-thick intertrappean section at Khar village, Khargaon District, Madhya Pradesh. These ostracod species are the following: *Frambocythere tumiensis anjarensis*, *Gomphocythere akalypton*, *G. paucisulcatus*, *G. strangulata*, *Paracypretta jonesi*, *Stenocypris cylindrica*, *Zonocypris gujaratensis*, *Cypridopsis hyperectyphos*, *Eucypris intervolcanus*, *E. pelasgicos*, *Cyclocypris amphibolos* and *Cypria cyrtonidion*. Based on palaeoecology, these genera were divided into two groups: i) non-swimmer, endobenthonic or epibenthonic walker/crawlers such as *Frambocythere* and *Gomphocythere*, and ii) swimmers, which are *Paracypretta*, *Stenocypris*, *Zonocypris*, *Cypridopsis*, *Eucypris*, *Cyclocypris* and *Cypria*. The recovered ostracods from the Khar intertrappean beds suggested an Upper Cretaceous (Maastrichtian) age for these intertrappean beds. Also, they indicated that a non-marine permanent water body (lacustrine environment) was present during the accumulation of these beds. Subsequently, Rathore (2018) collected 9 genera belonging to 12 species from the thinly white/yellow laminated shale at the intertrappean locality exposed at Pinjaurni, Chandrapur District, Maharashtra. The collected species are the following: *Stenocypris subarcuata*, *Zonocypris gujaratensis*, *Z. labyrinthicos*, *Z. spirula*, *Cypridopsis hyperectyphos*, *C. wynnei*, *Eucypris intervolcanus*, *Cyclocypris amphibolos*, *Cypria cyrtonidion*, *Eucandona kakamorpha*, *Limnocythere deccanensis* and *Frambocythere tumiensis anjarensis*.

Recently, Kapur et al. (2019) described five ostracod species from the 1.50-m-thick calcareous shale of the intertrappean beds exposed at Manawar, Dhar District, Madhya Pradesh. These species are the following: *Eucypris intervolcanus*, *Cypridopsis hyperectyphos*, *Gomphocythere strangulata*, *Frambocythere tumiensis anjarensis* and *Zonocypris gujaratensis*. The absence of broken ostracod valves suggested a lack of turbulence in the water (Szczechura 1978). The genus *Frambocythere*, *Zonocypris* and *Gomphocythere* are heavily ornamented, indicating an increase in alkalinity in the water body. The genus *Cypridopsis* contains 17 species that are thought to have migrated from India and survived in Mongolia and Alaska. The genus *Gomphocythere*, which includes nine endemic species, has survived largely in North China, Africa and Alaska. The genus *Eucypris*, which contains 11 endemic species in India, has spread throughout Mongolia, China and Europe. *Eucypris* is the first ostracod genus discovered in the Upper Cretaceous (Cenomanian) Bayan Shireh Formation of Mongolia and China (Khand, 2000). The genus *Zonocypris* was previously reported from the Aptian/Albian of Brazil (Colin and Depeche 1997). *Frambocythere* had already been reported from Upper Cretaceous (Maastrichtian) to Middle Eocene deposits in Iran, China and Europe (Hou et al. 1978; Tambareau 1984; Tambareau et al. 1991; Colin 2011).

Kshetrimayum et al. (2021) described and illustrated 12 ostracod species, including 2 new species, from the intertrappean beds at Gujri, Dhar District, Madhya Pradesh. The newly reported species are *Gomphocythere testudo* and *Candona Phaseolus*. Other species were *Cyclocypris sahnii*, *Cypria cyrtonidion*, *Paracypretta jonesi*, *Periosocypris megistus*, *Stenocypris cylindrica*, *Frambocythere tumiensis anjarensis*, *Gomphocythere strangulata*, *G. paucisulcatus*, *Zonocypris gujaratensis* and *Eucypris intervolcanus*. Kshetrimayum et al. (2021) inferred a Maastrichtian age for the fossiliferous horizon based on the ostracod fauna. They suggested the presence of non-marine ostracods in association with terrestrial vertebrates, gastropods and charophytes favoured a freshwater depositional environment.

2.4 Foraminiferans

Bhalla (1967) described 18 species of foraminiferans from the intertrappean sections exposed in the Pangadi area, Andhra Pradesh. He reported the following species of foraminiferans from the different clayey units: *Bathysiphon eocenicus*, *Quinqueoloculina* sp., *Triloculina decipiens*, *Fissurina levigata*, *Globulina inaequalis*, *Vaginulina icenii*, *Virgulina* sp. cf. *Virgulina dubia*, *Epistominella dubia*, *Protephidium adamsi*, *P. duddukuruense*, *Rosalina* sp. cf. *Rosalina depressa*, *R. subvilardeboana*, *R. spharuligera*, *Nonion kingi*, *Discorbis toddae*, *Globorotalia* (*Globorotalia*) sp., *Globorotalia* (*Turborotalia*) sp. cf. *G.* (*T.*) *trails* and *Cibicides reinholdi*. On the basis of recovered foraminiferans, Bhalla (1967) suggested the existence of alternating marine and estuarine conditions during the deposition of the intertrappean beds in the Palaeocene-Eocene time interval.

Recently, Keller et al. (2008) carried out a sedimentologic, microfacies and biostratigraphic study of the four intertrappean sections situated in the Duddukuru lake, Government, Balaji and Church quarries in the Rajahmundry Traps of Andhra Pradesh, south-east India. These authors recovered the following species of foraminiferans: *Globoconusa daubjergensis*, *Parvularugoglobigerina eugubina*, *Parasubbotina pseudobulloides*, *Praemurica compressa*, *Subbotina triloculinoides*, *Globigerina pentagona*, *Chiloguembelina* sp., *Globanomalina compressa*, *Globotruncana arca*, *G. dupeublei*, *G. rosetta*, *Contusotruncana contusa*, *C. plicata*, *Rugoglobigerina rotundata*, *R. rugosa*, *R. macrocephala*, *Pseudoguembelina palpebra*, *Laeviheterohelix glabrans* and *Globotruncanella petaloidea*. They suggested that these trap sediments were deposited during magnetic polarity C29N, and the biota recovered from them indicates that volcanism may have played a significant role in the Cretaceous-Palaeogene mass extinction.

Subsequently, Keller (2009a, b) undertook a biostratigraphic, sedimentologic and chemostratigraphic study of the Jhilmili intertrappean beds. They recovered several forms of foraminiferans belonging to *Eoglobigerina edita*, *E. eobulloides*, *Globanomalina compressa*, *Globigerina* (*Eoglobigerina*) *pentagona*, *Globigerinelloides aspera*, *Globoconusa daubjergensis*, *Guembelitria cretacea*, *Hedbergella holmdelensis*, *Parasubbotina pseudobulloides*, *Parvularugoglobigerina eugubina*, *Praemurica taurica*, *Subbotina triloculinoides* and *Woodringina hornerstownensis* from the pink, clayey limestone unit of the section. They demarcated the Cretaceous-Palaeogene transition in the Jhilmili intertrappean beds based on the presence of planktic foraminiferans. They also argued that the Jhilmili intertrappeans formed close to the main phase of the Deccan volcanic eruption under terrestrial, semi-humid to arid environments and that these beds also witnessed shallow marine incursions. The presence of planktic foraminiferans in terrestrial intertrappean sediments some 800 km away from the nearest ocean suggests the presence of a narrow seaway along the Narmada and Tapti rift zones.

Malarkodi et al. (2010) observed benthic and planktic foraminiferans in thin sections made from the different lithologies of intertrappean beds of the Government and Sunnamrayalu quarries of the Rajahmundry Traps (Andhra Pradesh). These workers found planktic foraminiferans assigned to *Parvularugoglobigerina eugubina*, *Parasubbotina pseudobulloides*, *Globoconusa daubjergensis*, *Guembelitria cretacea*, *Praemurica taurica*, *Hedbergella monmouthensis*, *Laeviheterohelix glabrans*, *L. pulchra*, *Rugoglobigerina rugosa*, *R. scotti*, *Globotruncanella havanensis*, *Globotruncana arca* and *Globotruncana* sp., as well as benthic foraminiferans – *Cibicides* spp., *Nonion kingi*, *Discorbis* spp., *D. ubiqua*, *Gyroidina* spp., *Valvulina* spp., *Dentalina basiplanata*, *Triloculina* spp., *Quinqueloculina* spp., *Spiroloculina* spp., *Protelphidium brotzeni* and *Nodosaria* spp. The benthic foraminiferans were considered to indicate deposition of sediments that probably took place in shallow inner shelf to brackish environments. The presence of ostracods and foraminiferans in the Rajahmundry Trap sediments together was used to confirm that depositional conditions were initially restricted to shallow marine to estuarine, lagoonal and, later, open marine conditions developed, followed by abrupt emersion and palaeosol deposition prior to the arrival of the upper trap flows at or near the base of C29N.

Further, Keller et al. (2011) studied the core samples of the ONGC wells in the Krishna-Godavari basin, Rajahmundry area. They found the following foraminiferans: *Pseudoguembelina palpebra*, *P. hariaensis*, *P. costulata*, *P. kempensis*, *P. excolata*, *Heterohelix planata*, *H. navarroensis*, *H. globulosa*, *H. labellosa*, *H. rajagopalani*, *Rugoglobigerina rotundata*, *R. rugosa*, *R. scotti*, *Pseudotextularia elegans*, *P. deformis*, *Racemiguembelina fructicosa*, *Plummerita hantkeninoides*, *Globotruncana aegyptiaca*, *G. arca*, *Planoglobulina acervulinoides*, *Hedbergella monmouthensis*, *Globigerinelloides aspera*, *Contusotruncana contusa*, *Globotruncanita stuarti*, *Gublerina acuta*, *G. cuvillieri*, *Globotruncanella citae*, *Laeviheterohelix glabrans*, *Abathomphalus mayaroensis* and *Planoglobulina multicamerata*. In these wells, two phases of Deccan volcanism were recorded. The planktic foraminiferans were recovered with nanofossils from the sediments associated with the Rajahmundry Traps. These fossils were correlated with global warming and subsequent cooling that took place near the end of the Maastrichtian epoch.

Recently, Khosla (2015) presented a detailed account of the foraminiferal assemblage from the Jhilmili intertrappean beds. He described *Eoglobigerina edita*, *E. eobulloides*, *Globanomalina compressa*, *Globigerina* (*Eoglobigerina*) *pentagona*, *Globigerinelloides aspera*, *Globoconusa daubjergensis*, *Guembelitria cretacea*, *Hedbergella holmdelensis*, *Parasubbotina pseudobulloides*, *Parvularugoglobigerina eugubina*, *Praemurica taurica*, *Subbotina triloculinoides* and *Woodringina hornerstownensis* from the pink clayey limestone. The presence of planktic foraminiferans modified the age from the earlier assigned Maastrichtian to the Early Danian and favoured a brackish-marine depositional environment. Khosla (2015) suggested that the Jhilmili intertrappean beds were deposited in terrestrial, semi-humid to arid environments, with a short interval of freshwater ponds and lakes, which was followed by shallow coastal marine/estuarine conditions.

2.5 Fishes

Sphareodus rugulosus, the oldest pycnodont fish, was described by Egerton (1845) from the Upper Cretaceous deposits of the Cauvery basin in South India. *Enchodus forex*, a predatory fish, was discovered by Leidy (1855) in the Upper Cretaceous intertrappean beds of Nagpur, Maharashtra. Fedden (1884) documented the fish fauna of the intertrappean beds of Bambanbor and Ninama in Gujarat. Woodward (1908) discovered *Lepisosteus indicus* (Lepisosteidae), *Pycnodus lametae* (Pycnodontidae) and *Eosserranus hislopi* in the Lameta Formation of Maharashtra.

Hora (1938) reported six new families of fishes – Polycanthidae, Cyprinidae, Serranidae, Nandidae, Pristolepididae and Clupeidae – from the Upper Cretaceous intertrappean deposits of Central India. Jain and Sahni (1983) recovered fishes assigned to *Igdabatis sigmodon*, *Rhombodus* sp., *Indotrigonodon ovatus*, *Eotrigodon wardhaensis* and *Pisdurodon spatulats* from the infratrappean beds of Pisdura, Maharashtra. Gayet et al. (1984) reported dental remains and scales of *Pycnodus lametae* from the intertrappean deposits of Gitti Khadan, near Nagpur, Maharashtra. Prasad and Sahni (1987) described a fish assemblage from the Asifabad infratrappean beds, Telangana. This assemblage is represented by *Raja* sp., *Coupatezia woutersi*, *Rhombodus* cf. *R. levis*, *Rhombodus* sp., *Igdabatis sigmodon*, *Lepidotes* sp., *Lepisosteus indicus*, *Pycnodus lametae*, *Pycnodus* cf. *P. praecursor*, *Pycnodus* sp., *Belonostomus* cf. *B. cinctus*, *Eomuraena* cf. *E. sagittidens*, *Phareodus* sp., *Apateodus striatus*, *Enchodus* sp., *Stephanodus libycus*, *Stephanodus* sp., *Eotrigodon indicus*, *E. wardhaensis* and *Indotrigonodon ovatus*.

Prasad (1989) investigated the infra- and intertrappean beds of Asifabad, Naskal, Marepalli and Timsanpalli of Andhra Pradesh and Telangana. They recovered fish assemblages from these beds comprising *Lepisosteus*, *Phareodus*, osteoglossid otoliths, *Pycnodus*, *Rhombodus*, *Stephanodus* and *Igdabatis*. Prasad and Khajuria (1990) reported eight fish taxa – *Igdabatis* sp., *Lepisosteus indicus*, *Phareodus*, *Osteoglossidarum deccanensis*, *O. intertrappus*, *Notopteridarum nolfi*, *Apateodus striatus* and *Dapalis* sp. – from the intertrappean beds of Naskal, Telangana. Later, Prasad and Singh (1991) described a fish fauna from the infratrappean beds of Marepalli and Timsanpalli, Rangareddy District, Telangana: Dasyatidae indeterminate, *Igdabatis* sp., *Rhombodus*, *Lepisosteus indicus*, *Phareodus* sp., *Pycnodus lametae*, *P. bicresta*, *Eotrigonodon jonesi*, *Indotrigonodon ovatus*, *Stephanodus libycus* and *Apateodus striatus*. Prasad and Cappetta (1993) investigated the significance of batoid fish recovered from the Telangana infratrappean beds of Asifabad and Marepalli. Two new species, *Raja sudhakari* and *Igdabatis indicus*, as well as some other Rajiformes indeterminate, *Rhombodus* sp., 1 and *Rhombodus* sp. 2, were reported from these locations. Khosla et al. (2004) described a fish fauna consisting of *Lepisosteus* cf. *L. indicus*, *Igdabatis indicus*, Pycnodontidae and Osteoglossidae from the Kisalpuri intertrappean beds of the Dindori District, Madhya Pradesh.

Verma et al. (2017) investigated the fish fauna of the Pisdura, Nand-Dongargaon and Kisalpuri areas and provided a detailed account of *Igdabatis indicus* and Pycnodontidae. They hypothesised that the presence of *Igdabatis* indicates a series of shallow marine dispersals between Africa and India, possibly along the margins of the Kohistan-Ladakh island arc during the latest Cretaceous, and that the presence of pycnodont fish favours Gondwana dispersal events. *Igdabatis indicus*, *Lepisosteus indicus*, Osteoglossidae, Pycnodontidae and Siluriformes were discovered by Lourembam et al. (2017) in the intertrappean beds of Piplanarayanwar, Chhindwara District, Madhya Pradesh. They proposed a near-shore, deltaic or estuarine palaeoenvironment for the intertrappean beds, as well as a Late Cretaceous (Maastrichtian) age. The reported fish fauna resembled the intertrappean fish faunas of Asifabad and the infratrappean beds of Marepalli, implying a coastal plain environment of deposition. Very recently, Kapur et al. (2019) discovered *Phareodus* and an indeterminate teleost in a calcareous shale unit of the Manawar intertrappean beds in the Dhar District of Madhya Pradesh.

References

Andreu B, Colin JP, Singh J (2007) Cretaceous (Albian to Coniacian) ostracodes from the subsurface of the Jaisalmer Basin, Rajasthan, India. J Micropaleontol 53(5):345–370

Andreu B, Colin JP, Singh J (2012) Middle and upper Jurassic ostracods from Western Kachchh, Gujarat, India: biostratigraphy and paleobiogeography. Gond Res 22(3–4):1110–1124

Babinot JF, Colin JP, Randrianasolo A (2009) Les ostracodes de l'Albien-Turonien moyen de la régiond' Antsiranana (Nord Madagascar): systématique, paléoécologie et paléobiogéographie. Carnets de Géologie/Notebooks on Geology, Brest, Article 2009/01 (CG2009_A01)

Bajpai S, Whatley RC (2001) Late Cretaceous non-marine ostracods from the Deccan Intertrappean Beds, Kora: Western Kachchh, India. Rev Esp de Micropaleontol 33(1):91–111

Bajpai S, Holmes J, Bennett C, Mandal N, Khosla A (2013) Palaeoenvironment of northwestern India during the Late Cretaceous Deccan volcanic episode from trace-element and stable-isotope geochemistry of intertrappean ostracod shells. Glob Planet Change 107:82–90

Ballent SC, Whatley RC (2007) The distribution of the gondwanine ostracod *Rostrocytheridea* Dingle: palaeozoogeographical implications. Geol J 42(2):113–125

Bhalla SN (1965) A note on the Ostracoda from the inter-trappean beds near Pangadi, Andhra Pradesh, vol 1. Annals of Geology Department, Aligarh Muslim University, Aligarh, p 57

Bhalla SN (1967) Foraminifera from the inter-trappean beds (lower Eocene) of the Pangadi area, India. Micropaleontology 13(3):351–368

Bhandari A, Colin JP (1999) Limnic ostracods of the intertrappean sediments (terminal Maastrichtian-basal Paleocene) of the Anjar region (Kachchh, Gujarat state), India: systematics, paleoecology, and paleobiogeographic affinities. J Micropaleontol 42(1):3–20

Bhatia SB (1984) Ostracod faunas of the Indian subcontinent – their palaeozoogeographic and palaeoecologic implications. J Palaeontol Soc Ind 20:1–8

Bhatia SB (1992) Recent advances in studies on fossil Charophyta. Palaeobotanist 40:413–419

Bhatia SB, Mannikeri MS (1976) Some Charophyta from the Deccan intertrappean beds near Nagpur, Central India. Geophytology 6(1):75–81

Bhatia SB, Rana RS (1984) Palaeogeographic implications of the Charophyta and Ostracoda of the intertrappean beds of peninsular India. Mém Soc Geol Fr 147:29–35

Bhatia SB, Prasad GVR, Rana RS (1990a) Deccan volcanism, a Late Cretaceous event: conclusive evidence of ostracodes. In: Sahni A, Jolly A (eds) Cretaceous event stratigraphy and the correlation of the Indian nonmarine strata. A Seminar cum Workshop IGCP 216 and 245, Chandigarh, pp 47–49

Bhatia SB, Riveline J, Rana RS (1990b) Charophyta from the Deccan intertrappean beds near Rangapur, Andhra Pradesh, India. The Palaeobotanist 37:316–323

Bhatia SB, Prasad GVR, Rana RS (1996) Maastrichtian non-marine ostracodes from peninsular India: Palaeobiogeographic and age implications. Mem Geol Soc Ind 37:297–311

Brenner P, Oertli HJ (1976) Lower Cretaceous ostracodes (Valanginian to Hauterivian) from the Sundays River Formation, Algoa Basin, South Africa. Bull Centre Rese, Pau-SNPA 10:471–533

Brookfield ME, Sahni A (1987) Palaeoenvironment of the Lameta Beds (Late Cretaceous) at Jabalpur, M.P., India: soils and biota of a semi-arid alluvial plain. Cretac Res 8:1–14

Carter HJ (1852) Geology of the Island of Bombay. J Bombay Br R Asiat Soc 21:161–215

Case JA (2002) A new biogeographic model for dispersal of Late Cretaceous vertebrates into Madagascar and India. J Vertebr Paleontol 22(3):42A

Chanda S, Deb U, Mukherjee TN (1989) Study of fossil Charophyta from the intertrappean beds at Kateru area, Krishna-Godavari Basin, Andhra Pradesh, South India. First International Symposium on Extant and Fossil Charophytes, Montpellier, pp 1–11
Chaudhary M (2017) The distribution of the Gondwanine ostracode genus *Makatinella* dingle from the Bagh Formation (Upper Cretaceous) of Narmada Basin. Indian J Appl Res 7(9):9–11
Chaudhary M, Nagori ML (2019) Ostracods from the Bagh Formation (Upper Cretaceous) of the Narmada Basin, India: their age and palaeobiogeographical significance. Micropaleontology 65(6):513–543
Chaudhary M, Nagori ML, Bhanat N (2017) *Cytherelloidea* (ostracoda) from the Upper Cretaceous Bagh formation of Madhya Pradesh, India: its affinity and palaeogeographic implications. J Palaeontol Soc Ind 62(2):217–224
Chaudhary M, Nagori ML, Khosla SC (2019) Ostracod genus *Rostrocytheridea* from the Bagh Formation (Upper Cretaceous) of Narmada Basin, Madhya Pradesh and Gujarat, India. J Palaeontol Soc Ind 64(1):73–82
Colin JP (2011) From light to darkness: *Frambocythere* Colin 1980 to *Kovalevskiella* Klein, 1963 (Limnocytheridae: Timiriasevinae). J Geol Paläontol 11:44–47
Colin JP, Dépêche F (1997) Faunes d'ostracodes lacustres des bassins intra-cratoniques d' âge albo-aptienen Afrique de l'Ouest (Cameroun, Tchad) et au Brésil: considerations d'ordre paléoécologique et paléobiogéographique. Afr Geosci Rev 4:431–450
Compagno LJV (1973) Interrelationships of living elasmobranchs. Zool J Linnean Soc 53:15–61
Dingle RV (1969) Marine Neocomian Ostracoda from South Africa. Trans R Soc S Afr 38:139–164
Egerton DPM (1845) Description of the mouth of a *Hybodus* found by Mr. Boscawen Ibbetson in the Isle of Wight. Quart J Geol Soc Lond 1(1):197–199
Fedden F (1884) The geology of the Kathiawar peninsula in Gujarat. Mem Geol Surv Ind 21:1–64
Gayet M, Rage JC, Rana RS (1984) Nouvelles ichthyofaune et herpétofaune de Gitti Khadan, le plus anciengisement connu du Deccan (Crétacé/Paléocène) à microvertébrés. Implications paléogéographiques. Mém Soc Géol Fr 147:55–65
Goswami A, Prasad GVR, Upchurch P, Boyer DM, Seiffert ER, Verma O, Gheerbrant E, Flynn JJ (2011) A radiation of arboreal basal eutherian mammals beginning in the Late Cretaceous of India. Proc Natl Acad Sci USA 108:16333–16338
Goswami A, Prasad GVR, Verma O, Flynn JJ, Benson RBJ (2013) A troodontid dinosaur from the latest Cretaceous of India. Nat Commun 4:1–5. https://doi.org/10.1038/ncomms2716
Guha DK (1970) Observation on the Cenozoic and some Mesozoic Ostracoda of India, vol 7. Publication of the Centre of Advanced Study in Geology, Panjab University, Chandigarh, pp 205–212
Guha AK, Ghosh BK (1970) Fossil Bryozoa and Ostracoda from the Bagh beds, MP. Proc 57th Sess Indian Sci Cong, Part III (Abs.), pp 194
Hora SL (1938) On some fossil fish-scales from the inter-trappean beds at Deothan and Kheri, central provinces. Rec Geol Surv Ind 73(2):267–294
Hou Y, Junde H, Chunhui Y (1978) The Cretaceous-Tertiary ostracods from the marginal region of the Yangtze-Han River plain in Central Hubei, vol 9. Mem Nanjing Institute of Geology and Paleontol, pp 129–206
Jain SP (1961) Discovery of Ostracoda and smaller foraminifera from the Upper Cretaceous Bagh Beds, MP. Curr Sci 30(9):341–342
Jain SP (1975) Ostracoda from the Bagh Beds (Upper Cretaceous) of Madhya Pradesh. Geophytology 5(2):188–212
Jain SP (1978) Ostracodes des "inter-trappean beds" (Eocene Inferieur) de Kateru, Rajahmundry, Côte est de l' Inde. Rev de Microplaeontol 21(2):51–58
Jain SL, Sahni A (1983) Some Upper Cretaceous vertebrates from central India and their palaeogeographic implications. In: Symposium on Cretaceous of India: palaeoecology, palaeogeography and time boundaries. Birbal Sahni Institue of Palaeobotany, Lucknow, pp 66–83
Jones T (1860) Notes on the fossil Cypride. Quart J Geol Soc London 16:154–189
Joyce WG, Bandyopadhyay S (2020) A revision of the pelomedusoid turtle *Jainemys pisdurensis* from the Late Cretaceous (Maastrichtian) Lameta Formation of India. Peer J 8:e9330. https://doi.org/10.7717/peerj.9330

Kania S, Khosla A, Verma O, Prashanth M (2022) Charophyte assemblage in the Cretaceous–Palaeogene boundary of the Deccan intertrappeans from Chhindwara District (Madhya Pradesh), Central India. Him Geol 43(2):383–396

Kapur VV, Khosla A (2019) Faunal elements from the Deccan volcano-sedimentary sequences of India: a reappraisal of biostratigraphic, palaeoecologic, and palaeobiogeographic aspects. Geol J 54(5):2797–2828

Kapur VV, Khosla A, Tiwari N (2019) Paleoenvironmental and paleobiogeographical implications of the microfossil assemblage from the Late Cretaceous intertrappean beds of the Manawar area, district Dhar, Madhya Pradesh, Central India. Hist Biol 31(9):1145–1160

Keller G, Adatte T, Gardin S, Bartolini A, Bajpai S (2008) Main Deccan volcanism phase ends near the K-T boundary: evidence from the Krishna-Godavari Basin, SE India. Earth Planet Sci Lett 268:293–311

Keller G, Khosla SC, Sharma R, Khosla A, Bajpai S, Adatte T (2009a) Early Danian planktic Foraminifera from Cretaceous-Tertiary intertrappean beds at Jhilmili, Chhindwara District, Madhya Pradesh, India. J Foram Res 39(1):40–55

Keller G, Adatte T, Bajpai S, Mohabey DM, Widdowson M, Khosla A, Sharma R, Khosla SC, Gertsch B, Fleitmann D, Sahni A (2009b) K-T transition in Deccan traps of Central India marks major marine seaway across India. Earth Planet Sci Lett 282:10–23

Keller G, Bhowmick PK, Upadhyay H, Dave A, Reddy AN, Jaiparkash BC, Adatte T (2011) Deccan volcanism linked to the Cretaceous-Tertiary boundary mass extinction: new evidence from ONGC wells in the Krishna-Godavari Basin. J Geol Soc India 78:399–428

Khand Y (2000) The origins of modern nonmarine ostracod faunas: evidence from the Late Cretaceous and Early Palaeogene of Mongolia. Hydrobiologia 419:119–124

Khosla A (2014) Upper Cretaceous (Maastrichtian) charophyte gyrogonites from the Lameta Formation of Jabalpur, Central India: palaeobiogeographic and palaeoecological implications. Acta Geol Pol 64(3):311–323

Khosla A (2015) Palaeoenvironmental, palaeoecological and palaeobiogeographical implications of mixed fresh water and brackish marine assemblages from the Cretaceous-Palaeogene Deccan intertrappean beds at Jhilmili, Chhindwara District, Central India. Rev Mex de Cien Geol 32(2):344–357

Khosla A (2021) Paleobiogeographical inferences of Indian Late Cretaceous vertebrates with special reference to dinosaurs. Hist Biol:33(9):1431–1442 https://doi.org/10.1080/08912963.2019.1702657

Khosla A, Lucas SG (2020a) Introduction of Indian Late Cretaceous dinosaur eggs and eggshells of Peninsular India. In: Khosla A, Lucas SG (eds) Late Cretaceous dinosaur eggs and eggshells of Peninsular India: oospecies diversity and taphonomical, palaeoenvironmental, biostratigraphical and palaeobiogeographical inferences, Topics in Geobiology, vol 51. Springer Nature International Publishing, Cham, pp 1–30

Khosla A, Lucas SG (2020b) Historical background of Late Cretaceous dinosaur studies and associated biota in India. In: Khosla A, Lucas SG (eds) Late Cretaceous dinosaur eggs and eggshells of Peninsular India: oospecies diversity and taphonomical, palaeoenvironmental, biostratigraphical and palaeobiogeographical inferences, Topics in Geobiology, vol 51. Springer Nature International Publishing, Cham, pp 31–56

Khosla A, Lucas SG (2020c) Geology and stratigraphy of dinosaur eggs and eggshell-bearing infra- and intertrappean beds of peninsular India. In: Khosla A, Lucas SG (eds) Late Cretaceous dinosaur eggs and eggshells of Peninsular India: oospecies diversity and taphonomical, palaeoenvironmental, biostratigraphical and palaeobiogeographical inferences, Topics in Geobiology, vol 51. Springer Nature International Publishing, Cham, pp 57–115

Khosla A, Lucas SG (2020d) Indian Late Cretaceous dinosaur nesting sites and their systematic studies. In: Khosla A, Lucas SG (eds) Late Cretaceous dinosaur eggs and eggshells of Peninsular India: oospecies diversity and taphonomical, palaeoenvironmental, biostratigraphical and palaeobiogeographical inferences, Topics in Geobiology, vol 51. Springer Nature International Publishing, Cham, pp 117–205

Khosla A, Lucas SG (2020e) Discussion: Oospecies diversity, biomineralization aspects, taphonomical, biostratigraphical, palaeoenvironmental, palaeoecological and palaeobiogeographical Inferences of the dinosaur-bearing Lameta Formation of Peninsular India. In: Khosla A, Lucas SG (eds) Late Cretaceous dinosaur eggs and eggshells of Peninsular India: Oospecies diversity and taphonomical, palaeoenvironmental, biostratigraphical and palaeobiogeographical inferences, Topics in Geobiology, vol 51. Springer, Cham, pp 207–271

Khosla A, Lucas, SG (2021) End-Cretaceous Extinctions. In: Elias S, Alderton, D (eds) Encyclopedia of Geology, 2nd edition. Earth Systems and Environmental Sciences (Academic Press, Elsevier), pp. 665–678. https://doi.org/10.1016/B978-0-12-409548-9.124

Khosla SC, Nagori ML (2007a) Ostracoda from the Intertrappean beds of Mohgaon-Haveli, Chhindwara District, Madhya Pradesh. J Geol Soc Ind 69:209–221

Khosla SC, Nagori ML (2007b) A revision of the Ostracoda from the inter-trappean beds of Takli, Nagpur District, Maharashtra. J Palaeontol Soc Ind 52(1):1–15

Khosla A, Sahni A (2000) Late Cretaceous (Maastrichtian) ostracodes from the Lameta Formation, Jabalpur cantonment area, Madhya Pradesh, India. J Palaeontol Soc India 45:57–78

Khosla A, Sahni A (2003) Biodiversity during the Deccan volcanic eruptive episode. J Asi Earth Sci 21(8):895–908

Khosla A, Verma O (2015) Paleobiota from the Deccan volcano-sedimentary sequences of India: paleoenvironments, age and paleobiogeographic implications. Hist Biol 27(7):898–914

Khosla A, Kapur VV, Sereno PC, Wilson JA, Dutheil D, Sahni A, Singh MP, Kumar S, Rana RS (2003) First dinosaur remains from the Cenomanian-Turonian of the Nimar Sandstone (Bagh Beds), district Dhar, Madhya Pradesh, India. J Palaeontol Soc Ind 48:115–127

Khosla A, Prasad GVR, Verma O, Jain AK, Sahni A (2004) Discovery of a micromammal-yielding Deccan intertrappean site near Kisalpuri, Dindori District, Madhya Pradesh. Curr Sci 87(3):380–383

Khosla SC, Nagori ML, Mohabey DM (2005) Effect of Deccan volcanism on non-marine Late Cretaceous Ostracode fauna: a case study from Lameta formation of Dongargaon area (Nand-Dongargaon basin), Chandrapur District, Maharashtra. Gond Geol Magaz 8:133–146

Khosla A, Sertich JJW, Prasad GVR, Verma O (2009) Dyrosaurid remains from the intertrappean beds of India and the Late Cretaceous distribution of Dyrosauridae. J Vert Paleontol 29(4):1321–1326

Khosla SC, Nagori ML, Jakhar SR, Rathore AS (2009a) Mixed marine, brackish water and non-marine microfaunal association in the inter-trappean beds (early Palaeocene) of Jhilmili, Chhindwara District, Madhya Pradesh. J Geol Soc Ind 73:724–732

Khosla SC, Nagori ML, Jakhar SR, Rathore AS, Kumari M (2009b) A restudy of the Ostracoda *Cypris cylindrica* Sowerby (in Malcolmson, 1840) from the Deccan intertrappean beds (Late Cretaceous) of Lakshmipur, Kachchh, Gujarat. J Geol Soc Ind 74:579–584

Khosla SC, Nagori ML, Jakhar SR, Rathore AS (2010) Stratigraphical and palaeoecological implications of the Late Cretaceous ostracods from the Lameta Formation of Pisdura, Chandrapur District, Maharashtra, India. Gond Geol Magaz 25:115–124

Khosla SC, Nagori ML, Jakhar SR, Rathore AS (2011a) Early Danian lacustrine–brackish water Ostracoda from the Deccan Intertrappean beds near Jhilmili, Chhindwara District, Madhya Pradesh, India. Micropaleontology 57(3):223–245

Khosla SC, Rathore AS, Nagori ML, Jakhar SR (2011b) Non-marine Ostracoda from the Lameta Formation (Maastrichtian) of Jabalpur, Madhya Pradesh and Nand-Dongargaon Basin, Maharashtra, India: their correlation, age and taxonomy. Rev Esp de Micropaleontol 43:209–260

Khosla SC, Rathore A, Nagori M, Jakhar S (2013) Palaeoecology and affinity of ostracod fauna from the classic localities of Lameta formation of Jabalpur, Madhya Pradesh, India. Spec Publ Geol Soc Ind 1:401–415

Khosla A, Chin K, Alimohammadin H, Dutta D (2015) Ostracods, plant tissues, and other inclusions in coprolites from the Late Cretaceous Lameta Formation at Pisdura, India: Taphonomical and palaeoecological implications. Palaeogeog Palaeoclimat Palaeoecol 418:90–100

Khosla A, Chin K, Verma O, Alimohammadin H, Dutta D (2016) Paleobiogeographical and paleoenvironmental implications of the freshwater Late Cretaceous ostracods, charophytes and distinctive residues from coprolites of Lameta Formation at Pisdura, Chandrapur District (Maharashtra), Central India. In: Khosla A, Lucas SG (eds) Cretaceous period: biotic diversity and biogeography, vol 71. New Mexico Museum of Natural History and Science, Albuquerque, pp 173–184

Khosla A, Kania S, Lucas SG, Verma O (2022) Charophytes from the Cretaceous–Palaeogene transition in the Jhilmili intertrappean beds of Central India. Geol J 57(11):4412–4438. https://doi.org/10.1002/gj.4528

Kshetrimayum DS, Parmar V, Lourembam RS, Prasad GVR (2021) A diversified Ostracoda (Crustacea) assemblage from the Upper Cretaceous intertrappean beds of Gujri, Dhar District, Madhya Pradesh, India. Cretac Res 124. https://doi.org/10.1016/j.cretres.2021.104784

Leidy J (1855) Contributions towards a knowledge of the marine invertebrate fauna of the coasts of Rhode Island and New Jersey. Merrihewand Thompson

Lourembam RS, Prasad GVR, Grover P (2017) Ichthyofauna (Chondrichthyes, Osteichthyes) from the Upper Cretaceous intertrappean beds of Piplanarayanwar, Chhindwara District, Madhya Pradesh, India. Island Arc 26(1):e12180. https://doi.org/10.1111/iar.12180

Mahalakshmi P, Hussain SM (2015) Distribution of ostracods from Pazhaverkadu (Pulicat lagoon), Tamil Nadu, India. In: Vasudevan S (ed) Lakes and wetlands. Partridge, India, pp 268–287

Malarkodi N, Keller G, Fayazudeen PJ, Mallikarjuna UB (2010) Foraminifera from the early Danian intertrappean beds in Rajahmundry quarries, Andhra Pradesh. J Geol Soc Ind 75:851–863

Malcolmson JG (1840) On the fossils of the eastern portion of the Great Basaltic District of India. Trans Geol Soc London S2(5):537–575

Mallikarjuna UB, Nagaraja HM (1996) Ostracodes from the Ariyalur group (Late Cretaceous), Cauvery Basin, southern India. J Geol Soc Ind 48(2):189–201

Mathur AK, Verma KK (1988) Freshwater ostracodes from the intertrappean beds of southeastern Rajasthan. Spec Publ Geol Surv Ind 11(2):169–174

Matley CA (1921) On the stratigraphy, fossils and geological relationships of the Lameta beds of Jubbulpore. Rec Geol Surv Ind 53:142–164

McKenzie KG (1971) Entomostraca of Aldabra, with special reference to the genus *Heterocypris* (Crustacea, Ostracoda). Philos Trans R Soc London B 260:257–297

Medlicott HB (1860) On the geological structure of the central portion of the Nerbudda District. Mem Geol Surv Ind 2:96–278

Mohabey DM (1996) Depositional environment of Lameta Formation (Late Cretaceous) of Nand-Dongargaon inland basin, Maharashtra: the fossil and lithological evidences. Mem Geol Soc Ind 37:363–386

Mohabey DM, Udhoji SG (1990) Fossil occurrences and sedimentation of Lameta Formation of Nand area, Maharashtra: Palaeoenvironmental, palaeoecological and taphonomical implications. In: Sahni A, Jolly A (eds) Cretaceous event stratigraphy and the correlation of the Indian nonmarine strata. A Seminar cum Workshop IGCP 216 and 245, Chandigarh, pp 75–77

Mohabey DM, Udhoji SG, Verma KK (1993) Palaeontological and sedimentological observations on non-marine Lameta formation (Upper Cretaceous) of Maharashtra, India: their palaeontological and palaeoenvironmental significance. Palaeogeog Palaeoclimat Palaeoecol 105:83–94

Naqvi SM (2005) Geology and evolution of the Indian plate (from Hadean to Holocene – 4 Ga to 4 Ka). Capital Publishing Company, New Delhi

Piovesan EK, Nicolaidis DD, Fauth G, Viviers MC (2013) Ostracodes from the Aptian–Santonian of the Santos, Campos and Espírito Santo basins, Brazil. J S Am Earth Sci 48:240–254

Prasad GVR (1989) Vertebrate fauna from the infra- and intertrappean beds of Andhra Pradesh: age implications. J Geol Soc Ind 34(2):161–173

Prasad GVR (2012) Vertebrate biodiversity of the Deccan volcanic province of India: a review. Bull de la Soc Géol de France 183:597–610

Prasad GVR, Cappetta H (1993) Late Cretaceous selachians from India and the age of the Deccan traps. Palaeontology 36(1):231–248

Prasad GVR, Khajuria CK (1990) A record of microvertebrate fauna from the inter-trappean beds of Naskal, Andhra Pradesh. J Palaeontol Soc Ind 35:151–161
Prasad GVR, Sahni A (1987) A coastal-plain microvertebrate assemblage from the terminal Cretaceous of Asifabad, Andhra Pradesh. J Palaeontol Soc Ind 32:5–19
Prasad GVR, Singh V (1991) Microvertebrates from the inter-trappean beds of Rangareddi District, Andhra Pradesh and their biostratigraphic significance. Bull Ind Geol Assoc 24(1):1–20
Prasad GVR, Verma O, Sahni A, Khosla A (2021) Cretaceous mammals of India–Stratigraphic distribution, diversity and intercontinental affinities. J Palaeosci 70:173–192
Prasad GVR, Verma O, Sahni A, Krause DW, Khosla A, Parmar V (2007a) A new Late Cretaceous Gondwanatherian mammal from Central India. Proc Indian Nat Sci Acad 73(1):17–24
Prasad GVR, Verma O, Sahni A, Parmar V, Khosla A (2007b) A Cretaceous hoofed mammal from India. Science 318:937
Prasad GVR, Verma O, Gheerbrant E, Goswami A, Khosla A, Parmar V, Sahni A (2010) First mammal evidence from the Late Cretaceous of India for biotic dispersal between India and Africa at the K/T transition. Comp Rend Palevol 9:63–71
Prasad GVR, Verma O, Flynn JJ, Goswami A (2013) A new Late Cretaceous vertebrate fauna from the Cauvery basin, South India: implications for Gondwana palaeobiogeography. J Vertebr Paleontol 33(6):1260–1268
Prasad GVR, Sharma A, Verma O, Khosla A, Singh LR, Priyadarshin R (2015) Testudoid and crocodiloid eggshells from the Upper Cretaceous Deccan Intertrappean Beds of Central India. Comp Rend Palevol 14:513–526
Puckett TM (2012) Paleogeographic significance of muscle scars in global populations of Late Cretaceous ostracodes. Micropaleontology 58(3):259–271
Rage J-C, Prasad GVR, Verma O, Khosla A, Parmar V (2020) Anuran Lissamphibian and squamate reptiles from the Upper Cretaceous (Maastrichtian) Deccan Intertrappean Sites in Central India, with a review of Lissamphibian and squamate diversity in the northward drifting Indian plate. In: Prasad GVR, Patnaik R (eds) Biological Consequences of Plate Tectonics: New Perspectives on Post-Gondwanal and Break-up—A Tribute to Ashok Sahni, Vertebrate Paleobiology and Paleoanthropology. Springer, Switzerland, pp 99–101 https://doi.org/10.1007/978-3-030-49753-8_6
Rathore AS (2018) Late Cretaceous (Maastrichtian) non-marine Ostracod fauna from the Intertrappean beds of Pinjaurni, Chandrapur district, Maharashtra. JK Knowledge Intitiative 2(1):64–73
Rathore AS, Grover P, Verma V, Lourembam RS, Prasad GVR (2017) Late Cretaceous (Maastrichtian) non-marine ostracod fauna from Khar, a new intertrappean locality, Khargaon District, Madhya Pradesh, India. Paleontol Res 21(3):215–229
Roychowdhary M, Sastri VV (1962) On the revised classification of the Cretaceous and associated rocks of the Man River section of lower Narbada valley. Rec Geol Surv Ind 91:283–304
Sahni A, Khosla A (1994) A Maastrichtian ostracode assemblage (Lameta Formation) from Jabalpur Cantonment, Madhya Pradesh, India. Curr Sci 67(6):456–460
Sastri VV (1961) Foraminifera and Ostracoda from the inter-trappean beds near Rajahmundry. Indian Minerals 15(2):197–198
Sastri VV (1963) A note on the Foraminifera and Ostracoda from the inter-trappean beds near Rajahmundry. Rec Geol Surv Ind 92(2):299–310
Sastry MV, Mamgain VD, Rao BJ (1972) Ostracod fauna of the Ariyalur group (Upper Cretaceous), Tiruchirapalli District, Tamil Nadu. Mem Geol Surv Ind Palaeontol Indica 40:1–40
Seeling J, Colin JP, Fauth G (2004) Global Campanian (Upper Cretaceous) ostracod palaeobiogeography. Palaeogeog Palaeoclimat Palaeoecol 213(3–4):379–398
Sharma R, Khosla A (2009) Early Palaeocene Ostracoda from the Cretaceous-Tertiary (K-T) Deccan intertrappean sequence at Jhilmili, district Chhindwara, Central India. J Palaeontol Soc Ind 54(2):197–208
Sharma R, Bajpai S, Singh MP (2008) Freshwater Ostracoda from the Paleocene-age Deccan intertrappean beds of Lalitpur (Uttar Pradesh), India. J Paleontol Soc Ind 53(2):81–87

Shome S, Chandel RS (2013) Palaeontological studies of Papro formation (intertrappean) of Lalitpur District, Uttar Pradesh – its age, correlation and palaeoecology. Indian J Geosci 65:49–62

Singh P (1997) Ostracods from the subsurface Khuiala Formation (lower Eocene) of Manhera Tibba well-1, Jaisalmer, Rajasthan, India. Geosci J18:149–233

Sowerby J d C (1840) Explanations of the plates and wood-cuts. In: Malcolmson JC (ed), On the fossils of the eastern portion of the Great Basaltic District of India. Trans Geol Soc London Ser 2(5):511–567

Srinivasan S (1991) Geology and micropalaeontology of Deccan Trap associated sediments of northern Karnataka, peninsular India. Unpublished PhD Thesis, Panjab University, Chandigarh

Srinivasan S, Bajpai S, Sahni A (1994) Charophytes from Deccan intertrappean beds of peninsular India: implications for age and correlation of Deccan volcanics. Geobios 27:559–571

Srinivasan S, Sahni A, Bajpai S (1992) Fossil Charophyta from the Deccan intertrappean beds of Gurmatkal, Gulbarga District, Karnataka. Curr Sci 63:396–398

Srivastava AK, Mankar RS (2015) Lithofacies architecture and depositional environment of Late Cretaceous Lameta Formation, Central India. Arab J Geosci 8(1):207–226

Szczechura J (1978) Fresh-water ostracodes from the Nemegt Formation (Upper Cretaceous) of Mongolia. Palaeontol Pol 38:65–21

Tambareau YP (1984) Les ostracodes du "Montien Continental" de Hainin, Hainaut, Belgique. Rev de Micropalaeontol 27(2):144–156

Tambareau Y, Gruas-Cavagnetto C, Feist M, Villatte J (1991) Flores et faunes continentals ilerdiennes du versant sud de la Montagne Noire et de la Montagne d'Alaric. Rev Micropaleontol 34:69–89

Thakre D, Samant B, Mohabey DM, Sangode S, Srivastava P, Kapgate DK, Mahajan R, Upretii N, Manchester SR (2017) A new insight into age and environments of intertrappean beds of Mohgaon-Kalan, Chhindwara District, M.P. using palynology, megaflora, magnetostratigraphy and clay mineralogy. Curr Sci 112(11):2193–2197

Udhoji SG, Mohabey DM (1996) In: Pandey J, Azmi RJ, Bhandari A, Dave A (eds) Ostracoda and Charophyta from the Late Cretaceous Lameta formation of Maharashtra: paleobiogeographic and age implication. Contributions to XV Indian Colloquium on Micropalaeontology and Stratigraphy, Dehradun, pp 409–418

Valdiya KS (2016) The making of India: geodynamic evolution. Springer, Cham

Verma O (2015) Cretaceous vertebrate fauna of the Cauvery Basin, southern India: palaeodiversity and palaeobiogeographic implications. Palaeogeog Palaeoclimat Palaeoecol 431:53–67

Verma O, Khosla A (2018) Application of internet technology in assembling literature for palaeontological research. Iran J Sci Technol Trans A Sci 4:1715–1723

Verma O, Khosla A (2019) Developments in the stratigraphy of the Deccan Volcanic Province, peninsular India. Compt Rendus Geosci 351:461–476

Verma O, Prasad GVR, Khosla A, Parmar V (2012) Late Cretaceous Gondwanatherian mammals of India: distribution, interrelationships and biogeographic implications. J Paleontol Soc Ind 57:95–104

Verma O, Khosla A, Goin FJ, Kaur J (2016) Historical biogeography of the Late Cretaceous vertebrates of India: comparison of geophysical and paleontological data. In: Khosla A, Lucas SG (eds) Cretaceous period: biotic diversity and biogeography, vol 71. New Mexico Museum of Natural History and Science, Albuquerque, pp 317–330

Verma O, Khosla A, Kaur J, Prasanth M (2017) Myliobatid and pycnodont fish from the Late Cretaceous of Central India and their paleobiogeographic implications. Hist Biol 29(2):253–265

Verma O, Prashanth M, Greco R, Khosla A, Singh K (2022) Geological education scenario in India and role of open educational resources in the light of COVID-19 pandemic. Earth Sci Res J 26(2):239–254

Whatley RC (1992) The reproductive and dispersal strategies of Cretaceous non-marine Ostracoda: the key to pandemism. In: Mateer NJ, Chen PJ (eds) Aspects of non-marine Cretaceous Geology. China Ocean Press, Beijing, pp 177–192

Whatley RC (2012) The 'out of India' hypothesis: further supporting evidence from the extensive endemism of Maastrichtian non-marine Ostracoda from the Deccan volcanic region of peninsular India. Rev de Paleobiol 11:229–248

Whatley RC, Bajpai S (2000a) A new fauna of Late Cretaceous non-marine Ostracoda from the Deccan intertrappean beds of Lakshmipur, Kachchh (Kutch District), Gujarat, western India. Rev Esp de Micropaleontol 32(3):385–409

Whatley RC, Bajpai S (2000b) Further nonmarine Ostracoda from the Late Cretaceous intertrappean deposits of the Anjar region, Kachchh, Gujarat, India. Rev Micropaleontol 43(1):173–178

Whatley RC, Bajpai S (2000c) Zoogeographical relationships of the Upper Cretaceous Nonmarine Ostracoda of India. Curr Sci 79(6):694–696

Whatley RC, Bajpai S (2005) Some aspects of the paleoecology and distribution of non-marine ostracoda from Upper Cretaceous intertrappean deposits and the Lameta Formation of peninsular India. J Paleontol Soc Ind 50(2):61–76

Whatley RC, Bajpai S (2006) Extensive endemism among the Maastrichtian nonmarine Ostracoda of India with implications for palaeobiogeography and "out of India" dispersal. Rev Esp de Micropaleontol 38(2–3):229–244

Whatley RC, Bajpai S, Srinivasan S (2002a) Upper Cretaceous nonmarine Ostracoda from intertrappean horizons in Gulbarga district, Karnataka state, South India. Rev Esp de Micropaleontol 34(2):163–186

Whatley RC, Bajpai S, Srinivasan S (2002b) Upper Cretaceous intertrappean nonmarine Ostracoda from Mohgaonkala (Mohgaon-Kalan), Chhindwara District, Madhya Pradesh state, Central India. J Micropaleontol 21:105–114

Whatley RC, Bajpa S, Whittaker JE (2002c) New record and new species of Upper Cretaceous Ostracoda for Indian intertrappean deposits. Boll Soc Paleontol Ital 41(2–3):163–173

Whatley RC, Bajpai S, Whittaker J (2003a) Freshwater Ostracoda from the Upper Cretaceous intertrappean beds at Mamoni (Kota district), southeastern Rajasthan, India. Rev Esp de Micropaleontol 35(1):75–86

Whatley RC, Bajpai S, Whittaker JE (2003b) Indian intertrappean Ostracoda in the collections of the Natural History Museum, London. Cretac Res 24(1):73–88

Whatley RC, Bajpai S, Whittaker JE (2003c) The identity of the non-marine ostracod *Cypris subglobosa* Sowerby from the intertrappean deposits of Peninsular India. Palaeontology 46(6):1281–1296

Whatley RC, Khosla SC, Rathore AS (2012) *Periosocypris megistus* n. gen. and n. sp.: a new gigantic non-marine cyprid ostracod from the Maastrichtian Lameta formation of India. J Palaeontol Soc Ind 57(2):113–117

Woodward AS (1908) On some fish remains from the Lameta Beds at Dongargaon, Central Province. Mem Geol Sur India, Palaeontol Indica NS 3:1–6

Chapter 3
Geology and Stratigraphy of Microbiota-Bearing Intertrappean Beds of the Chhindwara District, Madhya Pradesh, India

3.1 Introduction

The Cretaceous-Palaeocene system has a special status in the geology of India. During the Jurassic, the Indian plate separated from eastern Gondwana. Subsequently, in response to the geotectonic displacement of the plate during the Cretaceous to Palaeocene interval, it witnessed intense volcanism and plutonism as well as basinal and regional geotectonic events that led to the development of distinctive rock units, ranging from marine to terrestrial sedimentary and igneous to metamorphic rocks (Acharyya and Lahiri 1991; Vaidyanadhan and Ramakrishnan 2010). The Late Cretaceous-Early Palaeocene is regarded as a period of major marine transgressions that resulted in marine sedimentation within the plate's continental areas (Naqvi 2005). During the Early Cretaceous and the Upper Cretaceous-Early Palaeocene, the Indian plate crossed the Kerguelen and Réunion hotspots, resulting in the formation of the Rajmahal and Deccan Traps in eastern and central western India, respectively. In contrast, an initial subduction of the Indian plate beneath the Asian plate, most likely in the Late Cretaceous, produced the Dras volcanics on the plate's north-western margin (Vaidyanadhan and Ramakrishnan 2010). The eruptions of the Deccan Traps ended Cretaceous sedimentation in India.

The Upper Mesozoic-Lower Cenozoic stratigraphy of India has continuously been updated. The earliest classic work on the Cretaceous geology of India was nicely documented by Pascoe (1959) and Krishnan (1968) and more recent work by Acharyya and Lahiri (1991), Sahni and Khosla (1994), Naqvi (2005) and Vaidyanadhan and Ramakrishnan (2010). The Cretaceous rock systems in India are well developed both in the Himalayan and the peninsular regions of the country (Fig. 3.1). The Himalayan region mostly contains marine sediments in the Tethyan belt extending from the north-west to the north-east Himalayas (Acharyya and Lahiri 1991; Sahni and Khosla 1994). The Cretaceous deposits are represented in

A. Khosla et al., *Microbiota from the Late Cretaceous-Early Palaeocene Boundary Transition in the Deccan Intertrappean Beds of Central India*, Topics in Geobiology 54, https://doi.org/10.1007/978-3-031-28855-5_3

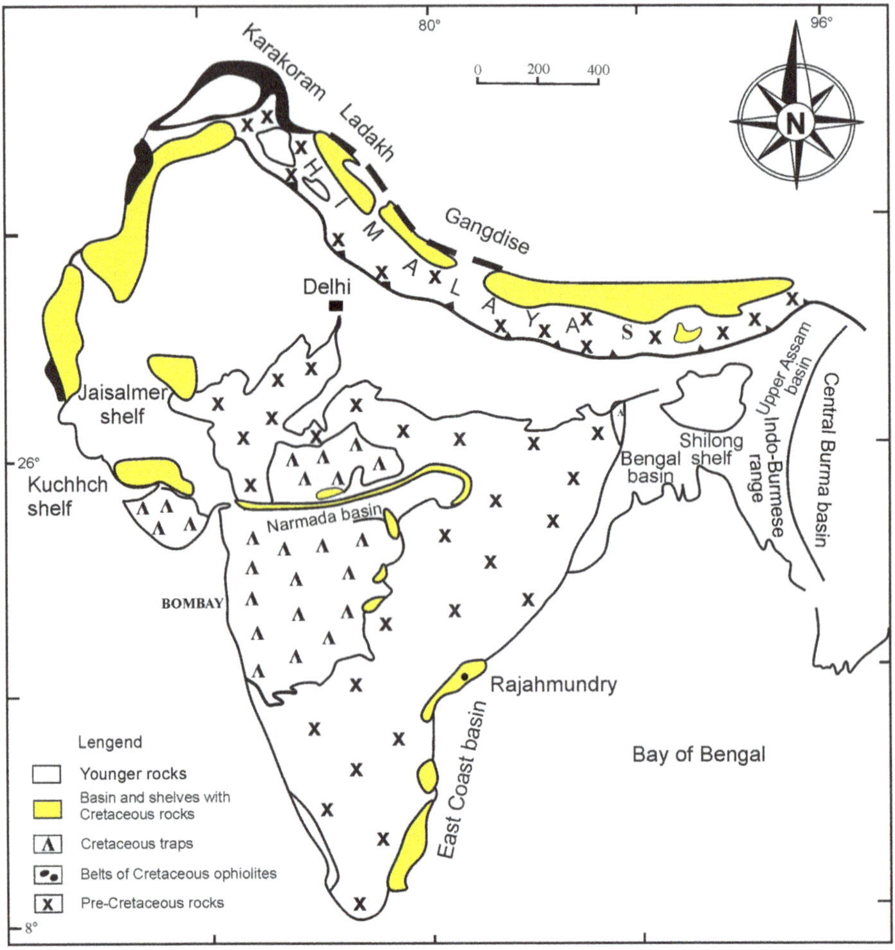

Fig. 3.1 Map of India showing Cretaceous successions. (Modified after Acharyya and Lahiri 1991)

Ladakh by the Khalsi and Dras volcanic formations of the Sangeluma Group and the Hundri Formation of the Shyok-Nubra belt. The Giumal and Chikkim formations are found in Himachal Pradesh between the Upper Jurassic Spiti Shale Formation and the Lower Palaeogene Kangi La Formation. The Cretaceous succession in Uttarakhand is represented by the Spiti Shale, Giumal Sandstone and Jhangu Formation of the Sancha Malla Group. The Mahadek Formation in the north-eastern Himalayas represents the Upper Cretaceous sediments in Assam and Meghalaya (Vaidyanadhan and Ramakrishnan 2010).

Peninsular India's Cretaceous successions are present in a variety of basins. The shallow marine Cretaceous (Albian to Maastrichtian) sequence is well preserved in the pericratonic Cauvery basin of Tamil Nadu, southern India. In ascending order, the sequence is classified into three groups based on lithology and fossil content:

Uttattur, Trichinopoly and Ariyalur (Blanford 1862; Sundaram et al. 2001). The Ariyalur Group's topmost Upper Maastrichtian Kallamedu Formation has yielded many remains of freshwater to terrestrial vertebrates such as fishes, crocodiles, dinosaurs, etc. (Verma et al. 2012a, 2017, 2022; Goswami et al. 2013; Prasad et al. 2010, 2013; Verma 2015). Lower Cretaceous marine deposits can be found in the pericratonic Kachchh basin, Gujarat, western India. The Cretaceous sequence begins here with the Umia Formation, which is located above the Middle Jurassic Katrol Formation (Biswas 1977, 1991). The strata in the pericratonic Saurashtra basin (Kathiawar peninsula), Gujarat, on the other hand, are composed of well-developed Lower Cretaceous fluvio-deltaic sediments known as the Dhrangadhra Group. The Dhrangadhra Group is divided into four formations – (in ascending order) Than, Surajdewal, Ranipat and Wadhwan – that lie directly on the Lower Proterozoic basement and beneath the Deccan basalts (Casshyap and Aslam 1992). Rajasthan's intracratonic Jaisalmer and Barmer rift basins, which formed adjacent to the Kachchh basin, contain well-preserved Lower Cretaceous deposits. Because of their proximity, the Cretaceous successions of the Kachchh, Jaisalmer and Barmer basins share striking similarities in lithological and faunal content. The Cretaceous succession of the Jaisalmer basin is divided into the Pariwar and Habur Formations. The Neocomian Pariwar Formation lies above the Upper Jurassic Bhadesar Formation, and the Aptian Habur Formation is overlain by the Palaeocene Sanu Formation (Vaidyanadhan and Ramakrishnan 2010). In the Barmer basin, Cretaceous deposits comprise the Fatehgarh Formation, which lies above the Jurassic Jaisalmer Formation and below the Akli Formation of Palaeocene age (Mathur et al. 2005).

Fluvio-marine Cretaceous successions can be found in the Narmada basin, which stretches from Barwaha in Madhya Pradesh to Rajpipla in Gujarat in the central and western peninsular region. These deposits, known as the Bagh and Lameta Formations, underlie the Deccan Traps and, in some places, overlie the Precambrian basement and the Vindhyan or Gondwana sediments. The Bagh Formation is a shallow marine succession that is divided into the Nimar Sandstone, Nodular Limestone, Deola-Chirakham marl and Coralline Limestone members and ranges in age from Cenomanian to Turonian (e.g. Roy Chowdhary and Sastry 1962; DasSarma and Sinha 1975; Singh and Srivastava 1981; Badve 1987; Joshi 1995; Khosla et al. 2003; Dhiman et al. 2021). The overlying Lameta Formation has an aerial extent of over 10,000 km^2 and is found in numerous isolated outcrops in Madhya Pradesh, Gujarat and Maharashtra. The Lameta Formation is famous for its diverse vertebrate fauna, particularly dinosaurs (Khosla and Sahni 2003; Khosla and Verma 2015; Khosla and Lucas 2020a, b, c, d, e). Numerous depositional environments have been inferred for the Lameta Formation, ranging from semi-arid, freshwater to marginal marine conditions with mixed tidal influence (e.g. Singh 1981; Brookfield and Sahni 1987; Tandon et al. 1990, 1995, 1998; Saha et al. 2010; Khosla and Verma 2015; Kapur and Khosla 2019; Khosla and Lucas 2020a, b, c, d, e; Khosla 2021; Kumari et al. 2021).

The Lower Cretaceous deposits also are present in the Upper Gondwana basins of peninsular and coastal regions of India. For example, the Ahmednagar

(Himmatnagar) sandstone near Ahmedabad, Gujarat; the Gangapur Formation of the Pranhita-Godavari basin, Telangana; the Golapalli, Raghavapuram and Tirupati formations in the Eluru area and the Budavada, Vemavaram and Pavalur formations in the Ongole area of the Krishna-Godavari basin, Andhra Pradesh; the Sriperumbudur and Satyavedu Formations of the Palar basin, Tamil Nadu; and the Athgarh Formation of the Mahanadi basin, Cuttack, Odisha, are Early Cretaceous in age (Acharyya and Lahiri 1991; Vaidyanadhan and Ramakrishnan 2010).

The Indian plate witnessed three continental flood basalt provinces during its northward journey: the Rajmahal, the Sylhet and the Deccan Traps. The Rajmahal-Sylhet basalt provinces are situated in eastern India, especially in the states of Jharkhand, West Bengal and Meghalaya, whereas the Deccan Traps are located in the central and western parts of India. It has been considered that the volcanic eruptions of the Rajmahal-Sylhet flood basalt provinces would have been associated with the early volcanic activity of the Kerguelen Plateau (Baksi 1995; Ray et al. 2005). These flood basalts are dated to be of Early Cretaceous (118–115 Ma) age. Interestingly, the Rajmahal Traps contain numerous intertrappean beds, but, to date, only a single fish skeleton has been reported from them (Misra and Saxena 1964; Tripathi et al. 2013).

3.2 Deccan Volcanic Province

The Deccan Volcanic Province is one of the world's large igneous provinces. It has been proposed that during the Cretaceous-Palaeocene drift, the Indian plate crossed over the Reunion mantle plume, resulting in the formation of the Deccan Volcanic Province, following the fragmentation of the Indian plate from the Gondwana landmasses (Morgan 1981; Cox 1983). It was a remarkable event in Indian geology, with numerous lava flows poured out primarily through fissure-type volcanic eruptions in a short span of time, covering a large area of peninsular India. These lava flows combined to form one of the world's largest igneous provinces, known as the Deccan Volcanic Province or Deccan Traps (Fig. 3.2). Because lava flows are primarily composed of basalts, they are frequently referred to as trap rocks. It is important to note, however, that the term "trap" is strictly used to refer to a step-like topographic feature and, thus, has no relation to any specific rock type (Rao 1969).

The Deccan Volcanic Province erupted numerous lavas flows over the Precambrian to Mesozoic rock surface of peninsular India. Precambrian rock units include the Dharwar, Aravalli, Bastar and Bundelkhand cratons, while Cretaceous rock units include the Bagh Formation, Lameta Formation and Dhrangadhra Group (Krishnan 1968; Naqvi 2005; Vaidyanadhan and Ramakrishnan 2010). The province is made up of several lava flows, with individual flow thickness ranging from a few metres to as much as 40 metres, and a total of 48 lava flows have been reported (West 1958). Most lava flows occur in the form of horizontal sheets, with each lava flow covering an area of about 1000 km^2. The province has a maximum thickness of about 2.5 km on the western side of the plateau, near Bombay in the Western Ghats, and is the thinnest on the eastern side. It occupies an area of about 500,000 km^2 in

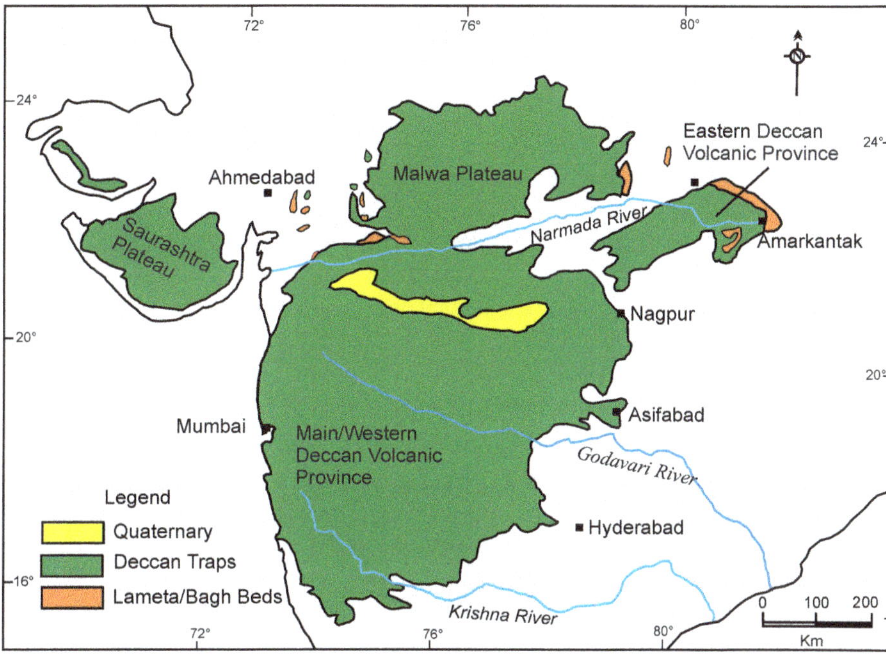

Fig. 3.2 Map showing the Deccan Volcanic Province in peninsular India

western and Central India, covering large portions of Maharashtra and Gujarat states and parts of Rajasthan, Madhya Pradesh, Andhra Pradesh, Telangana, Karnataka, Uttar Pradesh and Goa, Daman and Diu (Fig. 3.2).

The Deccan Volcanic Province is mostly made up of fine- to medium-grained, black to dark grey tholeiitic basalts. Other types of volcanic rocks found in the province include alkali-olivine basalts, rhyolites, trachytes, nepheline syenites, nephelinites, carbonatites, lamprophyres and picrites (Naqvi 2005; Vaidyanadhan and Ramakrishnan 2010). The province's lava flows are separated from one another by thin sedimentary and volcanic ash beds. It should be noted that volcanic ash beds associated with lava flows are a common feature of the traps. Intertrappean beds are sedimentary beds that are sandwiched between two successive lava flows, whereas infratrappean beds are also sedimentary beds that just lie below the first or the oldest lava flow. The intertrappean beds were formed during non-volcanic periods when drainage networks such as streams were blocked by lava flows and the streams created lakes and other water bodies (Khosla and Verma 2015). The intertrappean and infratrappean beds contain a variety of faunal and floral remains. The Deccan Volcanic Province is subdivided into four subprovinces (Ramakrishnan and Vaidyanadhan 2010; Fig. 3.2):

(a) *Main Deccan plateau*: It refers to the main Deccan Volcanic Province, which lies south of the Narmada River, comprising the western, central and south eastern regions of the Deccan Volcanic Province. It broadly occurs in Maharashtra state (Vaidyanadhan and Ramakrishnan 2010).

(b) *Malwa plateau*: It lies north of the Narmada River and occupies the areas around Indore, Bhopal and Sagar in Madhya Pradesh. The Malwa plateau is separated from the Western Deccan Volcanic Province by the Satpura hill (Verma and Khosla 2019).
(c) *Mandla lobe*: It is an isolated lava pile, situated in the eastern part of the main Deccan Volcanic Province in Central India. It covers areas around Chhindwara, Seoni and Jabalpur in Madhya Pradesh. This lava pile is present as an outlier near Mandla, so it is commonly known as the Mandla lobe or Eastern Deccan Volcanic Province (Verma and Khosla 2019).
(d) *Saurashtra plateau*: It is a square-shaped plateau located between the Khambhat graben in the east and the Son-Narmada fault in the south in Gujarat. This plateau also includes the curvilinear outcrops of the Kachchh rift (Verma and Khosla 2019).

Among these subprovinces, the main/Western Deccan Volcanic Province is one of the best studied subprovinces of the Deccan Volcanic Province. Historically, the main Deccan Volcanic Province was subdivided into three stratigraphic groups, the lower, middle and upper, with infratrappean beds (Lameta Formation) situated at the bottom (Pascoe 1964; Krishnan 1968; Table 3.1). According to this classification, the oldest lower traps covered the eastern and southern parts, the middle traps encompassed the central part, and the upper traps occupied the western parts of the Deccan Volcanic Province. However, this classification of the Deccan Volcanic Province is largely considered invalid because the oldest traps have now been recognised in the western part of the province (Radhakrishna and Vaidyanadhan 1994).

In 1968, the Geological Survey of India began systematic geological mapping of Western Maharashtra's Deccan Volcanic Province and proposed a new lithostratigraphic classification (Godbole et al. 1996). Based on the presence of megacryst horizons and the nature and type of lava flows, the Deccan Volcanic Province is divided into three subgroups and eight formations in this classification (Table 3.2). The North Sahyadri Group was named by the Geological Survey of India after the studied area in Western Maharashtra.

Table 3.1 Stratigraphy of the Deccan Volcanic Province

Traps	Distribution	Lithology
Nummulitics of Surat and Broach; Eocene of Kutch; laterite		
~~ Unconformity ~~~~~~~~~~~~~~~~~~~~~		
Upper traps (450 m thick)	Bombay and Saurashtra	Lava flows with numerous volcanic ash beds and fossiliferous sedimentary intertrappean beds
Middle traps (1200 m thick)	Malwa and Central India	Lava flows and ash beds forming the thickest part of the traps.
Lower traps (150 m thick)	Madhya Pradesh and Nagpur	Lava flows with a few ash beds and numerous fossiliferous intertrappean beds
~~ Slight unconformity ~~~~~~~~~~~~~~~~~~~~		
Lameta or infratrappean beds, Bagh beds and older rocks		

After Pascoe (1964) and Krishnan (1968)

Table 3.2 Lithostratigraphy of the Western Deccan Volcanic Province

Group	Subgroup	Formation		Characteristic feature	Thickness (m)
North Sahyadri Group	Wai	Mahabaleshwar		Simple and aa phyric flows	600
		Purandhargad		Simple and aa type flows	–
		Diveghat		Aphyric aa type flows	900
	Lonavala	Karla		Compound pahoehoe flows	–
		Indrayani		Simple flows of columnar jointed and aphyric types	700
		Ratangad	Upper	Compound flows of phyric type	–
			Lower		
	Kalsubai	Salher		Simple pahoehoe flows of phyric type	~1500

After Godbole et al. (1996) and Vaidyanadhan and Ramakrishnan (2010)

Table 3.3 Chemostratigraphy or flow stratigraphy of the Western Deccan Volcanic Province

Group	Subgroup	Formation	Magnetic chron
Deccan basalt	Wai (~500 m)	Desur	29N
		Panhala	29N
		Mahabaleshwar	29N
		Ambenali	29R
		Poladpur	29R
	Lonavala (525 m)	Bushe	29R
		Khandala	29R
	Kalsubai (2000 m)	Bhimashankar	29R
		Thakurvadi	29R
		Neral	29R
		Igatpuri	30N
		Jawhar	30N

Modified after Beane et al. (1986), Vaidyanadhan and Ramakrishnan (2010), Chenet et al. (2009) and Richards et al. (2015)

Beane et al. (1986) proposed a flow stratigraphy for the 500 km belt of Deccan basalts exposed along the Western Ghat. Based on the geochemical and isotopic characteristics of the lava flows, they divided the Western Deccan Volcanic Province into three subgroups, Kalsubai, Lonawala and Wai, and ten formations, where each formation is characterised by its own specific chemical composition (Verma and Khosla 2019; Table 3.3). Later, two youngest formations, the Panhala and Desur, were added into the classification (e.g. Subbarao and Hooper 1988; Peng et al. 1994; Bondre et al. 2006; Vaidyanadhan and Ramakrishnan 2010).

The Kalsubai Subgroup is the basal-most subgroup of the Western Deccan Volcanic Province and made up of five formations, the Jawhar, Igatpuri, Neral, Thakurvadi and Bhimashankar, from bottom to top. The subgroup is characterised by the predominance of amygdaloidal compound lava flows and has a high content of MgO with phenocrysts of olivine and clinopyroxene (Beane et al. 1986; Verma and Khosla 2018, 2019). The overlying Lonavala Subgroup consists of two formations, the Khandala and Bushe. The group is characterised by having simple individual lava flows and has high Ba/Ti and Ba/Sr values (Beane et al. 1986). The topmost Wai Subgroup lies above the Lonavala Subgroup and is made up of five formations, the Poladpur, Ambenali, Mahabaleshwar, Panhala and Desur, in ascending order. This subgroup is characterised by the presence of simple lava flows and minute phenocrysts of plagioclase, olivine and pyroxene (Beane et al. 1986). It has been proposed that the Cretaceous-Palaeogene boundary is within magnetic chron 29R, and it would have been near the onset of the volcanic eruption of the Wai Subgroup (Keller et al. 2012; Richards et al. 2015). It is important to mention that a chemostratigraphic framework is yet to be proposed for the Eastern Deccan Volcanic Province, so it is difficult to correlate the various flows of the Eastern Deccan Volcanic Province with the well-established chemostratigraphy of the Western Deccan Volcanic Province.

The majority of stratigraphic age data indicates that the Deccan volcanic eruptions took place from 69 to 64 million years ago in three phases (Chenet et al. 2007, 2008, 2009; Verma and Khosla 2018, 2019). The first phase is marked by the beginning of the eruption of lava flows of the Deccan Traps at 67.5 million years ago, followed by a quieter period of two million years. At the Cretaceous-Palaeogene boundary (66 million years ago), the second phase of volcanic eruptions took place. It is considered the main event, during which about 80% of the total lava flows of the Deccan Traps were erupted (Keller et al. 2008, 2009a, b, 2012; Punekar et al. 2014). The last and third phase began after the Cretaceous-Palaeogene boundary around 64 million years ago, and about 14% of lava flows of the total Deccan Traps were erupted. Thus, the major portion of the Deccan Traps in less than one million years during the second phase of volcanic eruption at the Cretaceous-Palaeogene boundary, and this phase is linked with the Cretaceous-Palaeogene mass extinctions (Keller et al. 2008, 2009a, b, 2012; Punekar et al. 2014). In addition, based on the fossils, especially foraminiferans, ostracods and plants discovered from the infratrappean and intertrappean beds, a Maastrichtian to Danian age has been proposed for the Deccan Traps (Khosla and Verma 2015; Verma and Khosla 2019).

3.2.1 Eastern Deccan Volcanic Province

The Eastern Deccan Volcanic Province is an isolated lava pile situated on the north eastern margin of the main Deccan Volcanic Province in Central India. The Eastern Deccan volcanism occurs as an outlier near the Mandla area; hence, it is also known as the Mandla lobe (Pattanayak and Shrivastava 1999). This Deccan lava pile is about 900 m thick and covers an area of around 29, 400 km^2 (Pathak et al. 2016). This lobe is about 344 km long in the east-to-west direction from Amarkantak to Chhindwara and about 156 km wide in the north-to-south direction from Kundam to Seoni and covers the Seoni, Chhindwara, Jabalpur, Dindori, Amarkantak and Mandla regions of Madhya Pradesh state (Fig. 3.3). The north-eastern part of the

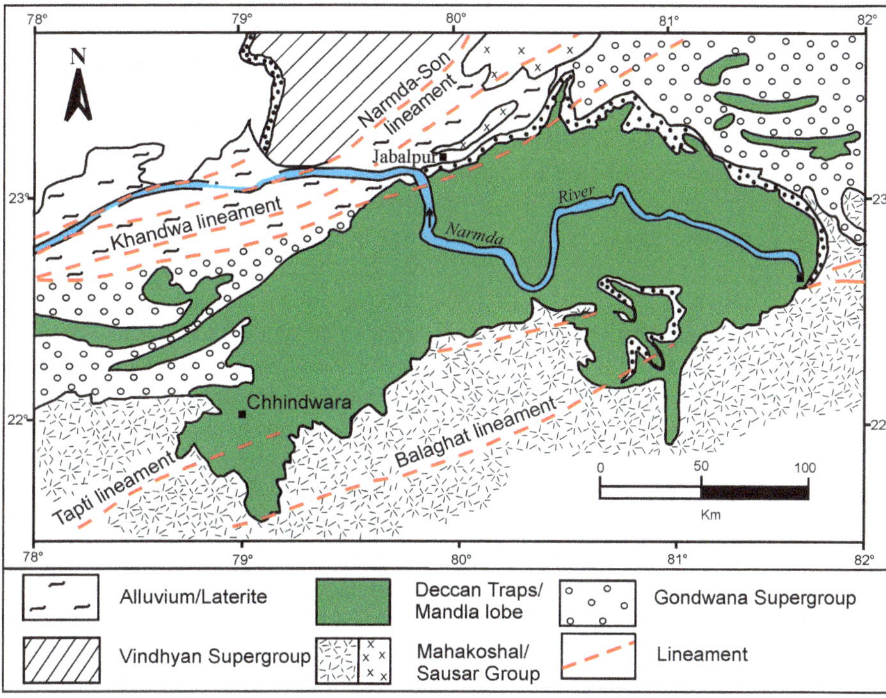

Fig. 3.3 Geological map of the Eastern Deccan Volcanic Province (Mandla lobe), Madhya Pradesh, India. (After Solanki et al. 1996)

lobe lies above the arc-shaped outcrops of the Lameta Formation that are present between Jabalpur to Amarkantak. In turn, the Lameta Formation lies above the sediments of the Gondwana Supergroup, which rest on the rocks of the Vindhyan Supergroup (Solanki et al. 1996). In the south and south-western parts, the base of the lobe directly lies above the Precambrian granitic gneiss basement from Amarkantak to Chhindwara. The rocks of the Gondwana Supergroup are present on the north-western extremity of the lobe from Chhindwara to Jabalpur (Solanki et al. 1996; Kania et al. 2022). Based on order of superposition of the lava flows and geographical distribution of the Mandla lobe, it has been observed that the Seoni-Jabalpur-Shahpura sector in the west contains the oldest flows, whereas the youngest flows are present in the Dindori-Amarkantak sector in the east (e.g. Pathak et al. 2016). A total of 37 lava flows have been recorded in the Eastern Deccan Volcanic Province (Pattanayak and Shrivastava 1999; Pathak et al. 2016).

It is worth noting that the Mandla lobe's presence within the Narmada-Tapti rift system and at the Cretaceous-Palaeogene transition makes its study very important from a chemostratigraphic, magnetostratigraphic, palaeomagnetic and palaeontologic standpoint (Deshmukh et al. 1996; Khosla et al. 2004, 2009, 2022; Prasad et al. 2007a, b; Keller et al. 2009a, b; Pathak et al. 2016; Kania et al. 2022). The deep-seated Narmada-Son and Tapti lineaments, respectively, bounded the northern and southern margins of the Mandla lobe. The Khandwa lineament, which is thought to be a southern extension of the Narmada-Son lineament, bounds the northern and southern margins of the lobe, and it is bounded by a major lineament that is most likely associated with the Tapti tectonic zone (Solanki et al. 1996; Fig. 3.3). These tectonic zones were most likely formed during the Proterozoic, and their reactivation during the Late Cretaceous resulted in the upwelling of the lobe's basaltic magma (e.g. Solanki et al. 1996). Kaila (1988) conducted deep seismic sounding investigations on the Deccan Volcanic Province and concluded that the thickness of the province gradually decreases from 1500 m along India's western coast to around 100 m near the village Khajuria Kalan near Bhopal in Madhya Pradesh. Furthermore, Kaila (1988) noted that the province's thickness increases eastward and reaches 900 m in the form of an isolated lava pile, the Mandla lobe, after a small gap. The main conclusion from deep seismic sounding investigations is that the Mandla lobe is an entity separate from the main Deccan Volcanic Province and has a different origin (Kaila 1988). Other workers have also proposed a similar interpretation (e.g. Pattanayak and Shrivastava 1999; Shrivastava and Ahmad 2005; Pathak et al. 2016). The lava flows of the Mandla lobe are less understood in terms of petrology, chemostratigraphy, magnetostratigraphy and palaeomagnetism than the main Deccan Volcanic Province, and the data currently available are insufficient to test the above hypothesis.

Roychowdhary et al. (1968) provided a general stratigraphic framework of the Eastern Deccan Volcanic Province (Table 3.4). Subsequently, Deshmukh et al. (1996), Nair et al. (1996), Pattanayak and Shrivastava (1996, 1999), Solanki et al. (1996) and Yedekar et al. (1996) proposed a lava flow stratigraphy of the Mandla lobe. Pattanayak and Shrivastava (1996) carried out the first detailed investigations on the stratigraphy of the 900-m-thick lava pile of the lobe. Following that, Pattanayak and Shrivastava (1999) identified the presence of 37 lava flows in this lava sequence. Desmukh et al. (1996) proposed a chemostratigraphic framework of

Table 3.4 Generalised stratigraphy of the Eastern Deccan Volcanic Province

Formation/supergroup	Lithology	Age	Thickness
Laterite	Laterite with associated bauxite and clay	Upper Tertiary to Quaternary	20–50 m
Deccan Trap	Basaltic flows, dykes and intertrappean beds	Upper Cretaceous to Eocene	200–300 m
Lameta	Cherty limestone, sandstone and clay	Upper Cretaceous	30 m
~~ Unconformity ~~~~~~~~~~~~~~~~~~~~~~			
Gondwana	Sandstone, shale and coal seams	Upper Carboniferous to Permian	900 m
~~ Unconformity ~~~~~~~~~~~~~~~~~~~~~~			
Archaean schist, gneisses with granite intrusion			

After Roychowdhary et al. (1968)

the lava flows that occur in the Eastern Deccan Volcanic Province based on geological and chemical studies of the lava flows. Furthermore, the presence of various formations of the Wai Subgroup of the Western Deccan Volcanic Province in the Mandla lobe was recognised by these authors. Solanki et al. (1996) classified a 500-m-thick lava pile surrounding the Mandla lobe as the Mandla, Shahpura, Pipardahi and Rai formations based on lithofacies, flow type and long-distance continuity of flows. However, these authors had made no attempt to correlate these formations with those of the main Deccan Volcanic Province.

Yedekar et al. (1996) investigated three sections of the traps exposed in the Chhindwara, Jabalpur-Seoni and Jabalpur-Piparia areas of the Mandla lobe. They classified the lava sequences of the Chhindwara, Jabalpur, Mandla and Seoni areas into four informal formations based on variations in incompatible element ratios and compared and correlated these informal formations with the various formations of the Western Deccan Volcanic Province. These authors concluded that it would be difficult to make a comprehensive correlation between the formations of the Eastern and Western Deccan Volcanic Provinces based on the chemical signature of the flows. Keller et al. (2009a, b) discovered a formation associated with the Wai Subgroup of the Western Deccan Volcanic Province in the Mandla lobe exposed in the Jhilmili area of Chhindwara. Pattanayak and Shrivastava (1999) and Pathak et al. (2016) investigated the number of lava flows in the Mandla lobe using petrography, major-oxide chemistry, palaeomagnetism and magnetostratigraphy and concluded that these lava flows are compositionally similar to the Bushe, Poladpur, Ambenali and Mahabaleshwar formations of the Western Deccan Volcanic Province.

Importantly, numerous fossiliferous and non-fossiliferous intertrappean beds can be found in the Mandla lobe's lava flows (e.g. Bande et al. 1988; Solanki et al. 1996; Khosla et al. 2004, 2009; Keller et al. 2009a; Verma et al. 2012b; Khosla 2015; Rage et al. 2020; Prasad et al. 2021; Kania et al. 2022; Khosla et al. 2022). The Mandla lobe of the Chhindwara area contains a good number of intertrappean beds exposed at Paladaun, Mohgaon-Kalan and Jhilmili. In addition, numerous intertrappean sequences are exposed in wells (Fig. 3.4). These intertrappean beds range in thickness from 2 to 14 m and have a limited vertical extent, but some of them can be traced for 2 to 3 km horizontally. They are composed of clayey limestones, marls, siltstones, claystones, mudstones and cherts. These fossiliferous intertrappeans are very important from a palaeontological point of view because they contain important biotic remains that can help us to understand biodiversity patterns during the Deccan volcanic event and the role of the Indian plate as a geographic centre for the origin, evolution and dispersal of biota during northward drift of the plate and to document environmental consequences of the volcanic eruptions on the contemporaneous biotas (e.g. Khosla and Sahni 2003; Khosla et al. 2004; Keller et al. 2009a, b; Pal et al. 2013; Khosla 2015; Khosla and Verma 2015; Verma et al. 2016; Kapur and Khosla 2019; Khosla and Lucas 2020a; Wilson et al. 2022; Khosla et al. 2022).

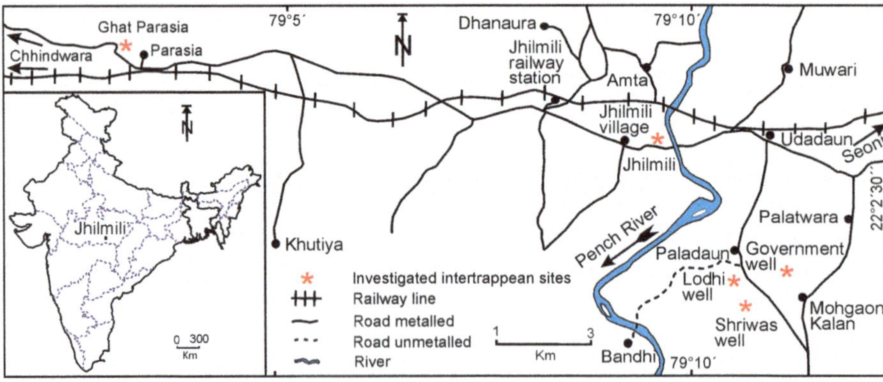

Fig. 3.4 Map showing investigated intertrappean sites (marked by stars) in the Chhindwara District, Madhya Pradesh. Inset is the map of India showing location of the Jhilmili village and other intertrappean sites, Chhindwara District, Madhya Pradesh

3.3 Geology of the Investigated Intertrappeans of Chhindwara Region

Lava flows of the Mandla lobe exposed in the Chhindwara District of Madhya Pradesh, Central India, contain fossiliferous intertrappean beds (Keller et al. 2009a, b; Sharma and Khosla 2009; Khosla 2015; Khosla and Verma 2015; Kania et al. 2022; Khosla et al. 2022). For the recovery of microfossils, extensive palaeontological field investigations were conducted on the intertrappean beds exposed on the surface at Jhilmili (22° 02′ 44″:79° 09′ 22″), Ghat Parasia (22° 03′ 53.74″: 79° 02′ 45.57″) and in two wells, at Government well (22° 01′ 06.61″: 79° 11′ 10.69″) and Shriwas (=Shiraj) well (22° 01′ 17.8″: 79° 11′ 06.3″) (Fig. 3.4). These intertrappean beds have yielded the remains of charophytes, ostracods, gastropods, foraminiferans, fishes and trace fossils.

3.3.1 Geology of Jhilmili Intertrappean Site

The Jhilmili (22° 02′ 44" N; 79° 09′ 22″ E) intertrappean section is exposed on a small hill and situated along the right bank of the Pench River near the village Jhilmili on the left side of the Chhindwara-Seoni road (Figs. 3.4 and 3.5). This section is located nearly 33 km to the east of Chhindwara and 5 km to the north-west of the famous plant-bearing locality Mohgaon-Kalan. It is an approximately 14-m-thick succession that consists of claystones, palaeosols, siltstones, marlstones and limestones and has been divided into six lithological units, including two units of lava flows (Figs. 3.5 and 3.6). Unit 1 represents the lowest weathered lava flows of the Deccan basalts. The overlying unit 2 is 6 m thick and consists of

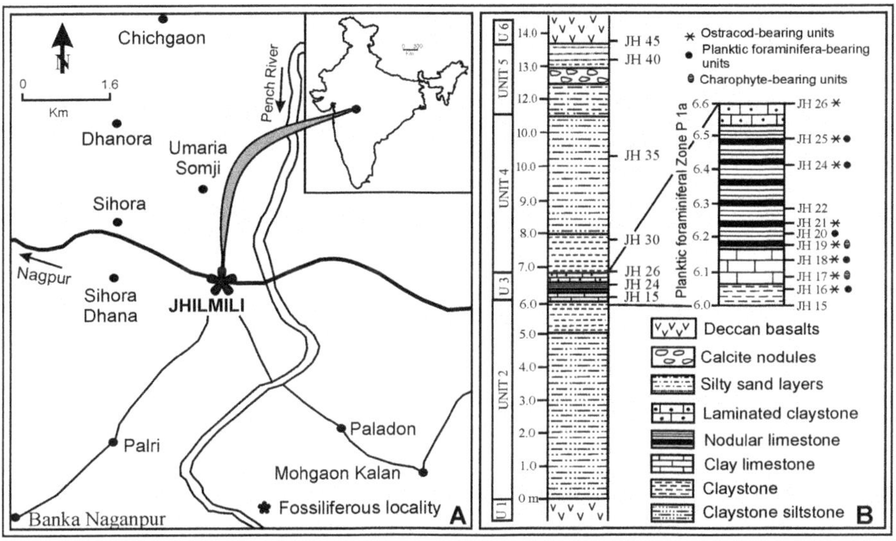

Fig. 3.5 Jhilmili intertrappean beds. (**A**) Map showing the location of the Jhilmili intertrappean section, Chhindwara District, Madhya Pradesh, India (modified after Sahni and Rode 1937; Khosla 2015), and (**B**) lithostratigraphic succession of the intertrappean beds at Jhilmili showing the microfossil-yielding levels. (Modified after Khosla 2015)

coarse-grained sandstone, purple siltstone, red-clayey siltstone and red claystone. The palaeosols are dominant at the bottom of the unit. It also contains substantial amounts of carbonate nodules, slickensides, pseudomorphic calcite and manganese streaks (Keller et al. 2009a, b; Pal et al. 2013). A palustrine to floodplain environment has been inferred for this unit (Keller et al. 2009a, b; Khosla 2015; Khosla et al. 2022). Unit 3 is 60 cm thick. Its basal part (18 cm thick) consists of yellow and pink claystone and yellow clayey limestone. Following this, there is a 30-cm-thick layer of nodular limestone. The topmost part of the unit consists of strongly laminated claystone. Geologically, unit 3 is very important because it has yielded diverse assemblages of planktic foraminiferans, freshwater and brackish water ostracods, charophytes and marine benthic calcareous chlorophytes that, together, mark the Cretaceous-Palaeogene transition (Keller et al. 2009a, b; Khosla 2015; Kundal et al. 2018; Khosla et al. 2022). Freshwater, lacustrine to brackish marine environments have been inferred for deposition of this unit based on the presence of the microfossils (Keller et al. 2009a, b; Khosla 2015; Kundal et al. 2018). The charophytes, ostracods and foraminiferans described here were recovered from this unit (Kania et al. 2022; Khosla et al. 2022).

Unit 4 consists of red clay and siltstone with fine sand layers yielding clasts of clay and carbonate. Unit 5 encompasses green-grey siltstone with rare layers of fine sand. A palustrine environment has been recognised for the deposition of both units 4 and 5 (Keller et al. 2009a, b; Khosla 2015). The topmost unit 6 consists of weathered flows of the Deccan traps.

Fig. 3.6 Field photographs of the Jhilmili section. (**A**) Panoramic view of the Jhilmili intertrappean beds displaying a thin interval (clayey and nodular limestone marked by arrow) of lacustrine to brackish marine unit 3 (60 cm thick), which has produced charophytes, planktic foraminiferans and ostracods. (**B**) Enlarged view of the Jhilmili intertrappean beds displaying a thin interval of clayey and nodular limestone of lacustrine to brackish marine unit 3 (60 cm thick). (**C**) Enlarged view of the fossiliferous white, light-yellow and pink-brownish coloured clays with hard clayey limestone bands yielding microfossils. Scale = 14 cm

The intertrappean beds of the Jhilmili section were deposited during a quiescent interval of the Deccan volcanic activity between two successive lava flows and fall close to the Cretaceous-Palaeogene boundary (Khosla 2015; Kania et al. 2022; Khosla et al. 2022). The lithostratigraphy, chemostratigraphy, magnetostratigraphy and biostratigraphy of the Jhilmili area have been established by Keller et al. (2009a, b), Sharma and Khosla (2009), Khosla et al. (2011) and Khosla (2015). The Jhilmili intertrappean beds have been interpreted as the deposits of rivers, lakes, ponds and a brackish river system. The richly fossiliferous unit 3 has been dated biostratigraphically as latest Cretaceous to Early Danian in age, largely based on the foraminiferal assemblages (Keller et al. 2009a, b; Sharma and Khosla 2009; Khosla 2015; Khosla et al. 2022). It may be noted that coarse sandstone is mainly exposed in a nearby area located adjacent to the classic Jhilmili section and has yielded isolated teeth of a gar, *Lepisosteus indicus*. Nearly 800 kg of samples were collected from the 60-cm-thick unit 3 (i.e. clay limestone) encompassing the Cretaceous-Palaeogene boundary. After processing and sorting of the samples at laboratory, a rich assemblage consisting of charophytes (*Platychara perlata* Peck and Reker 1947, *Platychara raoi* Bhatia and Mannikeri 1976, *Platychara sahnii* Bhatia and Mannikeri 1976, *Platychara compressa* Peck and Reker 1948, *Platychara* sp., *Peckichara cf. varians* Grambast 1957, *Nemegtichara* cf. *grambasti* Bhatia et al. 1990b, ?*Grambastichara* sp., *Microchara shivarudrappai* sp. nov. and *Chara chhindwaraensis* sp. nov.), ostracods (*Buntonia whittakerensis* sp. nov., *Neocyprideis raoi* Jain 1978, *Limnocythere deccanensis* Khosla et al. 2005, *Limnocythere martensi* sp. nov., *Frambocythere tumiensis anjarensis* Bhandari and Colin 1999, *Gomphocythere strangulata* Jones 1860, *Gomphocythere paucisulcatus* Whatley et al. 2002b, *Gomphocythere dasyderma* Whatley et al. 2002a, *Paracypretta subglobosa* Sowerby 1840, *Paracypretta jonesi* Bhatia and Rana 1984, *Paracypretta verruculosa* Whatley et al. 2002a, *Strandesia jhilmiliensis* Khosla et al. 2011a, *Stenocypris cylindrica* Sowerby in Malcolmson 1840, *Periosocypris megistus* Whatley et al. 2012, *Zonocypris spirula* Whatley and Bajpai 2000a, *Zonocypris viriensis* Khosla and Nagori 2005, *Zonocypris penchi* sp. nov., *Cypridopsis astralos* Whatley et al. 2002a, *Cypridopsis hyperectyphos* Whatley and Bajpai 2000a, *Eucypris pelasgicos* Whatley and Bajpai 2000a, *Cyclocypris amphibolos* Whatley et al. 2002a, *Cypria cyrtonidion* Whatley and Bajpai 2000a and *Talicypridea pavnaensis* Khosla et al. 2005) and foraminiferans (*Subbotina triloculinoides* Plummer 1926, *Globanomalina compressa* Plummer 1926, *Woodringina hornerstownensis* Olsson 1960, *Woodringina claytonensis* Loeblich and Tappan 1957b, *Hedbergella holmdelensis* Olsson 1964, *Guembelitria cretacea* Cushman 1933, *Parasubbotina pseudobulloides* Plummer 1926, *Globigerinelloides aspera* Ehrenberg 1854, *Globigerina* (*Eoglobigerina*) *pentagona* Morozova 1961) was recovered from unit 3 of the Jhilmili section.

It is worth mentioning that various studies such as biostratigraphy based on planktic foraminiferans, ostracods and marine benthic calcareous algae as well as chemostratigraphic and magnetostratigraphic methods were carried out to deduce the age of the Jhilmili intertrappean beds (Sharma and Khosla 2009; Keller et al. 2009a, b; Khosla 2015). These studies indicate that this intertrappean locality is of Late Cretaceous-Early Palaeocene age (Khosla 2015; Kania et al. 2022; Khosla et al. 2022). The charophyte flora (JH 17, 19 and 25), which consists of *Platychara perlata, Platychara raoi*, *Platychara sahnii*, *Platychara compressa*, *Platychara sp.*, *Peckichara cf. varians*, *Nemegtichara* cf. *grambasti*, ?*Grambastichara* sp., *Microchara shivarudrappai* sp. nov. and *Chara chhindwaraensis* sp. nov., indicates a Late Cretaceous to Early Palaeocene age. Together with charophytes, very small planktic foraminiferans (*Globoconusa daubjergensis*, *Eoglobigerina eobulloides*, *E. edita*, *Woodringina hornerstownensis*, *Parvularugoglobigerina extensa* and *P. eugubina*), ranging in size from 40 to 100 μm, were recovered from clayey limestone and a nodular limestone band (JH 16, 17, 19 and JH 25) of middle unit 3. Some larger planktic foraminiferans, which are more than 150 μm in diameter, were recovered using sieving methods and include *Praemurica taurica*, *Globigerina* (*Eoglobigerina*) *pentagona*, *Subbotina triloculinoides* and *Parasubbotina pseudobulloides*, further suggesting an Early Danian age for the Jhilmili intertrappean beds (Keller et al. 2009a, b; Khosla 2015 and this study).

Apart from the Early Danian assemblage, planktic foraminiferans, mostly of Cretaceous age, have also been recorded, for instance, *Guembelitria cretacea*, *Hedbergella* cf. *holmdelensis* and *Globigerinelloides aspera*. The stratigraphic ranges of these characteristically Cretaceous species extend into the Danian, so they are commonly known as Cretaceous-Palaeogene survivors (Keller et al. 2002, 2009a, b; Khosla 2015). Sharma and Khosla (2009) and Keller et al. (2009a, b) assigned an Early Danian P0–P1a (1) age to the lower part of the 6-m-thick palaeosols (red clayey siltstone), while the upper part of the intertrappeans is also 6 m thick and lithologically consists of red shales (palaeosols) and has been assigned to the upper part of zone P1a (2). The presence of *Parvularugoglobigerina eugubina* indicates that the deposition of the foraminiferan-bearing horizon took place around 100–150 ka later than the Cretaceous-Palaeogene boundary (Keller et al. 2009a, b).

The intertrappean beds of Jhilmili also contain ostracod assemblages of Maastrichtian to Early Palaeocene age. The freshwater ostracods of Late Cretaceous age are in the basal part of unit 3 (JH 17–21): *Limnocythere deccanensis*, *Cypria cyrtonidion*, *Paracypretta verruculosa*, *Paracypretta jonesi*, *Zonocypris viriensis*, *Zonocypris* sp., *Darwinula torpedo* and *Cypridopsis hyperectyphos*. The upper part of unit 3 (JH 25–26) also contains some freshwater ostracods such as *Zonocypris spirula*, *Eucypris pelasgicos*, *Strandesia jhilmiliensis*, *Heterocypris* sp., *Frambocythere tumiensis anjarensis*, *Stenocypris cylindrica*, *Gomphocythere strangulata*, *Gomphocythere paucisulcatus*, *Centrocypris megalopos* and *Cyclocypris amphibolos* (Sharma and Khosla 2009; Khosla et al. 2011; Khosla 2015). Apart from the freshwater assemblage, two brackish water ostracods (JH 17–21, *Neocyprideis raoi* and *Buntonia whittakerensis* sp. nov.) are abundant and can be used to identify the beginning of the Palaeocene and, further, indicate an Early Danian age. Overall, the recovered microbiota comprising ostracods and planktic foraminiferans suggests a Late Cretaceous to Early Palaeocene age.

Furthermore, the Jhilmili intertrappean beds correspond to the shallow marine intertrappean beds exposed in the Rajahmundry quarries, which have also produced

Early Danian planktic foraminiferal assemblages between the lower and upper basalt traps of the C29R and C29N magnetic polarity zones (Keller et al. 2008; Khosla 2015). Keller et al. (2009a, b) reported magnetostratigraphic data for Jhilmili that show that the Cretaceous-Palaeogene boundary is bordered by lower traps in C29R and upper traps in C29R/C29N, which correspond to the upper Ambenali and lower Mahabaleshwar formations, respectively (Jay and Widdowson 2008; Keller et al. 2008). As a result, the lower volcanic flows at Jhilmili placed the Cretaceous-Palaeogene boundary near the end of the main Deccan volcanic eruptions, as evidenced by Early Danian planktic foraminiferal assemblages overlying the last part of the lower traps (Keller et al. 2009a; Khosla 2015).

3.3.2 Geology of Intertrappean Beds in the Government Well

The Government well (22° 01′ 06.61″N: 79° 11′ 10.69″E) is located approximately 0.4 km north-east of Mohgaon-Kalan village and approximately 34 km east of Chhindwara town in Madhya Pradesh, Central India (Fig. 3.4). This well section is rich in fossiliferous material, about 100 cm thick, and is located between two lava flows (Figs. 3.7 and 3.8A). In ascending order, it is composed of calcareous shale, black laminated calcareous shale and green clay with an intercalation of calcareous shale. The most basal unit is carbonaceous calcareous shale, which is about 22 cm thick and has yielded the remains of gastropods, plants and palynomorphs. It is covered by a 38-cm-thick unit of black laminated carbonaceous calcareous shale with gastropods and wood fragments. The topmost unit is 40 cm thick and is made up of green clay intercalated with calcareous shale (Fig. 3.8A).

Nearly 450 kg (150 kg from each unit) of samples were taken from these intertrappean beds. Ostracods (*Frambocythere tumiensis lakshmiae* Whatley and Bajpai 2000a, *Gomphocythere strangulata* Jones 1860, *Gomphocythere paucisulcatus* Whatley et al. 2002b, *Gomphocythere* sp. 1, *Cypridopsis elachistos* Whatley et al.

Fig. 3.7 Field photographs of the Government well, Chhindwara District, Madhya Pradesh, Central India. (**A**) Side view of the well and (**B**) Internal view of the well

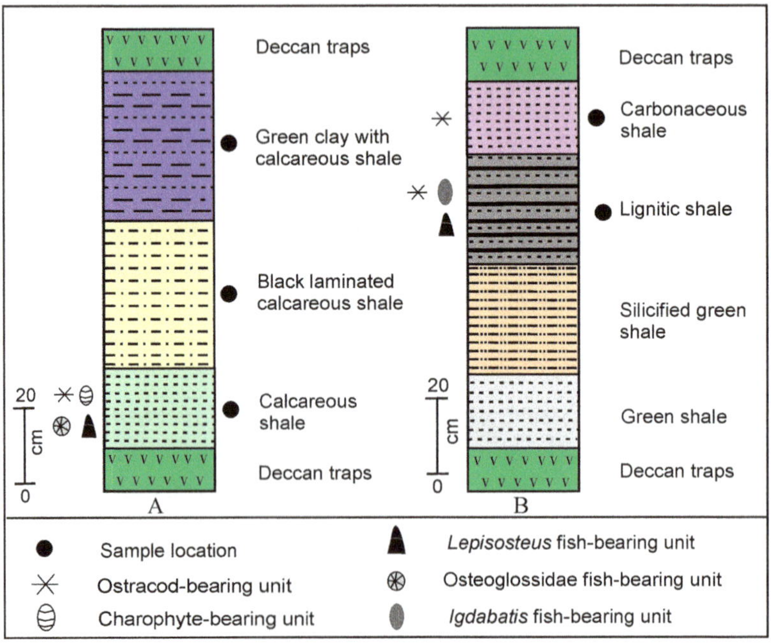

Fig. 3.8 Lithologic units of the investigated well sections. (**A**) Government well and (**B**) Shriwas well, Chhindwara District, Madhya Pradesh

2002b,?*Eucypris verruculosa* Whatley et al. 2002a and *Cyprois rostellum* Whatley and Bajpai 2000a), gastropods (*Lymnaea* sp.) and fishes (*Lepisosteus indicus* Woodward 1908 and Osteoglossidae genus and species indeterminate) were recovered from the calcareous shale.

3.3.3 Geology of Intertrappean Beds in the Shriwas (=Shiraj) Well

The Shriwas well (22° 01′ 17.8″N: 79° 11′ 06.3″E) is situated about 0.5 km southwest of Mohgaon-Kalan village and nearly 34 km east of Chhindwara town, Madhya Pradesh, Central India (Figs. 3.8B and 3.9A, B). The intertrappean beds of this section are around 1 m thick and sandwiched between two basaltic lava flows. Lithologically, the Shriwas well section consists of green shale, silicified green shale, lignitic shale and carbonaceous shale. The green shale forms the basal-most unit, which is about 20 cm thick. It is overlain by a 30-cm-thick band of gastropod-bearing silicified green shale. Following this is a 30-cm-thick layer of lignitic shale containing remains of molluscs, plants and palynoflora (Kumaran et al. 1997). A 20-cm-thick layer of carbonaceous shale having traces of pyrite lies at the top of the section. Based on unidentified egg shells of avian dinosaurs together with

Fig. 3.9 Field photographs of the Shriwas (= Shiraj) well, Chhindwara District, Madhya Pradesh. (**A**) Partial top view of the well showing the Deccan Traps and (**B**) Internal view of the well

ostracods and palynoflora, a Late Maastrichtian age has been assigned to these intertrappeans (Srinivasan 1996; Kumaran et al. 1997; Kar and Srinivasan 1998; Thakre et al. 2017). About 100 kg of samples from two units, the lignitic shale and carbonaceous shale, were collected for micropalaeontological investigations. Ostracods (*Frambocythere tumiensis lakshmiae* Whatley and Bajpai 2000a, *Gomphocythere strangulata* Jones 1860, *Gomphocythere paucisulcatus* Whatley et al. 2002b, *Zonocypris labyrinthicos* Whatley et al. 2002b, *Zonocypris gujaratensis* Bhandari and Colin 1999, *Cypridopsis elachistos* Whatley et al. 2002b and *Cyprois rostellum* Whatley and Bajpai 2000a), gastropods (*Lymnaea* sp.) and fishes (*Igdabatis indicus* Prasad and Cappetta 1993 and *Lepisosteus indicus* Woodward 1908) from the lignitic shale and ostracods (*Zonocypris labyrinthicos* Whatley et al. 2002b and *Zonocypris gujaratensis* Bhandari and Colin 1999) from the carbonaceous shale were recovered.

3.3.4 Geology of Ghat Parasia Intertrappean Site

The Ghat Parasia (22° 03′ 53.74″N: 79° 02′ 45.57″E) intertrappean site is located approximately 7 km east of Chhindwara and is exposed on both sides of the Chhindwara-Seoni road. The thickness of the section is 170 cm. It consists of reddish chert, fossiliferous greenish limestone, greenish chertified clay, fossiliferous black chert and hard clayey limestone (Figs. 3.10 and 3.11). This section has produced a diverse fauna consisting of invertebrates and vertebrates and a flora that includes charophytes and vascular plant remains. Reddish chert, fossiliferous clayey limestone, greenish chertified clay, fossiliferous black chert and hard clayey limestone have all yielded ostracods, gastropods, charophytes and fish teeth and scales.

The recovered biota from the Ghat Parasia intertrappean beds is assigned to *Platychara closasi* sp. nov. (charophyte); *Limnocythere deccanensis* Khosla et al. 2005, *Frambocythere tumiensis anjarensis* Bhandari and Colin 1999, *Gomphocythere*

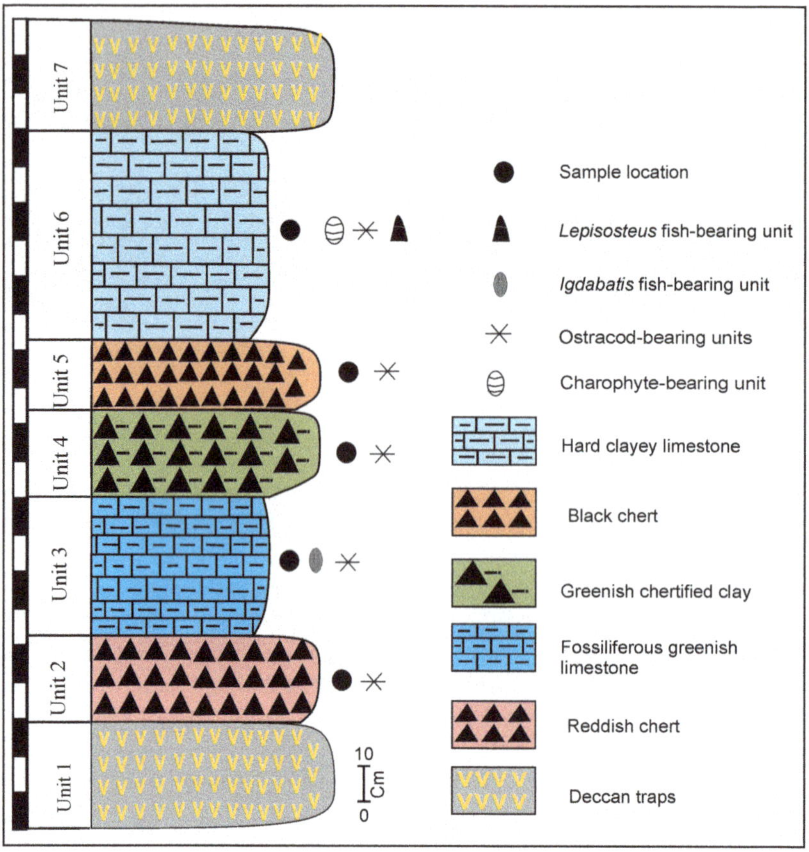

Fig. 3.10 Lithologic units of the Ghat Parasia intertrappean section, Chhindwara District, Madhya Pradesh

Fig. 3.11 Field photographs of the Ghat Parasia intertrappean section, Chhindwara District, Madhya Pradesh. (**A**) Panoramic view and (**B**) close-up view of the section

paucisulcatus Whatley et al. 2002b, *Periosocypris megistus* Whatley et al. 2012, *Candona* sp., *Eucypris* sp. 1, *Cyclocypris amphibolos* Whatley et al. 2002a, *Cyprois rostellum* Whatley and Bajpai 2000a, *Cyprois* sp. and *Darwinula* sp. (ostracods); and *Igdabatis indicus* Prasad and Cappetta 1993, *Lepisosteus indicus* Woodward 1908 and Osteoglossidae genus and species indeterminate (fishes). The presence of this biota in the Ghat Parasia intertrappean beds supports assigning it a Late Cretaceous (Maastrichtian) to ? Early Palaeocene age.

3.4 Conclusions

In this chapter, the detailed geological context of the microbiota-bearing sections of the intertrappean beds of Chhindwara District, Madhya Pradesh, India, is presented.

As a part of the field investigations for this study, four stratigraphic sections namely, Jhilmili, Ghat Parasia, Shriwas (=Shiraj) well and Government well in the Chhindwara area, were selected. The intertrappean beds in the Chhindwara area rest directly on the Archaean basement and Gondwana succession. Lithologically, the sections around Chhindwara area are composed of clayey limestones, marl, siltstones, claystones, mudstones and cherts. The intertrappean beds have yielded 10 taxa of charophytes including 3 new forms, 34 taxa of ostracods including 3 new forms, 10 taxa of planktic foraminiferans and 3 taxa of fishes of Late Cretaceous-Early Palaeocene age.

References

Acharyya SK, Lahiri TC (1991) Cretaceous palaeogeography of the Indian subcontinent: a review. Cretac Res 12:3–26

Badve RM (1987) A reassessment of stratigraphy of Bagh beds, Barwaha area, Madhya Pradesh, with description of trace fossils. J Geol Soc India 30:106–120

Baksi AK (1995) Petrogenesis and timing of volcanism in the 75 Ma Kerguelen hotspot reconstruction places it much Rajmahal flood basalt province, northeastern India. Chem Geol 121:73–90

Bande MB, Chandra A, Venkatachala BS, Mehtotra RC (1988) Deccan intertrappean floritisics and their stratigraphic implications. In: Maheshwari HK (ed) Proceedings of Symposium on Palaeocene of India: Limits and Subdivision. Indian Association of Palynostratigraphers, pp 83–123

Beane JE, Turner CA, Hooper PR, Subbarao KV, Walsh JN (1986) Stratigraphy, composition and form of the Deccan basalts, Western Ghats, India. Bull Volcanol 48:61–83

Bhandari A, Colin JP (1999) Limnic ostracods of the intertrappean sediments (terminal Maastrichtian-basal Paleocene) of the Anjar region (Kachchh, Gujarat state), India: Systematics, paleoecology, and paleobiogeographic affinities. J Micropaleontol 42(1):3–20

Bhatia SB, Mannikeri MS (1976) Some Charophyta from the Deccan intertrappean beds near Nagpur, Central India. Geophytology 6(1):75–81

Bhatia SB, Rana RS (1984) Palaeogeographic implications of the Charophyta and Ostracoda of the intertrappean beds of peninsular India. Mém Soc Geol Fr 147:29–35

Bhatia SB, Riveline J, Rana RS (1990b) Charophyta from the Deccan intertrappean beds near Rangapur, Andhra Pradesh, India. The Palaeobotanist 37:316–323
Biswas SK (1977) Mesozoic rock stratigraphy of Kachchh. Quart J Geol Mining Metall Soc India 49:1–52
Biswas SK (1991) Stratigraphy and sedimentary evolution of the Mesozoic basin of Kutch, western India. In: Tandon SK, Pant CC, Casshyap SM (eds) Stratigraphy and Sedimentary Evolution of Western India. Nainital Gyanodaya Prakashan, India, pp 74–103
Blanford HF (1862) On the Cretaceous and other rocks of the South Arcot and Trichinopoly district, South India. Mem Geol Surv India 4(1):1–217
Bondre NR, Hart WK, Sheth HC (2006) Geology and geochemistry of the Sangamner dyke swarm, Western Deccan Volcanic Province: implication for regional stratigraphy. J Geol 114:155–170
Brookfield ME, Sahni A (1987) Palaeoenvironment of the Lameta Beds (Late Cretaceous) at Jabalpur, M.P., India: soils and biota of a semi-arid alluvial plain. Cretac Res 8:1–14
Casshyap SM, Aslam M (1992) Deltaic and shoreline sedimentation in Saurashtra Basin, western India: An example of infilling in an Early Cretaceous failed rift. J Sediment Petrol 62:972–991
Chenet AL, Quidelleur X, Fluteau F, Courtillot V (2007) $^{40}Ar/^{39}Ar$ dating of the main Deccan large igneous province: further evidence of KTB age and short duration. Earth Planet Sc Lett 263:1–15
Chenet AL, Fluteau F, Courtillot V, Gerard M, Subbarao KV (2008) Determination of rapid Deccan eruptions across the Cretaceous-Tertiary boundary using palaeomagnetic secular variation: results from a 1200m thick section in the Mahabaleshwar. J Geophy Res 113:B04101. https://doi.org/10.1029/2006JB004635
Chenet AL, Courtillot V, Flutea F, Gerard M, Quidelleur X, Khadri SFR, Subbarao KV, Thordarson T (2009) Determination of rapid Deccan eruptions across the Cretaceous–Tertiary boundary using paleomagnetic secular variation: 2. Constraints from analysis of eight new sections and synthesis for a 3500m thick composite section. J Geophy Res 114:B06103. https://doi.org/10.1029/2008JB005644
Cox KG (1983) The Deccan Traps and the Karoo: stratigraphic implications of possible hot-spot origins. Hamburg, IAVCEI Abstr, p 96
Cushman JA (1933) Some new foraminiferal genera. Contrib Cushman Lab Foram Res 8:4–7
Dassarma DC, Sinha NK (1975) Marine Cretaceous formations of Narmada valley (Bagh Beds), Madhya Pradesh and Gujarat. Mem Geol Surv India Palaeontol Indica 42:1–123
de Sowerby JC (1840) Explanations of the plates and wood-cuts. In: Malcolmson JC (ed), On the fossils of the eastern portion of the Great Basaltic District of India. Trans Geol Soc London 2(5):511–567
Desmukh SS, Sano T, Fujii T, Nair KKK, Yedekar DB, Umino S, Iwamori H, Aramaki S (1996) Chemical stratigraphy and geochemistry of the basalt flows from the Central and eastern parts of the Deccan Volcanic Province of India. Gond Geol Magaz 2:145–170
Dhiman H, Dutta S, Kumar S, Verma V, Prasad GVR (2021) Discovery of proteinaceous moieties in Late Cretaceous dinosaur eggshell. Palaeontology 64(5):585–595
Ehrenberg CG (1854) Mikrogeologie. Das Erden und Felsen schaffende Wirken des unsichtbar kleinen selbstständigen Lebens auf der Erde. Leopold Voss, Leipzig, p xxviii + 374
Godbole SM, Rana RS, Natu SR (1996) Lava stratigraphy of Deccan basalts of Western Maharashtra. Gond Geol Magaz 2:125–134
Grambast L (1957) Ornamentation de la gyrogonite et systematique chez les charophytes fossiles. Rev Gén Bot 64:339–362
Goswami A, Prasad GVR, Verma O, Flynn JJ, Benson RBJ (2013) A troodontid dinosaur from the latest Cretaceous of India. Nat Commun 4:1–5. https://doi.org/10.1038/ncomms2716
Jain SP (1978) Ostracodes des "inter-trappean beds" (Eocene Infeerieur) de Kateru, Rajahmundry, Côte est de l' Inde. Rev de Microplaeontol 21(2):51–58
Jay AE, Widdowson M (2008) Stratigraphy, structure and volcanology of the south-eastern Deccan continental flood basalt province: implications for eruptive extent and volumes. J Geol Soc Lond 165:177–188
Jones T (1860) Notes on the fossil Cypride. Quart J Geol Soc London 16:154–189

Joshi AV (1995) New occurrence of dinosaur eggs from Lameta rocks (Maestrichtian) near Bagh. Madhya Pradesh J Geol Soc India 46(4):439–443

Kaila KL (1988) Mapping the thickness of Deccan Trap flows in India from DSS studies and inferences about a hidden Mesozoic Basin in the Narmada-Tapti region. In: Subbarao KV (ed), Deccan Flood Basalts. Mem Geol Soc India 10:91–116

Kania S, Khosla A, Verma O, Prashanth M (2022) Charophyte assemblage in the Cretaceous–Palaeogene boundary of the Deccan intertrappeans from Chhindwara District (Madhya Pradesh), Central India. Him Geol 43(2):383–396

Kapur VV, Khosla A (2019) Faunal elements from the Deccan volcano-sedimentary sequences of India: a reappraisal of biostratigraphic, palaeoecologic, and palaeobiogeographic aspects. Geol J 54(5):2797–2828

Kar RK, Srinivasan S (1998) Late Cretaceous palynofossils from the Deccan intertrappean beds of Mohgaon-Kalan, Chhindwara District, Madhya Pradesh. Geophytology 27:17–22

Keller G, Adatte T, Burns SJ, Tantaway A (2002) High stress paleoenvironment during the late Maastrichtian to early Paleocene in Central Egypt. Palaeogeogr Palaeoclimatol Palaeoecol 187:35–60

Keller G, Adatte T, Gardin S, Bartolini A, Bajpai S (2008) Main Deccan volcanism phase ends near the K-T boundary: evidence from the Krishna-Godavari Basin, SE India. Earth Planet Sci Lett 268:293–311

Keller G, Khosla SC, Sharma R, Khosla A, Bajpai S, Adatte T (2009a) Early Danian planktic foraminifera from Cretaceous-Tertiary intertrappean beds at Jhilmili, Chhindwara District, Madhya Pradesh, India. J Foram Res 39(1):40–55

Keller G, Adatte T, Bajpai S, Mohabey DM, Widdowson M, Khosla A, Sharma R, Khosla SC, Gertsch B, Fleitmann D, Sahni A (2009b) K-T transition in Deccan Traps of central India marks major marine seaway across India. Earth Planet Sci Lett 282:10–23

Keller G, Adatte T, Bhowmick PK, Upadhyay H, Dave A, Reddy AN, Jaiprakash BC (2012) Nature and timing of extinctions in Cretaceous–Tertiary planktic foraminifera preserved in Deccan intertrappean sediments of the Krishna–Godavari Basin, India. Earth Planet Sci Lett 341–344:211–221

Khosla A (2015) Palaeoenvironmental, palaeoecological and palaeobiogeographical implications of mixed fresh water and brackish marine assemblages from the Cretaceous–Palaeogene Deccan intertrappean beds at Jhilmili, Chhindwara District, Central India. Rev Mex de Cien Geol 32(2):344–357

Khosla A (2021) Paleobiogeographical inferences of Indian Late Cretaceous vertebrates with special reference to dinosaurs. Hist Biol 33:1431–1442

Khosla A, Lucas SG (2020a) Introduction of Indian Late Cretaceous dinosaur eggs and eggshells of Peninsular India. In: Khosla A, Lucas SG (eds) Late Cretaceous dinosaur eggs and eggshells of Peninsular India: Oospecies diversity and taphonomical, palaeoenvironmental, biostratigraphical and palaeobiogeographical inferences, Topics in Geobiology, vol 51. Springer Nature International Publishing, Cham, pp 1–30

Khosla A, Lucas SG (2020b) Historical background of Late Cretaceous dinosaur studies and associated biota in India. In: Khosla A, Lucas SG (eds) Late Cretaceous dinosaur eggs and eggshells of Peninsular India: Oospecies diversity and taphonomical, palaeoenvironmental, biostratigraphical and palaeobiogeographical inferences, Topics in Geobiology, vol 51. Springer Nature International Publishing, Cham, pp 31–56

Khosla A, Lucas SG (2020c) Geology and stratigraphy of dinosaur eggs and eggshell-bearing infra- and intertrappean beds of Peninsular India. In: Khosla A, Lucas SG (eds) Late Cretaceous dinosaur eggs and eggshells of Peninsular India: Oospecies diversity and taphonomical, palaeoenvironmental, biostratigraphical and palaeobiogeographical inferences, Topics in Geobiology, vol 51. Springer Nature International Publishing, Cham, pp 57–115

Khosla A, Lucas SG (2020d) Indian Late Cretaceous dinosaur nesting sites and their systematic studies. In: Khosla A, Lucas SG (eds) Late Cretaceous dinosaur eggs and eggshells of Peninsular India: Oospecies diversity and taphonomical, palaeoenvironmental, biostratigraphical and palaeobiogeographical inferences, Topics in Geobiology, vol 51. Springer Nature International Publishing, Springer Cham, pp 117–205

Khosla A, Lucas SG (2020e) Discussion: Oospecies diversity, biomineralization aspects, taphonomical, biostratigraphical, palaeoenvironmental, palaeoecological and palaeobiogeographical inferences of the dinosaur-bearing Lameta Formation of Peninsular India. In: Khosla A, Lucas SG (eds) Late Cretaceous dinosaur eggs and eggshells of Peninsular India: Oospecies diversity and taphonomical, palaeoenvironmental, biostratigraphical and palaeobiogeographical inferences, Topics in Geobiology, vol 51. Springer Nature International Publishing, Cham, pp 207–271

Khosla SC, Nagori ML (2005) A restudy of ostracode fauna from the intertrappean beds of the Anjar, Kachchh District, Gujarat. J Geol Soc India 66(5):573–580

Khosla A, Sahni A (2003) Biodiversity during the Deccan volcanic eruptive episode. J Asi Earth Sci 21(8):895–908

Khosla A, Verma O (2015) Paleobiota from the Deccan volcano-sedimentary sequences of India: paleoenvironments, age and paleobiogeographic implications. Hist Biol 27(7):898–914

Khosla A, Kapur VV, Sereno PC, Wilson JA, Dutheil D, Sahni A, Singh MP, Kumar S, Rana RS (2003) First dinosaur remains from the Cenomanian-Turonian of the Nimar Sandstone (Bagh Beds), District Dhar, Madhya Pradesh, India. J Palaeontol Soc India 48:115–127

Khosla A, Prasad GVR, Verma O, Jain AK, Sahni A (2004) Discovery of a micromammal-yielding Deccan intertrappean site near Kisalpuri, Dindori District, Madhya Pradesh. Curr Sci 87(3):380–383

Khosla A, Sertich JJW, Prasad GVR, Verma O (2009) Dyrosaurid remains from the intertrappean beds of India and the Late Cretaceous distribution of Dyrosauridae. J Vertebr Paleontol 29(4):1321–1326

Khosla SC, Nagori ML, Mohabey DM (2005) Effect of Deccan volcanism on non-marine Late Cretaceous ostracode fauna: a case study from Lameta Formation of Dongargaon area (Nand-Dongargaon basin), Chandrapur District, Maharashtra. Gond Geol Mag 8:133–146

Khosla SC, Nagori ML, Jakhar SR, Rathore AS (2011a) Early Danian lacustrine–brackish water Ostracoda from the Deccan Intertrappean beds near Jhilmili, Chhindwara District, Madhya Pradesh, India. Micropaleontology 57(3):223–245

Khosla A, Kania S, Lucas SG, Verma O (2022) Charophytes from the Cretaceous–Palaeogene transition in the Jhilmili intertrappean beds of Central India. Geol J 57(11):4412–4438. https://doi.org/10.1002/gj.4528

Krishnan MS (1968) Geology of India and Burma. The Madras Law Journal Office, Madras

Kumaran KPN, Bonde SD, Kanikar MD (1997) An *Aquilapollenites* associated palynoflora from Mohgaon-Kalan and its stratigraphic implications for age and stratigraphic correlation of Deccan intertrappean beds. Curr Sci 72:590–592

Kumari A, Singh S, Khosla, A (2021) Palaeosols and palaeoclimate reconstructions of the Maastrichtian Lameta Formation, Central India. Cret Res 117:104632 https://doi.org/10.1016/j.cretres.2020.104632

Kundal P, Humane S, Humane SK, Petkar SP (2018) Discovery of marine benthic chlorophycean algae in Early Danian Deccan intertrappean at Jhilmili, Central India: new insights into existence of marine seaway close to Cretaceous-Paleogene boundary. J Paleontol Soc India 63(2):203–211

Loeblich AR, Tappan H (1957b) *Woodringina,* a new foraminiferal genus (Heterohelicidae) from the Paleocene of Alabama. J Wash Acad Sci 47:39–40

Malcolmson JG (1840) On the fossils of the Eastern portion of the Great Basaltic District of India. Trans Geol Soc London 52(5):537–575

Mathur SC, Gour SD, Loyal RS, Tripathi A, Sisodia MS (2005) Spherules from the Late Cretaceous phosphorite of the Fatehgarh Formation, Barmer Basin, India. Gond Res 8(4):579–584

Misra KS, Saxena RS (1964) A new fossil fish, *Jhingrania roonwali*, from the Rajmahal Hills, India. J Palaeontol Soc India 4:30–34

Morgan WJ (1981) Hot-spot tracks and the opening of Atlantic and Indian oceans. In: Emiliani C (ed) The Sea, vol 7. Wiley-Interscience, New York, pp 443–1487

Morozova VG (1961) Datsko-Montskie planktonnye foraminifery yuga SSSR (Danian-Montian Planktonic Foraminifera of the Southern USSR). Paleontol Z 2:8–19

Nair KKK, Chatterjee AK, Sano T (1996) Stratigraphy and geochemistry of the Deccan basalts along Toranmal section, western Satpura Range. Gond Geol Magaz 2:23–148
Naqvi SM (2005) Geology and evolution of the Indian plate (From Hadean to Holocene – 4 Ga to 4 Ka). Capital Publishing Company, New Delhi
Olsson RK (1960) Foraminifera of latest Cretaceous and earliest Tertiary age in the New Jersey coastal plain. J Paleontol 34:1–58
Olsson RK (1964) Late Cretaceous planktonic foraminifera from New Jersey and Delaware. Micropaleontology 10:157–188
Pal S, Srivastava S, Shrivastava JP (2013) Mineral chemistry of clays associated with the Jhilmili intertrappean bed in the eastern Deccan Volcanic Province: Palaeoenvironmental inferences and KTB transition. J Geol Soc India 82:38–52
Pascoe EH (1959) A manual of the Geology of India and Burma. Govt. of India Press, Calcutta
Pascoe EH (1964) A Manual of the Geology of India and Burma. Delhi. Geol Sur India. 3
Pathak V, Patil SK, Shrivastava JP (2016) Tectonomagmatic setting of lava packages in the Mandla lobe of the eastern Deccan volcanic province, India: palaeomagnetism and magnetostratigraphic evidence. Geol Soc London Spec Publ 445. https://doi.org/10.1144/SP445.3
Pattanayak SK, Shrivastava JP (1996) Petrominerography and major oxide geochemical variations in the basaltic lava flow sequence from eastern Deccan Volcanic province: a stratigraphic and genetic model. Gond Geol Magaz 2:537
Pattanayak SK, Shrivastava JP (1999) Petrography and major-oxide geochemistry of basalts from the Eastern Deccan Volcanic Province, India. Mem Geol Soc India 43:233–270
Peck RE, Reker CC (1947) Cretaceous and Lower Cenozoic Charophyta from Peru. Am Mus Nat Hist Novit 1369:1–6
Peck RE, Reker CC (1948) Eocene Charophyta from North America. J Paleontol 22(1):85–90
Peng ZX, Mahoney JJ, Hooper PR, Harris C, Beane JE (1994) Role for lower continental crust in flood basalt genesis? isotopic and incompatible element study of the lower six formations of the Western Deccan traps. Geochim Cosmochim Acta 58(1):267–288
Plummer HJ (1926) Foraminifera of the Midway Formation in Texas. Univ Tex Bull 2644:1–206
Prasad GVR, Cappetta H (1993) Late Cretaceous selachians from India and the age of the Deccan Traps. Palaeontology 36(1):231–248
Prasad GVR, Verma O, Sahni A, Krause DW, Khosla A, Parmar V (2007a) A new Late Cretaceous gondwanatherian mammal from Central India. Proc Indian Natl Sci Acad 73(1):17–24
Prasad GVR, Verma O, Sahni A, Parmar V, Khosla A (2007b) A Cretaceous hoofed mammal from India. Science 318:937
Prasad GVR, Verma O, Gheerbrant E, Goswami A, Khosla A, Parmar V, Sahni A (2010) First mammal evidence from the Late Cretaceous of India for biotic dispersal between India and Africa at the K/T transition. C R Palevol 9:63–71
Prasad GVR, Verma O, Flynn JJ, Goswami A (2013) A new Late Cretaceous vertebrate fauna from the Cauvery basin, south India: implications for Gondwana palaeobiogeography. J Vertebr Paleontol 33(6):1260–1268
Prasad GVR, Verma O, Sahni A, Khosla A (2021) Cretaceous mammals of India– Stratigraphic distribution, diversity and intercontinental affinities. J Palaeosci 70:173–192
Punekar J, Keller G, Khozyem H, Hamming C, Adatte T, Tantawy AA, Spangenberg JE (2014) Late Maastrichtian–early Danian high–stress environments and delayed recovery linked to Deccan volcanism. Cret Res 49:63–82
Radhakrishna BP, Vaidyanadhan R (1994) Geology of Karantaka. Geological Society of India, Bangalore
Ramakrishnan M, Vaidyanadhan R (2010) Geology of India, Geol Soc India Bangalore 1:232–233
Rage JC, Prasad GVR, Verma O, Khosla A, Parmar V (2020) Anuran lissamphibian and squamate reptiles from the Upper Cretaceous (Maastrichtian) Deccan intertrappean sites in Central India, with a review of lissamphibian and squamate diversity in the northward drifting Indian plate. In: Prasad GVR, Patnaik R (eds) Biological Consequences of Plate Tectonics. Vertebrate Paleobiology and Paleoanthropology, Springer, Cham, pp 99–121

Rao SU (1969) Problems of the Deccan Traps. Mineral Wealth 7(1):1–8
Ray JS, Pattanayak SK, Pande K (2005) Rapid emplacement of the Kerguelen plume-related Sylhet Traps, eastern India: Evidence from ^{40}Ar-^{39}Ar geochronology. Geophys Res Lett 32:1–4
Richards MA, Alvarez W, Self S, Karlstrom L, Renne PR, Manga M, Sprain CJ, Smit J, Vanderkluysen L, Gibson SA (2015) Triggering of the largest Deccan eruptions by the Chicxulub impact. Geol Soc Am Bull. https://doi.org/10.1130/B31167.1
Roychowdhary M, Sastri VV (1962) On the revised classification of the Cretaceous and associated rocks of the Man River section of lower Narbada Valley. Rec Geol Surv India 91:283–304
Roychowdhary MK, Venkatesh V, Paul DK (1968) Report on detailed investigation of some bauxite deposits of Amarkantak area, M.P. Bull Geol Surv India 28:1–208
Saha O, Shukla UK, Rani R (2010) Trace fossils from the Late Cretaceous Lameta Formation, Jabalpur Area, Madhya Pradesh: paleoenvironmental implications. J Geol Soc India 76: 607–620
Sahni A, Khosla A (1994) A Maastrichtian ostracode assemblage (Lameta Formation) from Jabalpur Cantonment, Madhya Pradesh, India. Curr Sci 67(6):456–460
Sahni B, Rode BP (1937) Fossil plants from the Deccan intertrappean beds of Mohgaon-Kalan with a sketch on the geology of Chhindwara District. Proc Nat Acad Sci India 7:165–174
Sharma R, Khosla A (2009) Early Palaeocene Ostracoda from the Cretaceous-Tertiary (K-T) Deccan intertrappean sequence at Jhilmili, District Chhindwara, Central India. J Palaeontol Soc India 54(2):197–208
Shrivastava JP, Ahmad M (2005) A review of research on Late Cretaceous volcanic–sedimentary sequences of the Mandla Lobe: implications for Deccan volcanism and the Cretaceous/Palaeogene boundary. Cretac Res 26:145–156
Singh IB (1981) Palaeoenvironment and palaeogeography of Lameta Group sediments (Late Cretaceous) in Jabalpur area, India. J Palaeontol Soc India 26:38–53
Singh SK, Srivastava HK (1981) Lithostratigraphy of the Bagh Beds and its correlation with Lameta Beds. J Palaeontol Soc India 26:77–85
Solanki JN, Bhattacharya DD, Jain AK, Mukherjee A (1996) Stratigraphy and tectonics of the Deccan Traps of Mandla. Gond Geol Magaz 2:101–114
Srinivasan S (1996) Late Cretaceous egg shells from the Deccan volcano-sedimentary sequence of Central India. Mem Geol Soc India 37:321–336
Subbarao KV, Hooper PR (1988) Reconnaissance map of the Deccan Basalt Group in the Western Ghats, India. Mem Geol Soc India 10:1–393
Sundaram R, Henderson A, Ayyasami K, Stilwell D (2001) A lithostratigraphic revision and palaeoenvironment assessment of the Cretaceous System exposed in the onshore Cauvery Basin, southern India. Cretac Res 22:743–762
Tandon SK, Andrews J, Sood A, Mittal S (1998) Shrinkage and sediment supply control on multiple calcrete profile development: a case study from the Maastrichtian of central India. Sed Geol 119:25–45
Tandon SK, Sood A, Andrews JE, Dennis PF (1995) Palaeoenvironments of the dinosaur-bearing Lameta Beds (Maastrichtian), Narmada valley, Central India. Palaeogeogr Palaeoclimatol Palaeoecol 117:153–184
Tandon SK, Verma VK, Jhingran V, Sood A, Kumar S, Kohli RP, Mittal S (1990) The Lameta Beds of Jabalpur, Central India: deposits of fluvial and pedogenically modified semi- arid fan-palustrine flat systems. In: Sahni A, Jolly A (eds) Cretaceous event stratigraphy and the correlation of the Indian nonmarine strata. A Seminar cum Workshop IGCP 216 and 245, Chandigarh, pp 27–30
Thakre D, Samant B, Mohabey DM, Sangode S, Srivastava P, Kapgate DK, Mahajan R, Upretii N, Manchester SR (2017) A new insight into age and environments of intertrappean beds of Mohgaon-Kalan, Chhindwara District, M.P. using palynology, megaflora, magnetostratigraphy and clay mineralogy. Curr Sci 112(11):2193–2197
Tripathi A, Jana BN, Verma O, Singh RK, Singh AK (2013) Early Cretaceous palynomorphs, dinoflagellates and plant megafossils from the Rajmahal basin, Jharkhand, India. J Palaeontol Soc India 56(1):125–134

Vaidyanadhan R, Ramakrishnan M (2010) Geology of India, vol II. Geological Society of India, Bangalore, India

Verma O (2015) Cretaceous vertebrate fauna of the Cauvery Basin, southern India: palaeodiversity and palaeobiogeographic implications. Palaeogeogr Palaeoclimatol Palaeoecol 431:53–67

Verma O, Khosla A (2018) Application of internet technology in assembling literature for palaeontological research. Iran J Sci Technol Trans A Sci 4:1715–1723

Verma O, Khosla A (2019) Developments in the stratigraphy of the Deccan Volcanic Province, peninsular India. Compt Rend Geosci 351:461–476

Verma O, Prasad GVR, Goswami A, Parmar V (2012a) *Ptychodus decurrens* Agassiz (Elasmobranchii: Ptychodontidae) from the Upper Cretaceous of India. Cretac Res 33:183–188

Verma O, Prasad GVR, Khosla A, Parmar V (2012b) Late Cretaceous gondwanatherian mammals of India: Distribution, interrelationships and biogeographic implications. J Paleontol Soc India 57:95–104

Verma O, Khosla A, Goin FJ, Kaur J (2016) Historical biogeography of the Late Cretaceous vertebrates of India: Comparison of geophysical and paleontological data. In: Khosla A, Lucas SG (eds) Cretaceous period: Biotic diversity and biogeography, vol 71. New Mexico Museum of Natural History & Science, Albuquerque, pp 317–330

Verma O, Khosla A, Kaur J, Prasanth M (2017) Myliobatid and pycnodont fish from the Late Cretaceous of Central India and their paleobiogeographic implications. Hist Biol 29(2):253–265

Verma O, Prashanth M, Greco R, Khosla A, Singh K (2022) Geological education scenario in India and role of open educational resources in the light of COVID-19 pandemic. Earth Sci Res J 26(2):239–254

West WD (1958) The petrography and petrogenesis of forty-eight flows of Deccan Traps penetrated by borings in western India. Trans Natl Inst Sci India 4:1–56

Whatley RC, Bajpai S (2000a) A new fauna of Late Cretaceous non-marine Ostracoda from the Deccan intertrappean beds of Lakshmipur, Kachchh (Kutch District), Gujarat, western India. Rev Esp de Micropaleontol 32(3):385–409

Whatley RC, Bajpai S, Srinivasan S (2002a) Upper Cretaceous nonmarine Ostracoda from intertrappean horizons in Gulbarga district, Karnataka state, South India. Rev Esp de Micropaleontol 34(2):163–186

Whatley RC, Bajpai S, Srinivasan S (2002b) Upper Cretaceous intertrappean nonmarine Ostracoda from Mohgaonkala (Mohgaon-Kalan), Chhindwara District, Madhya Pradesh state, Central India. J Micropaleontol 21:105–114

Whatley RC, Khosla SC, Rathore AS (2012) *Periosocypris megistus* n. gen. and n. sp.: A new gigantic non-marine cyprid ostracod from the Maastrichtian Lameta Formation of India. J Palaeontol Soc India 57(2):113–117

Wilson GP, Renne PR, Samant B, Mohabey DM, Dhobale A, Tholt AJ, Tobin TS, Widdowson M, Anantharaman S, Dassarma DS, Wilson JA (2022) New mammals from the Naskal intertrappean site and the age of India's earliest eutherians. Palaeogeog Palaeoclimat Palaeoecol 591(3):110857. https://doi.org/10.1016/j.palaeo.2022.110857

Woodward AS (1908) On some fish remains from the Lameta Beds at Dongargaon, Central Province. Mem Geol Sur India, Palaeontol Indica NS 3:1–6

Yedekar DB, Aramaki S, Fujii T, Sano T (1996) Geochemical signature and stratigraphy of the Chindwara-Jabalpur-Seoni-Mandla sector of the eastern Deccan volcanic province and problems of its correlation. Gond Geol Magaz 2:49–68

Chapter 4
Indian Late Cretaceous-Early Palaeocene Deccan Microbiota from the Intertrappean Beds of the Chhindwara District, Madhya Pradesh and Their Systematic Palaeontology

4.1 Introduction

The microbiota documented here, consisting of charophytes, ostracods, planktic foraminiferans and fishes, was recovered from four Deccan intertrappean beds, Jhilmili, Ghat Parasia, Government well and Shriwas (=Shiraj) well, situated in the Chhindwara District, Madhya Pradesh, Central India. The terminologies of Grambast (1957, 1962, 1974) and Peck and Reker (1947, 1948) have been followed for describing the charophyte gyrogonites, Kesling (1951), Moore et al. (1961), Howe (1962) and Neale (1969) for examining ostracod carapaces, and Loeblich and Tappan (1964) and Olsson et al. (1999, 2006) for studying foraminiferans.

4.2 Charophytes

4.2.1 Species Platychara perlata (Peck and Reker 1947) (Figs. 4.1, 4.2A–P, 4.3A–F; Table 4.1)

Division	Charophyta (Migula 1897)
Class	Charophyceae (Smith 1938) (emend. Schudack 1993)
Order	Charales (Lindley 1836)
Family	Characeae (Richard ex C. Agardh 1824)

Indian Late Cretaceous-Early Palaeocene Deccan Microbiota from the Intertrappean Beds of Chhindwara District, Madhya Pradesh and their Systematics Palaeontology

A. Khosla et al., *Microbiota from the Late Cretaceous-Early Palaeocene Boundary Transition in the Deccan Intertrappean Beds of Central India*, Topics in Geobiology 54, https://doi.org/10.1007/978-3-031-28855-5_4

Subfamily	Charoideae (Al. Braun in Migula 1897)
Genus	*Platychara* (Grambast 1962)
Species	***Platychara perlata*** (Peck and Reker 1947)

1924 *Chara elliptica* (Fritzsche), pp. 28, Pl. 2, pp. 165, Fig. 3 (non-Chara *elliptica* Hislop 1860).
1939 ?*Chara oehlerti* (Dollfus): Rao and Rao, pp. 8, Pl.1, Fig. 9, Pl. 3, Fig. 2.
1947 *Chara perlata*: Peck and Reker, pp. 3, Figs. 19–21.
1951 *Chara perlata*: Horn af Rantzien, pp. 661
1951 ?*Aclistochara cruciana* sp. nov.: Horn af Rantzien, pp. 672.
1967 *Platychara perlata* (Peck and Reker): Grambast et al. pp. 2.
1972 ?*Platychara cruciana* (Horn af Rantzien): Musacchio, pp. 229, Pl.1, Figs. 1, 4, 7, Pl. 2, Figs.15–17.
1979 *Platychara perlata* (Peck and Reker): Grambast et al. pp. 230, P1. 2, Figs. 5–8.
1984 *Platychara perlata* (Peck and Reker) (Grambast): Bhatia and Rana, pp. 30, P1. 1, Figs. 2, 3.
1990b *Platychara perlata* (Peck and Reker): Bhatia et al. pp. 318–319, P1.1, Figs. 1–5.
1992 *Platychara perlata* (Peck and Reker): Srinivasan et al. pp. 396–398, Pl.1, Figs. 2b.
1994 *Platychara perlata* (Peck and Reker): Srinivasan et al. pp. 562–564, P1.1, Figs. 1–5.
2022 *Platychara perlata* (Peck and Reker): Khosla et al. pp. 07–10, Figs. 3a–g, Fig. 4.

Material More than 200 well-preserved gyrogonite specimens.

Horizon, Age and Locality Upper Cretaceous to Lower Palaeocene clayey limestone of unit 3 (JH 17 and JH 19) of the Jhilmili intertrappean beds, Chhindwara District, Madhya Pradesh, India.

Description Small- to medium-sized gyrogonites, 616–690 μm length (mean 652 μm) and 720–823 μm width (mean 769 μm). Isopolarity index ranges from 85 to 91 (mean 85). Gyrogonites are prolate, subglobular to subprolate in shape, characteristically wider than long, and rounded at the top and base. The apical part is pointed, and the apical rosettes are usually well-developed and about 430–450 μm in diameter.

The lime spirals are 7–8 in number (80–130 μm width visible in lateral view). They are smooth, usually convex, flat, thin intercellular ridges without ornamentation and are inclined from right to left and are visible in lateral view. The gyrogonites become thinner along the equator towards the base. The basal part is flattened and arc shaped with a small basal pore (50 μm wide) in the centre.

Table 4.1 Dimensions of gyrogonite specimens of *Platychara perlata* (Peck and Reker 1947)

	Dimensions (μm)		
Catalogue no.	LPA	LED	ISI
MPL/SK/JML/101	616	720	86
MPL/SK/JML/102	661	761	87
MPL/SK/JML/103	641	823	78
MPL/SK/JML/104	690	787	88
MPL/SK/JML/105	687	753	91
MPL/SK/JML/106	618	770	81
Mean	652	769	85

Remarks The present specimens are comparable to *Chara elliptica* Fritzsche (1924), which was originally described from the lower Upper Cretaceous (Campanian) deposits of Tres Cruces and Negra Muerta, Province Jujuy in northern Argentina, South America (Doweld 2015). It was also recorded as *Chara perlata* from the Upper Cretaceous deposits of Peru (Peck and Reker 1947). This species was also named as ?*Aclistochara cruciana* by Horn af Rantzien (1951). Later, Musacchio (1972) re-assigned *Chara perlata* as *Platychara perlata*. Peck and Forester (1979) recorded *P. perlata* from the Upper Cretaceous deposits of the Western Hemisphere (Argentina, Peru, Bolivia and Mexico).

It is noted that *Platychara perlata* was frequently recorded in the Upper Campanian deposits of the Peruvian Andes (Jaillard et al. 1993). The Jhilmili specimens studied here are similar to *P. perlata* from the Western Hemisphere (Peck and Forester 1979) in size, shape, ISI, number and width of lime spirals, and some features of the apex and the base. It should be noted that *Platychara perlata* was frequently found in the Peruvian Andes' Late Campanian deposits (Jaillard et al. 1993). The Jhilmili specimens studied here are similar to *P. perlata* from the Western Hemisphere in size, shape, ISI, number and width of lime spirals and some apex and base features (Peck and Forester 1979).

In peninsular India, *Platychara perlata* has hitherto been recorded from the numerous Deccan intertrappean localities, that is, Nagpur (Bhatia and Rana 1984), Rajahmundry (Chanda et al. 1989), Rangapur (Bhatia et al. 1990b), Kachchh (Bajpai et al. 1990a; Srinivasan et al. 1992) and Gurmatkal, Chandarki and Yanagundi (Srinivasan et al. 1992, 1994). The present species is closely similar in overall shape, lime spirals and number of convolutions to specimens of *P. perlata* described from the intertrappean beds of Takli, Nagpur (Bhatia and Rana 1984). The gyrogonites of the Jhilmili specimens are similar in shape and convolution numbers to *P. perlata* described by Bhatia et al. (1990b) from the Upper Cretaceous intertrappean beds of Rangapur (Telangana).

The specimens from Rangapur differ from those from Jhilmili in being considerably larger (LPA ranging = 640–925 μm and LED ranging = 760–1050 μm). *Platychara perlata* was recorded from the intertrappean sections of Karnataka, such as Yanagundi, Chandarki and Gurmatkal (Srinivasan et al. 1992, 1994), and is slightly larger than the Jhilmili specimens (LPA ranging = 700–886 μm and LED ranging = 750–966 μm). The Jhilimili specimens are slightly smaller in size, but they are thought to be of the same species. The gyrogonite specimens from Manawar (District Dhar), Madhya Pradesh (Kapur et al. 2019), are slightly smaller in size than those from Jhilmili (LPA ranging = 372–494 m and LED ranging = 451–546 m). *Platychara sahnii* has tapering bases, whereas the current specimens have almost flat bases.

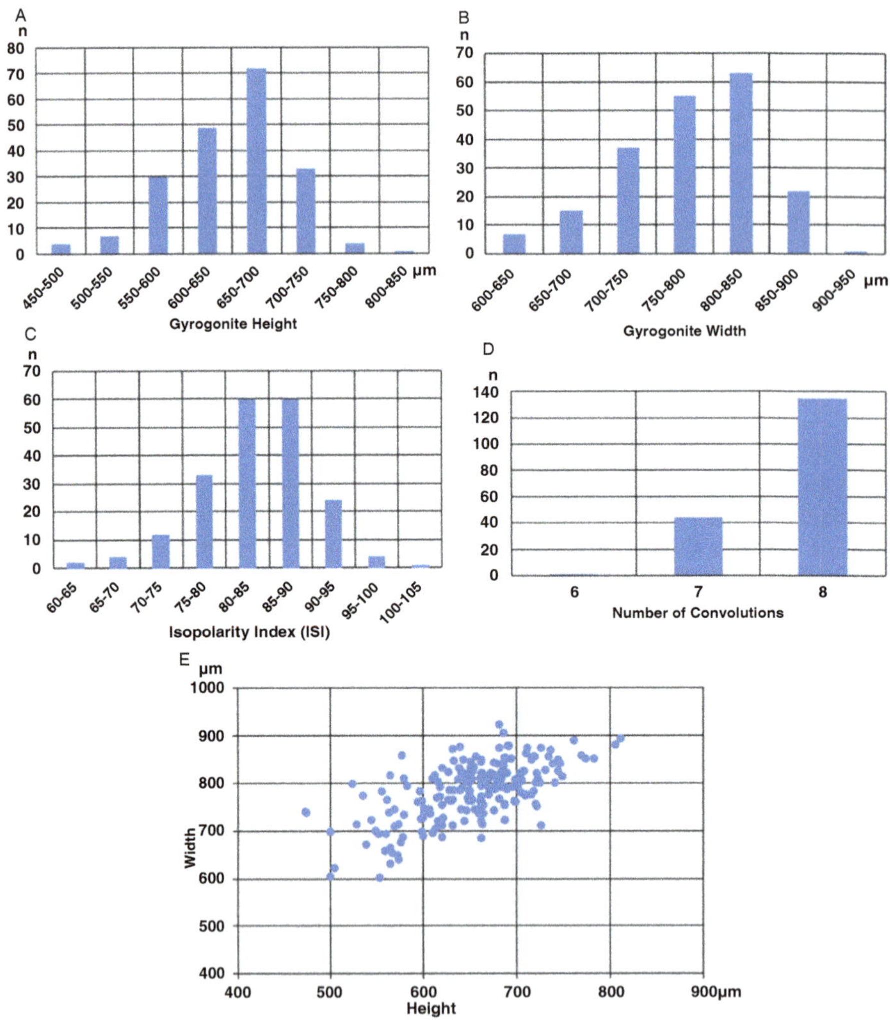

Fig. 4.1 Biometric data of *Platychara perlata* (200 gyrogonites) from the Jhilmili intertrappean beds, Chhindwara, Madhya Pradesh. (**A**) Histogram showing height, (**B**) Histogram showing width. (**C**) Histogram showing isopolarity index (ISI), (**D**) Histogram showing number of convolutions and (**E**) Dispersion graph of width/height

Distribution The recovered species *Platychara perlata* is widely distributed in the Upper Cretaceous intertrappean beds of peninsular India. It was reported from Rangapur, Telangana (Bhatia et al. 1990b), Kora, Gujarat (Bajpai et al. 1990a), Rajahmundry in southeastern India (Chanda et al. 1989), the Yanagudi, Gurmatkal and Chandakri sections in Karnataka (Srinivasan et al. 1994) and Manawar in District Dhar, Madhya Pradesh (Kapur et al. 2019). Elsewhere, the species is widely known from the Upper Cretaceous deposits of Argentina, Bolivia, Peru and Mexico (Peck and Reker 1947; Peck and Forester 1979; Musacchio 1972; Doweld 2015).

Fig. 4.2 (**A–P**) Gyrogonites of ***Platychara perlata*** (Peck and Reker 1947) (**A**) Lateral view (MPL/SK/JML/101); (**B**) Lateral view (MPL/SK/JML/102); (**C**) Lateral view (MPL/SK/JML/107); (**D**) Apical view (MPL/SK/JML/108); (**E**) Apical view (MPL/SK/JML/104); (**F**) Apical view (MPL/SK/JML/103); (**G**) Apical view (MPL/SK/JML/109); (**H**) Basal view (MPL/SK/JML/110); (**I**) Basal view (MPL/SK/JML/111); (**J**) Basal view (MPL/SK/JML/112); (**K**) Basal view (MPL/SK/JML/113); (**L**) Basal view (MPL/SK/JML/105); (**M**) Basal view (MPL/SK/JML/106); (**N**) Basal view (MPL/SK/JML/114); (**O**) Basal view (MPL/SK/JML/115); and (**P**) Basal view (MPL/SK/JML/116). (Figure **A–E**, **J**, and **M**, Reproduced from Khosla et al. 2022 with permission from Wiley)

Fig. 4.3 (**A**–**F**) Gyrogonites of ***Platychara perlata*** (Peck and Reker 1947). (**A**) Basal view (MPL/SK/JML/117); (**B**) Basal view (MPL/SK/JML/118); (**C**) Basal view (MPL/SK/JML/119); (**D**) Basal view (MPL/SK/JML/120); (**E**) Basal view (MPL/SK/JML/121); (**F**) Basal view (MPL/SK/JML/122); (**G**–**M**) Gyrogonites of ***Platychara raoi*** (Bhatia and Mannikeri 1976). (**G**) Lateral view (MPL/SK/JML/201); (**H**) Lateral view (MPL/SK/JML/205); (**I**) Lateral view (MPL/SK/JML/202); (**J**) Apical view (MPL/SK/JML/203); (**K**) Apical view (MPL/SK/JML/201); (**L**) Basal view (MPL/SK/JML/201); (**M**) Basal view (MPL/SK/JML/204); (**N**–**P**) Gyrogonites of ***Platychara sahnii*** Rao and Rao (Bhatia and Mannikeri 1976). (**N**) Lateral view (MPL/SK/JML/302); (**O**) Lateral view (MPL/SK/JML/301); and (**P**) Apical view (MPL/SK/JML/303). (Figure **N**, Reproduced from Khosla et al. 2022 with permission from Wiley)

4.2.2 Species Platychara raoi (Bhatia and Mannikeri 1976) (Figs. 4.3G–M, 4.4; Table 4.2)

1976 *Platychara raoi*: Bhatia and Mannikeri, Pl. 1, Figs. 11–12.

1994 *Platychara raoi*: (Bhatia and Mannikeri 1976): Srinivasan et al. pp. 564, P1.1, Fig. 11–12.

Table 4.2 Dimensions of gyrognite specimens of *Platychara raoi* (Bhatia and Mannikeri 1976)

	Dimensions (μm)		
Catalogue no.	LPA	LED	ISI
MPL/SK/JML/201	586	551	106
MPL/SK/JML/202	658	607	108
MPL/SK/JML/203	665	610	109
MPL/SK/JML/204	605	546	110
MPL/SK/JML/205	599	544	110
MPL/SK/JML/206	597	561	106
MPL/SK/JML/207	646	635	102
Mean	622	579	107

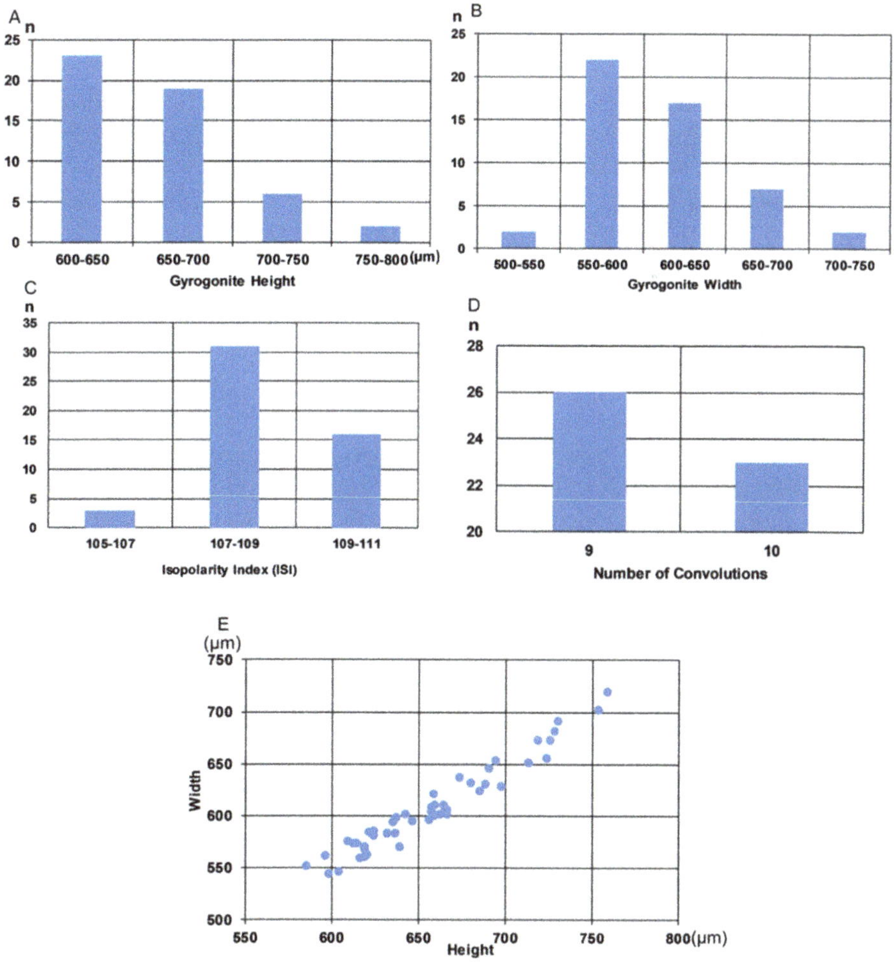

Fig. 4.4 Biometric data of *Platychara raoi* (50 gyrogonites) from the Jhilmili intertrappean beds, Chhindwara, Madhya Pradesh. (**A**) Histogram showing height, (**B**) Histogram showing width, (**C**) Histogram showing isopolarity index (ISI), (**D**) Histogram showing number of convolutions and (**E**) Dispersion graph of width/height

2022 *Platychara raoi*: (Bhatia and Mannikeri 1976): Khosla et al. pp. 10–12, Figs. 3n–r, Fig. 6.

Material More than 50 well-preserved gyrogonite specimens.

Horizon, Age and Locality Upper Cretaceous to Lower Palaeocene clayey limestone of unit 3 (JH 17 and JH 19) of the Jhilmili intertrappean beds, Chhindwara District, Madhya Pradesh, India.

Description Gyrogonites are medium to large in size, measuring 586–665 μm high (mean: 622 μm) and 551–635 μm wide (mean: 579 μm). The isopolarity index ranges between 106 and 110 (mean: 107). They are oblate spheroidal to suboblate in shape with a width that is greater than the length and a longitudinal flattening. Eight convolutions visible in lateral view. The spiral cells are generally concave or plane cells 75–100 μm wide separated by narrow intercellular ridges. The lime spirals make two sinistral turns at an equatorial angle of 10–16 degrees. The apical end cells become wider, thin and narrow at the periphery, and enlarge in the centre. The apex is flat and rounded. The base narrows and tapers to a point, with a pentagonal basal pore ~25–40 μm.

Remarks This species was first reported from the Upper Cretaceous deposits of the intertrappean beds at Gitti Khadan, Nagpur in Maharashtra (Bhatia and Mannikeri 1976; Bhatia and Rana 1984). It was later reported in the intertrappean beds of Gurmatkal, Chandakri and Yanagudi in Karnataka and Kachchh in Gujarat (Srinivasan et al. 1992, 1994). The present specimens are similar in shape, lime spirals and convolution numbers to *Platychara raoi* described by Bhatia and Mannikeri (1976) and Srinivasan et al. (1994) from the Upper Cretaceous intertrappean beds of Gitti Khadan, Maharashtra and Kachchh (Gujarat). However, the gyrogonite size is different from the specimens recorded from three of the localities. Thus, the Jhilmili specimens are considerably larger (LPA = 586–665 μm and LED = 544–635 μm) than specimens from Gitti Khadan (LPA = 440–500 μm and LED = 600–625 μm) and Kachchh (LPA = 425–500 μm and LED = 600–625 μm). However, though the Jhilmili specimens are rather larger in size in comparison to those from the abovementioned localities, we regard them as representing the same species.

In general outline and shape, *Platychara raoi* (Bhatia and Mannikeri 1976) compares closely with *Platychara caudata* (Grambast 1971) described from the Upper Cretaceous (Maastrichtian) continental sediments of south-east France in Europe. However, the Indian species is much larger and has less swollen and more broad-ended apical cells and a greater number of convolutions in lateral view in comparison to the French species.

Distribution *Platychara raoi* (Bhatia and Mannikeri 1976) was found in the Upper Cretaceous intertrappean localities of Kachchh in Gujarat; Yanagudi, Gurmatkal and Chandakri in Karnataka, and Nagpur in Maharashtra (Bhatia and Mannikeri 1976; Bhatia and Rana 1984; Srinivasan et al. 1994).

4.2.3 Species *Platychara sahnii* (Bhatia and Mannikeri 1976) (Figs. 4.3N–P, 4.5A–B, 4.6; Table 4.3)

1939 *Chara sahnii*: Rao and Rao, p. 10, Pl. 1, Fig. 12, Pl. 3, Fig. 5.
1976 *Platychara sahnii*: Bhatia and Mannikeri 1976: pp. 76–77, Figs. 8–13.
1992 *Platychara sahnii*: Bhatia and Mannieri 1976: Srinivasan et al. pp. 397, 398, Pl. 1, Figs. 2e–f.
1994 *Platychara sahnii*: Bhatia and Mannikeri 1976: Srinivasan et al. pp. 564, Pl. 1, Figs. 13–16.
2014 *Platychara* cf. *sahnii*: Bhatia and Mannikeri 1976: Khosla, pp. 314–315, Pl. 1, Fig. 1–4.
2022 *Platychara sahnii*: Bhatia and Mannikeri 1976: Khosla et al. pp. 10, Pl. 1, Figs. 3h–m, Fig. 5.

Material More than 50 well-preserved gyrogonite specimens.

Horizon, Age and Locality Upper Cretaceous to Lower Palaeocene clayey limestone of unit 3 (JH 17 and JH 19) of the Jhilmili intertrappean beds, Chhindwara District, Madhya Pradesh, India.

Description Gyrogonites are medium to large in size; 586–745 μm high (mean: 645 μm) and 500–616 μm wide (mean: 548 μm). Their isopolarity index ranges between 114 and 122 (mean: 117). They are spheroidal, oblate and suboblate in shape with six to seven convolutions in lateral view. The spiral cells are flat to concave, 80–100 μm wide and lack ornamentation. The apical part has a flat to round shape and is rounded apically due to swollen apical cells. The basal part is pointed, elongated and terminates with a small columnar basal pore. The maximum width is above mid-height.

Table 4.3 Dimensions of gyrogonite specimens of *Platychara sahnii* (Bhatia and Mannikeri 1976)

	Dimensions (μm)		
Catalogue no.	LPA	LED	ISI
MPL/SK/JML/301	745	616	121
MPL/SK/JML/302	643	526	122
MPL/SK/JML/303	629	544	116
MPL/SK/JML/304	644	557	115
MPL/SK/JML/305	628	550	114
MPL/SK/JML/306	586	500	117
Mean	645	548	117

Fig. 4.5 (**A**, **B**) Gyrogonites of ***Platychara sahnii*** Rao and Rao (Bhatia and Mannikeri 1976). (**A**) Basal view (MPL/SK/JML/305) and (**B**) Basal view (MPL/SK/JML/304); (**C–F**) Gyrogonites of ***Platychara compressa*** (Peck and Reker 1948). (**C**) Lateral view (MPL/SK/JML/C/1551); (**D**) Lateral view (MPL/SK/JML/C/1552); (**E**) Lateral view (MPL/SK/JML/C/1553); and (**F**) Apical view (MPL/SK/JML/C/1554), (**G–I**) Gyrogonites of ***Platychara*** sp. (**G**) Lateral view (MPL/SK/JML/C/1112); (**H**) Apical view (MPL/SK/JML/C/1113) and (**I**) Basal view (MPL/SK/JML/C/1114). (**J–M**) Gyrogonites of ***Platychara closasi*** sp. nov. (**J**) Lateral view (MPL/SK/GP/C/1255); (**K**) Deformed apical view (MPL/SK/GP/C/1256); (**L**) Apical view (MPL/SK/GP/C/1257); and (**M**) Basal view (MPL/SK/GP/C/1258), (**N–P**) Gyrogonites of ***Peckichara* cf. *varians*** (Grambast 1957). (**N**) Lateral view (MPL/SK/JML/1307); (**O**) Lateral view (MPL/SK/JML/401) and (**P**) Apical view (MPL/SK/JML/C/1308). (Figure **B**, Reproduced from Khosla et al. 2022 with permission from Wiley; Figures **G–K**, Reproduced from Kania et al. 2022 with permission from the Editor of Himalayan Geology)

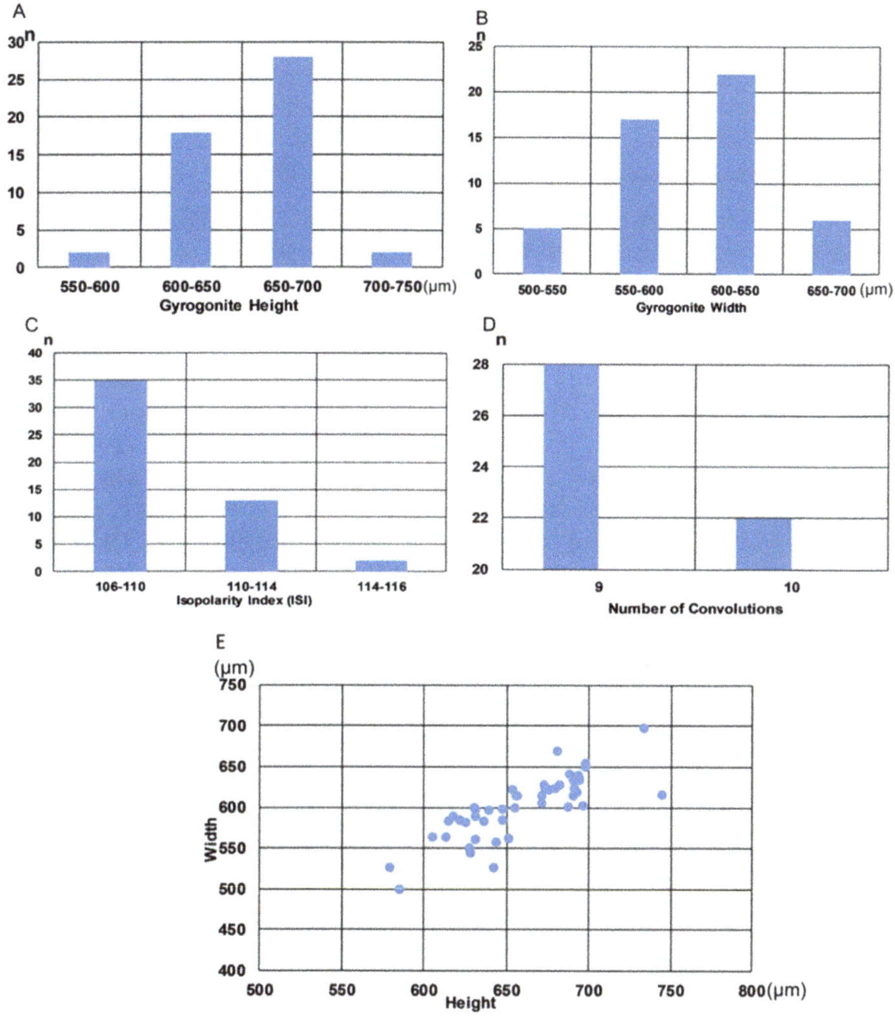

Fig. 4.6 Biometric data of *Platychara sahnii* (50 gyrogonites) from the Jhilmili intertrappean beds, Chhindwara, Madhya Pradesh. (**A**) Histogram showing height, (**B**) Histogram showing width, (**C**) Histogram showing isopolarity index (ISI), (**D**) Histogram showing number of convolutions and (**E**) Dispersion graph of width/height

Remarks This species was first included within the genus *Chara* (now a synonym, in part, of *Platychara*) by Rao and Rao (1939) and was recorded from the Upper Cretaceous intertrappean beds of Rajahmundry (East Godavari District), Andhra Pradesh. Later, *Chara sahnii* was re-assigned as *Platychara sahnii* based on material from the Upper Cretaceous intertrappean beds of Gitti Khadan, (Nagpur), Maharashtra, by Bhatia and Mannikeri (1976). The presently described specimens are similar to specimens of *Platychara sahnii* from the Gitti Khadan area in size, shape, ISI, number and width of lime spirals. They differ in having more or less similar or smaller dimensions (LPA = 586–745 μm) in comparison to the specimens from Gitti Khadan

(LPA = 550–800 μm). The widths of the Gitti Khadan specimens (LED = 500–700 μm) are slightly larger than the Jhilmili specimens (LED = 500–616 μm). The specimens described herein have higher ISI values (114–122 μm, most about 117 μm) than *P. perlata* (78–91 μm, most about 85 μm). *Platychara sahnii* was also reported from the intertrappean beds of Kachchh, Gujarat (Bajpai et al. 1990), and the Chandarki section in Gurmatkal, Karnataka (Srinivasan et al. 1992, 1994).

The specimens from Chandarki differ from the Jhilmili forms in being much larger in size (LPA = 900 μm and LED = 940 μm). They typically have a narrowing base, unlike other species of *Platychara*, for instance, *P. perlata*, which have a relatively flat base. Overall, the length and width of the gyrogonites from Jhilmili (LPA = 586–745 μm and LED = 500–616 μm) differ considerably from the other species of *Platychara* such as *P. perlata* (LPA = 616–690 μm; LED = 720–823 μm) and *P. raoi* (LPA = 586–665 μm; LED = 551–635 μm). The present forms are similar in shape and convolution numbers to *Platychara* cf. *sahnii* described by Khosla (2014) from the Upper Cretaceous Lameta Formation of Jabalpur (Madhya Pradesh), Central India. The Jabalpur specimens differ from the Jhilmili ones in being much smaller in size (LPA = 440–660 μm) but larger in width (LED = 610–660 μm). The Jhilmili specimens are larger in size in general, but we include them in the same species.

Distribution *Platychara sahnii* has been reported from the Upper Cretaceous intertrappean beds in the Chandarki section (Gurmatkal), Karnataka, South India (Srinivasan et al. 1992, 1994), Nagpur in Maharashtra (Bhatia and Mannikeri 1976) and in the Upper Cretaceous Lameta Formation of the Jabalpur Basin, Central India (Khosla 2014).

4.2.4 Species Platychara compressa (Peck and Reker 1948) (Figs. 4.5C–F)

1888 *Chara compressa*: Knowlton, pp. 156–157, Figs. 1–2; Groves 1933, pp. 14 (non-*Chara compressa* Kunth, 1815).
1948 *Aclistochara compressa*: Peck and Reker 1948, pp. 87, Pl. 21, Figs. 31–33 (non-Lankford, 1953, pp. 109, Pl. 12, Figs. 3–7).
1954 *Brachychara compressa*: Grambast and Grambast, pp. 3, 1954.
1955 *Tectochara compressa*: Madler, pp. 276; Peck 1959, pp. 95, Figs. 3–4.
1956 *Gyrogona compressa*: Grambast, pp. 280.
1959 *Nodosochara compressa*: Horn af Rantzien, pp. 81, 96.
1959 *Peckichara compressa*: Peck, pp. 118, Fig. 4.
1962 *Platychara compressa*: Grambast, pp. 76; Holifield 1964, pp. 31, Pl. 11, Figs. 11–18.
1979 *Platychara compressa*: Peck and Forster 1979.
1983 *Platychara* aff. *compressa*: Feist and Colmbo 1983, Figs. 13–14.
1987 *Platychara compressa*: Fouch et al. 1987, pp. 5.
1994 *Platychara compressa*: Srinivasan et al. pp. 564, Figs. 5–7.
1999 *Platychara compressa*: Difley et al. Fig. 8A.
2005 *Platychara compressa*: Ciccioli et al. Fig. 5A.

Material More than 20 well-preserved gyrogonite specimens.

Horizon, Age and Locality Upper Cretaceous to Lower Palaeocene clayey limestone of unit 3 (JH 17 and JH 19) of the Jhilmili intertrappean beds, Chhindwara District, Madhya Pradesh, India.

Description Gyrogonites are moderate to small in size with 493–683 μm length (mean: 465 μm) and 610–900 μm width (mean: 780 μm). Isopolarity index ranges from 58 to 94 (mean: 75). Gyrogonites are sub-spherical, sub-globular or oval in shape. There are 6–7 countable convolutions. The lime spirals cells are flat to concave without any ornamentation and inclined from right to left, discernable in lateral view. Apical part displays well-developed rosettes. Basal part has a small pentagonal or circular pore.

Remarks Originally, the species was named *Chara compressa* by Knowlton (1888), and later Peck and Reker (1948) named it as *Aclistochara compressa*. In three successive years, Grambast (1954), Madler (1955) and Grambast (1956) renamed this species as *Brachychara compressa*, *Tectochara compressa* and *Gyrogona compressa*, respectively. On the basis of morphology, Peck (1959) revised and named it as *Peckichara compressa*. This species is recognised on the basis of its typical compressed morphology.

Distribution Initially, this species was described as *Peckichara compressa* from the Lower Tertiary Wasatch Group of Wales Canyon, Utah, USA. Later, this species was recovered from Cretaceous and Lower Cenozoic deposits of Peru (South America), Palaeogene deposits of Mallorca (Spain), Miocene deposits near Oedonberg, Hungary (Europe) and Tertiary deposits of the Rocky Mountains, North America (Madler 1955; Peck 1959). This species was also documented from the Maastrichtian deposits of La Rioja Province (Argentina). In India, it has been reported from Upper Cretaceous deposits of Gurmatkal (Karnataka), Kateru area (Andhra Pradesh), and Kora (Gujarat).

4.2.5 Species Platychara sp. (Figs. 4.5G–I; Table 4.4)

Material Four well-preserved gyrogonite specimens.

Horizon, Age and Locality Upper Cretaceous to Lower Palaeocene clayey limestone of unit 3 (JH 17 and JH 19) of the Jhilmili intertrappean beds, Chhindwara District, Madhya Pradesh, India.

Description The large to moderate gyrogonites are 680–696 μm in length (mean: 687 μm) and 660–669 μm in width (mean: 665 μm). The isopolarity index ranges from 103 to 104 with a mean value of 103. Gyrogonites are sub-spherical, subrounded, ellipsoidal to oval in shape. They have 7–8 countable convolutions. The lime spirals are flat to concave and lack prominent ornamentation. In lateral view, they are inclined from right to left. The apical part possesses a well-developed rosette. The basal part contains a pore situated in the centre of the terminal column.

Table 4.4 Dimensions of three gyrogonites of *Platychara* sp.

	Dimensions (μm)		
Catalogue no.	LPA	LED	ISI
MPL/SK/JML/C/1112	686	666	103
MPL/SK/JML/C/1113	680	660	103
MPL/SK/JML/C/1114	696	669	104
Mean	687	665	103

Remarks Earlier, the genus *Platychara* was represented by three species, *Platychara perlata*, *P. raoi* and *P. sahnii*. *Platychara* sp. is likely a new species of this genus, but due to the lesser number of specimens, it has been identified up to the generic level. It differs from *P. perlata* by having a high height, less width and more elongation. It differs from *P. raoi* by having an arch shaped basal part, lime spirals that are comparatively much broader and a well-developed rosette. It differs from *P. sahnii* by having an arch shaped basal part, raised lime spirals and well-developed rosette.

Distribution This species is reported for the first time from the intertrappean beds of Jhilmili, Chhindwara District, Madhya Pradesh.

4.2.6 Species Platychara closasi sp. nov. (Figs. 4.5J–M; Table 4.5)

Holotype MPL/SK/GP/C/1255, well-preserved gyrogonite specimen.

Referred Material More than seven moderately preserved gyrogonite specimens.

Horizon, Age and Locality Upper Cretaceous to ? Lower Palaeocene hard clayey limestone horizon of the Ghat Parasia intertrappean beds, Chhindwara District, Madhya Pradesh, India.

Etymology Species named after Prof. Carles Martín-Closas to honour his great contributions to the charophytes of Europe.

Diagnosis Gyrogonites are medium in size, circular or oval in shape and possess 8–9 convolutions observable in lateral view.

Description Medium sized gyrogonites, 720–759 μm length (mean: 736 μm) and 809–829 μm width (mean: 818 μm). Isopolarity index ranges from 89 to 92 (mean: 90). Gyrogonites are circular or oval in shape. There are 8–9 countable convolutions. The lime spirals cells are sharp or concave without ornamentation and inclined from left to right, visible in lateral view. The apical part contains some rosettes. The basal part has a central pore.

Table 4.5 Dimensions of gyrogonites specimens of *Platychara closasi* sp. nov.

	Dimensions (μm)		
Catalogue no.	LPA	LED	ISI
MPL/SK/GP/C/1255	759	829	92
MPL/SK/GP/C/1256	720	809	89
MPL/SK/GP/C/1257	725	812	89
MPL/SK/GP/C/1258	739	821	90
Mean	736	818	90

Fig. 4.7 (**A**–**G**) Gyrogonites of ***Peckichara*** cf. ***varians*** (Grambast 1957). (**A**) Apical view (MPL/SK/JML/1303); (**B**) Basal view (MPL/SK/JML/404); (**C**) Basal view (MPL/SK/JML/C/1309) and (**D**) Basal view (MPL/SK/JML/C/1310); (**E**) Lateral view (MPL/SK/JML/402); (**F**) Apical (MPL/SK/JML/402) and (**G**) Apical view (MPL/SK/JML/403). (**H**–**M**) Gyrogonites of ***Nemegtichara*** cf. ***grambasti*** (Bhatia ct al. 1990b). (**H**) Lateral view (MPL/SK/JML/501); (**I**) Lateral view (MPL/SK/JML/1602); (**J**) Lateral view (MPL/SK/JML/505); (**K**) Apical view (MPL/SK/JML/1603); (**L**) Basal view (MPL/SK/JML/1603) and (**M**) Basal view (MPL/SK/JML/1606). (**N**–**P**) Gyrogonites of ?***Grambastichara*** sp. (**N**) Lateral view (MPL/SK/JML/C/1754); (**O**) Lateral view (MPL/SK/JML/C/1755) and (**P**) Apical view (MPL/SK/JML/C/1756). (Figure **B**, **E**, **F**, **G**, **H**, **K**, **L**, Reproduced from Khosla et al. 2022 with permission from Wiley)

Remarks The genus *Platychara* is now represented by six species: *Platychara perlata, P. raoi, P. sahnii, P. compressa, P.* sp. and *P. closasi* sp. nov. *Platychara closasi* sp. nov. differs from *P. perlata* in height and width. It has more height and a greater number of convolutions, and the rosette is not clearly visible, but the rosette is well-developed in *P. perlata*. *Platychara closasi* sp. nov. differs from *P. raoi* by having less height, a less observable apical part, an almost flat basal part and by the presence of a broad terminal pore. This species is different from *P. sahnii* in height, lime spirals and the basal part; it has less height, sharp lime spiral, a flat basal part

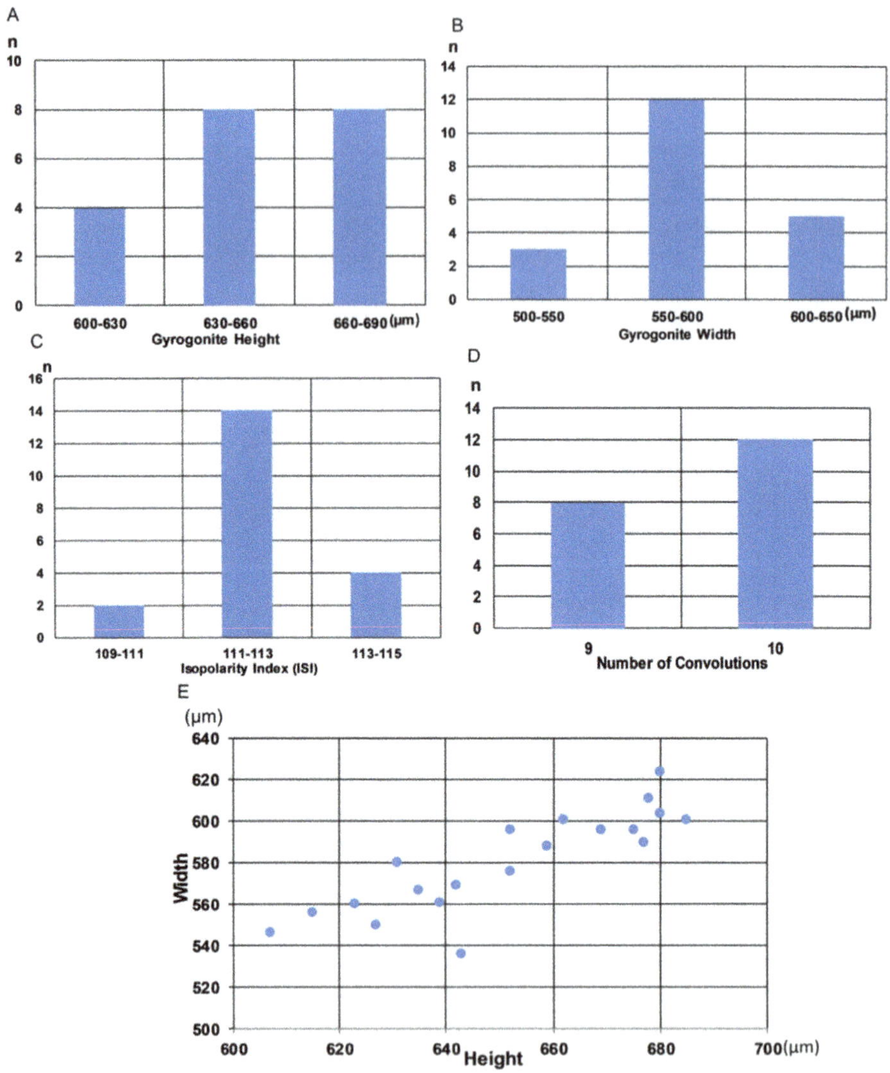

Fig. 4.8 Biometric data of *Peckichara* cf. *varians* (20 gyrogonites) from the Jhilmili intertrappean beds, Chhindwara, Madhya Pradesh. (**A**) Histogram showing height, (**B**) Histogram showing width, (**C**) Histogram showing isopolarity index (ISI), (**D**) Histogram showing number of convolutions and (**E**) Dispersion graph of width/height

and a broad basal pore. It differs from *P. compressa* by having high height and width. Finally, the lesser height, oval to sub-spherical shape, fewer lime spirals, flat basal part and broad to flat terminal pore of *Platychara closasi* sp. nov. differentiate it from *P.* sp.

Distribution *Platychara closasi* sp. nov. has been described for the first time from the hard clayey limestone unit of Upper Cretaceous Ghat Parasia intertrappean, Chhindwara District, Madhya Pradesh.

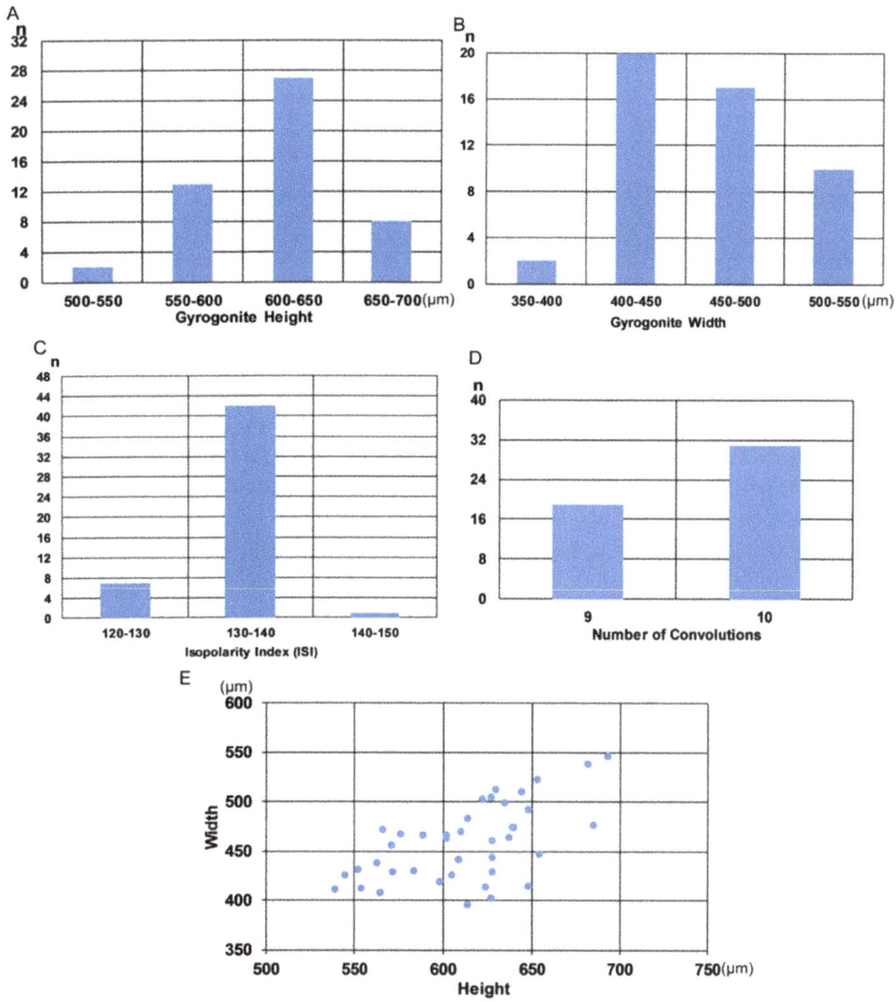

Fig. 4.9 Biometric data of *Nemegtichara* cf. *grambasti* (50 gyrogonites) from the Jhilmili intertrappean beds, Chhindwara, Madhya Pradesh. (**A**) Histogram showing height, (**B**) Histogram showing width, (**C**) Histogram showing isopolarity index (ISI), (**D**) Histogram showing number of convolutions and (**E**) Dispersion graph of width/height

4.2.7 *Species Peckichara cf. varians (Grambast 1957) (Figs. 4.5N–P, 4.7A–G, 4.8; Table 4.6)*

Genus	*Peckichara* Grambast (1957)
Species	***Peckichara* cf. *varians*** Grambast (1957)

1957 *Peckichara varians*: Grambast sp. nov. Grambast: pp. 14–15, pl. VIII, Figs. 1–8.
1992 *Peckichara varians*: Srinivasan et al. pp. 397–398, Figs. 2g–j.
1994 *Peckichara varians* cf. *P. varians meridionalis*: Srinivastan et al. pp. 564, Pl. 2, Figs. 1–5.
2016 *Peckichara varians*: Li et al. pp. 189–191, Figs. 6 A–L and Fig. 7, Table 2.
2022 *Peckichara* cf. *varians*: Khosla et al. pp. 10–13, Figs. 7a–h, Fig. 8.

Material More than 20 well-preserved gyrogonite specimens.

Horizon, Age and Locality Upper Cretaceous to Lower Palaeocene clayey limestone of unit 3 (JH 17 and JH 19) of the Jhilmili intertrappean beds, Chhindwara District, Madhya Pradesh, India.

Description Large to very large gyrogonites, measuring 623–685 μm length (mean: approximately 620 μm) and 516–611 μm width (mean: 568 μm). Isopolarity index ranges from 102 to 115 (mean: 109), spherical or ovoid in shape. Apex shows depression with a hole. Lime spirals flat and without ornamentation, with 8–9 convolutions in lateral view. Base is funnel-shaped. Basal part of spirals is of a convex shape and shows widening at basal pore. Base is pointed with a circular pore, about 5–35 μm in diameter.

Table 4.6 Dimensions of gyrognite specimens of *Peckichara* cf. *varians* (Grambast 1957)

	Dimensions (μm)		
Catalogue no.	LPA	LED	ISI
MPL/SK/JML/401	623	560	112
MPL/SK/JML/402	627	550	114
MPL/SK/JML/1303	642	610	105
MPL/SK/JML/404	531	516	102
MPL/SK/JML/1305	680	604	113
MPL/SK/JML/1306	678	611	111
MPL/SK/JML/1307	685	601	114
MPL/SK/JML/1308	652	596	110
MPL/SK/JML/1309	677	590	115
Mean	571	582	111

Fig. 4.10 **A** Gyrogonite of ?***Grambastichara*** sp. (**A**) Basal view (MPL/SK/JML/1757). (**B–F**) Gyrogonites of ***Microchara shivarudrappai*** sp. nov. (**B**) Lateral view (MPL/SK/JML/901); (**C**) Basal view (MPL/SL/GP/902); (**D**) Apical view (MPL/SK/JML/903); (**E**) Basal view (MPL/SK/JML/904) and (**F**) Latero-basal view (MPL/SL/GP/905); (**G–P**) Gyrogonites of ***Chara chhindwaraensis*** sp. nov. (**G**) Lateral view (MPL/SK/JML/1001); (**H**) Lateral view (MPL/SK/JML/1002); (**I**) Lateral view (MPL/SK/JML/C/1158); (**J**) Lateral view (MPL/SK/JML/1001); (**K**) Apical view (MPL/SK/JML/1003); (**L**) Apical view (MPL/SK/JML/1001); (**M**) Basal view (MPL/SK/JML/1003); (**N**) Basal view (MPL/SK/JML/1004); (**O**) Basal view (MPL/SK/JML/C/1160); and (**P**) Enlarged view of O (MPL/SK/JML/C/1160) showing a pentagonal depression. (Figure **I, M, O**, Reproduced from Kania et al. 2022 with permission from the Editor of the Himalayan Geology)

Remarks The species *Peckichara* cf. *varians* has been reported from the Upper Cretaceous intertrappean beds of Gurmatkal, Yanagundi and Chandarki (Gulbarga District), Karnataka and Kora, Gujarat (Srinivasan et al. 1992, 1994). In general outline, shape and number of lime spirals, the form described herein is quite similar to *P. varians*. The length and width of the gyrogonites from Gurmatkal (LPA = 715–925 μm and LED = 615–630 μm) are considerably larger than the Jhilmili specimens. However, the Jhilmili specimens are smaller in size, and it appears that they represent the same species. The present species is close to *Peckichara varians*, which was first reported from the Paris Basin (Palaeocene-

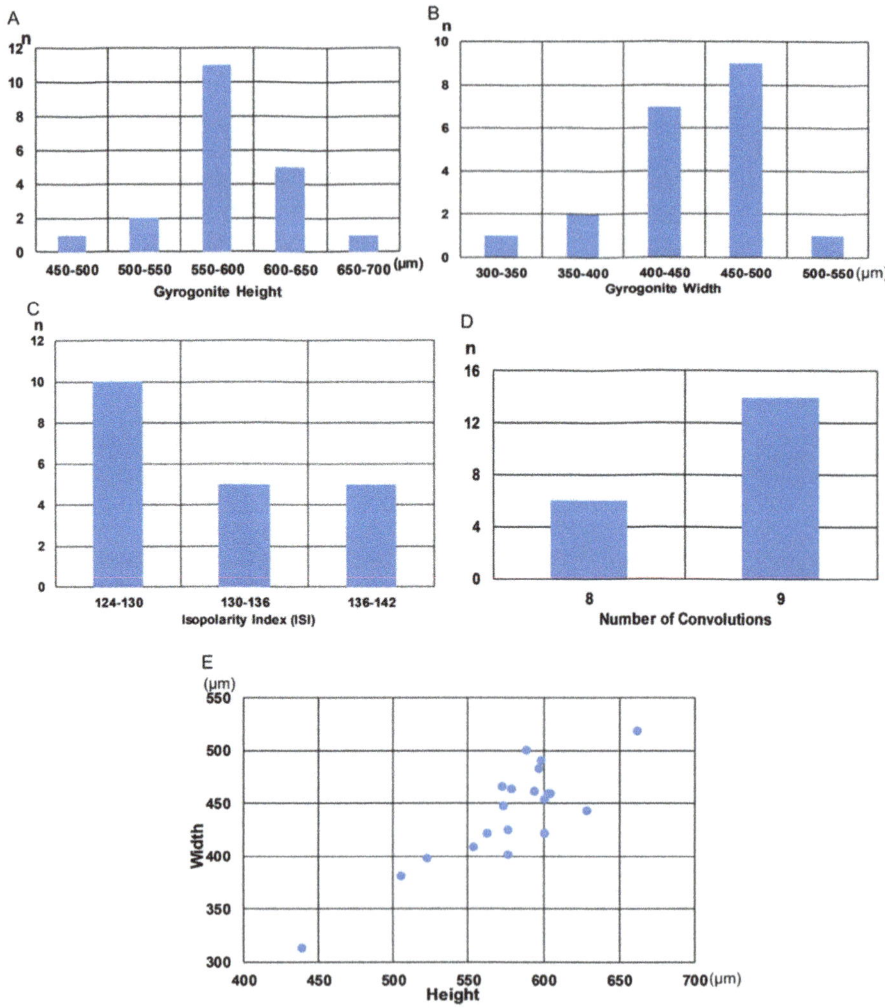

Fig. 4.11 Biometric data of *Microchara shivarudrappai* sp. nov. (20 gyrogonites) from the Jhilmili intertrappean beds, Chhindwara, Madhya Pradesh. (**A**) Histogram showing height, (**B**) Histogram showing width, (**C**) Histogram showing isopolarity index (ISI), (**D**) Histogram showing number of convolutions and (**E**) Dispersion graph of width/height

Early Eocene age), France (Grambast 1957). This species has also been reported from the Upper Cretaceous (Maastrichtian) deposits of Iran (Colin et al. 2012).

It has also been widely recorded from the Palaeocene-Eocene deposits of central and southern China (Wang 2004; Huang 1988) and Cretaceous-Palaeocene transition in the Pingyi Basin of eastern China (Li et al. 2016). Nevertheless, the Chinese gyrogonites are larger in size (LPA = 651–929 μm; LED = 631–931 μm, isopolarity index of 87–116 and 6–7 convolutions) in contrast to the Jhilmili species (623–685 μm, LED = 516–611 μm, ISI 111, and 8–9 convolutions).

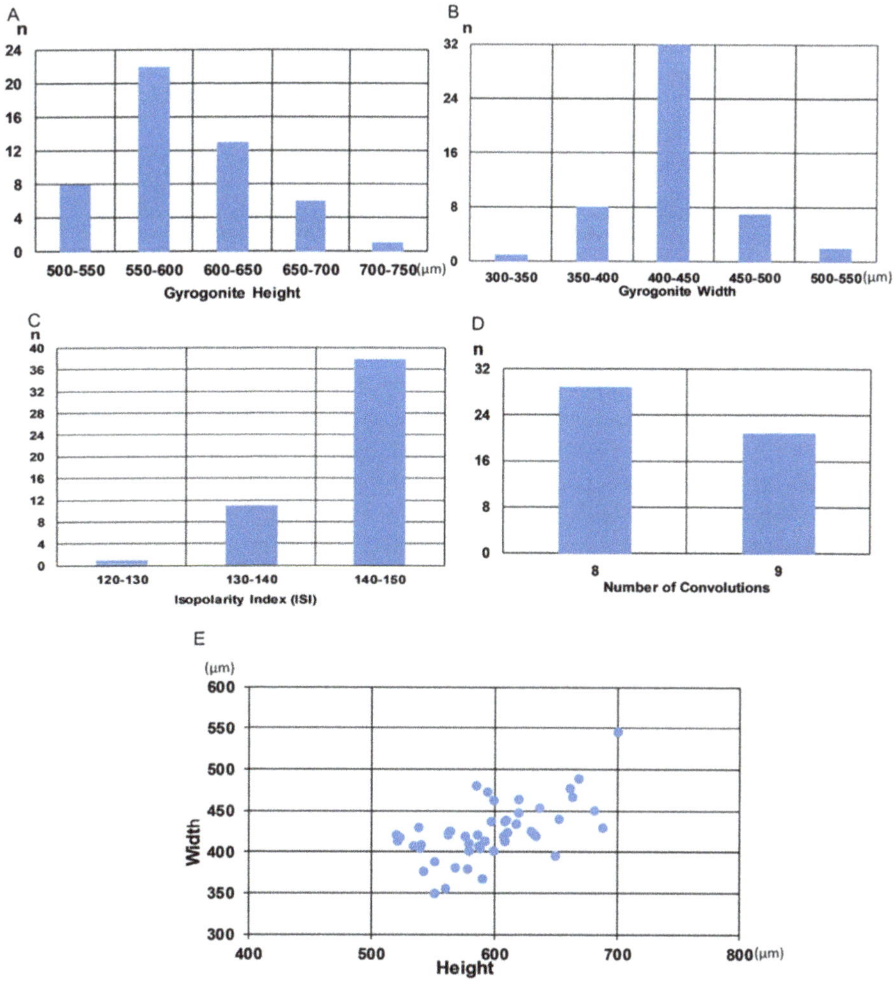

Fig. 4.12 Biometric data of *Chara chhindwaraensis* sp. nov. (50 gyrogonites) from the Jhilmili intertrappean beds, Chhindwara, Madhya Pradesh. (**A**) Histogram showing height, (**B**) Histogram showing width. (**C**) Histogram showing isopolarity index (ISI), (**D**) Histogram showing number of convolutions and (**E**) Dispersion graph of width/height

Distribution *Peckichara* cf. *varians* is widely distributed throughout the western and southern part of peninsular India. It was reported in the Upper Cretaceous intertrappean localities of Kora in Gujarat and the Yanagundi, Chandarki and Gurmatkal localities of Karnataka (Srinivasan et al. 1992, 1994). Outside India, *Peckichara varians* is extensively distributed in the Palaeocene-Early Eocene throughout Eurasia, including the Paris Basin, northern part of the Pyrenean basins, France (Grambast 1957; Massieux et al. 1981a, b), Spain (Feist and Colombo 1983), central and southern China (Wang 2004; Huang 1988) and the Cretaceous-Palaeocene transition in the Pingyi Basin, eastern China (Li et al. 2016). It has also been reported in the Late Cretaceous of Iran (Colin et al. 2012).

4.2.8 *Species Nemegtichara cf. grambasti (Bhatia et al. 1990b) (Figs. 4.7H–M, 4.9; Table 4.7)*

Genus	*Nemegtichara* (Karczewska and Ziembinska-Tworzydlo 1972)
Species	***Nemegtichara* cf. *grambasti*** (Bhatia et al. 1990b)

1990a *Nemegtichara grambasti*: Bhatia et al. pp. 118, Pl. 1, Fig. 12.
1990b *Nemegtichara grambasti* sp. nov: Bhatia et al. pp. 318–320, Pl. 1, Figs. 6–9.
1992 *Nemegtichara grambasti*: Srinivasan et al. pp. 397–398, Pl. 1, Figs. 2k–m.
1994 *Nemegtichara grambasti*: Srinivasan et al. pp. 564–566, Pl. 2, Figs. 8–10.
2014 *Nemegtichara grambasti*: Khosla, pp. 315, Pl.1, Figs. 5–7.
2022 *Nemegtichara grambasti*: Khosla et al. pp. 12–15, Figs. 7i–m, Fig. 9.

Material More than 65 moderately preserved gyrogonite specimens.

Horizon, Age and Locality Upper Cretaceous to Lower Palaeocene clayey limestone of unit 3 (JH 17 and JH 19) of the Jhilmili intertrappean beds, Chhindwara District, Madhya Pradesh, India.

Description Gyrogonites of small size, 601–686 μm high (mean: 652 μm) and 358–510 μm wide (mean: 449 μm). The isopolarity index ranges between 131 and 172 (mean: 146). Gyrogonites have long, ovoid and ellipsoidal shapes. The apex is either rounded or arc-shaped. The apical section shows very minor calcification of the lime spirals. The apical periphery has narrow lime spirals and is convex (50–90 μm high) with narrowing intercellular ridges. The lime spirals, as a whole, have eight to ten convolutions (often eight) and are inclined right to left in lateral view. The lime spirals are smooth, convex and without ornamentation. Gyrogonites protrude slightly in the middle. The basal portion is pointed and has a basal pore. It gradually tapers, and the basal pore opening is narrow, with a pentagonal pore (20–100 μm wide).

Table 4.7 Dimensions of gyrognite specimens of *Nemegtichara* cf. *grambasti* (Bhatia et al. 1990b)

	Dimensions (μm)		
Catalogue no.	LPA	LED	ISI
MPL/SK/JML/501	614	358	172
MPL/SK/JML/1602	624	414	151
MPL/SK/JML/1603	601	454	132
MPL/SK/JML/1604	685	477	143
MPL/SK/JML/505	669	510	131
MPL/SK/JML/1606	686	510	134
MPL/SK/JML/1607	636	430	147
Mean	652	450	144

Remarks This species was first described from the Upper Cretaceous intertrappean beds of the Rangapur locality of Telangana in peninsular India (Bhatia et al. 1990b). It was later recorded from the Upper Cretaceous intertrappean beds of Mamoni, Rajasthan (Bhatia et al. 1990a) and Gurmatkal, Karnataka (Srinivasan et al. 1992, 1994). In infratrappean beds, this species occurs in Upper Cretaceous deposits of Nand-Dongargaon in Maharashtra (Mohabey et al. 1993) and has also been reported from Jabalpur, Madhya Pradesh (Khosla 2014). In general outline, shape and number of lime spirals and the faint presence of calcified lime spirals in the apical part, the forms described herein are quite similar to *Nemegtichara grambasti* described from the Rangapur intertrappeans (Bhatia et al. 1990b).

The gyrogonites from Rangapur (LPA = 640–770 μm and LED = 487–564 μm) and the Yanagundi and Telangana areas (LPA = 730–785 μm and LED = 470–530 μm) of Gurmatkal, Karnataka (South India), are slightly longer and wider than the Jhilmili specimens (LPA = 601–686 μm and LED = 358–510 μm). In contrast, the gyrogonites from the infratrappeans of Jabalpur (Madhya Pradesh) have somewhat lesser biometric parameters (LPA = 510–580 μm; LED = 360–420 μm, ISI 139 and 8–10 convolutions) than the Jhilmili specimens. The gyrogonites from Jhilmili and Jabalpur are smaller in size than those from Yanagundi (Gurmatkal) and Rangapur, but they are thought to be the same species.

The present species is close to two species (*Nemegtichara prima* and *N. quarta*), which were first reported from the Nemegt Formation ("White Beds" of Palaeogene age) in the Nemegt Basin (Gobi Desert) of Mongolia (Karczewska and Ziembinska-Tworzydlo 1972). In northeast China, *Nemegtichara* was subsequently found in the Mingshui Formation in the Songliao Basin (Wang et al. 1985; Li et al. 2013, 2019). More recently, Li et al. (2019) described *N. prima* from the Cretaceous-Palaeocene boundary in the Songliao Basin of China. Nevertheless, the Chinese gyrogonites are smaller in size (LPA = 415–498 μm; LED = 347–514 μm, ISI 116–128, and 7–9 convolutions) in contrast to the *Nemegtichara* cf. *grambasti* (Bhatia et al. 1990b) from Jhilmili.

The Indian species is similar in shape and convolution numbers to the Chinese species *Nemegtichara prima* (Karczewska and Ziembinska-Tworzydlo 1972; Li et al. 2019) but differs in the size parameters such as smaller width. The length/width ratio of the Jhilmili specimens is quite close to *Nemegtichara quarta* from

Mongolia (Karczewska and Ziembinska-Tworzydlo 1972). *Nemegtichara grambasti* has a tapering base and maximum width in the upper third of the gyrogonite, whereas the Mongolian species displays a rounded base and maximum width in the middle part of the gyrogonite. More recently, Li et al. (2019) compared the two genera *Nemegtichara* and *Lamprothamnium* and concluded that the genus *Nemegtichara* is typified by narrow lime spirals in the apical periphery, resulting in a somewhat conical shape and pentagonal basal part, whereas *Lamprothamnium* exhibits a deep depression in its periapical part, and the basal part is not noticeable from the outer surface.

Distribution *Nemegtichara* cf. *grambasti* was reported in the Upper Cretaceous intertrappean beds of Rangapur, Telangana and Gurmatkal, Karnataka, south India (Bhatia et al. 1990b; Srinivasan et al. 1992, 1994), Mamoni in Rajasthan (Bhatia et al. 1990a), and in the Upper Cretaceous infratrappean beds of the Jabalpur and Nand-Dongargaon Basin, Central India (Mohabey et al. 1993; Khosla 2014). The Indian *N.* cf. *grambasti* shows close affinities with *N. prima* (Karczewska and Ziembinska-Tworzydlo 1972; Li et al. 2019). *Nemegtichara prima* was first described from the Palaeogene deposits of the Nemegt Basin of the Gobi Desert in Mongolia (Karczewska and Ziembinska-Tworzydlo 1972). Later, it was also reported in the Upper Cretaceous-Palaeocene in the Songliao Basin, northeastern China (Wang et al. 1985; Li et al. 2013, 2019).

4.2.9 Species ?Grambastichara sp. (Figs. 4.7N–P, 4.10A; Table 4.8)

Genus	*Grambastichara* (Horn Af Rantzian 1959)
Species	?***Grambastichara*** **sp.**

Material Four moderately preserved gyrogonite specimens.

Horizon, Age and Locality Upper Cretaceous to Lower Palaeocene clayey limestone of unit 3 (JH 17 and JH 19) of the Jhilmili intertrappean beds, Chhindwara District, Madhya Pradesh, India.

Description Large to very large gyrogonites, 582–729 μm length (mean: 588 μm) and 471– 520 μm width (mean: 503 μm). Isopolarity index ranges from 114 to 154 (mean: 126). Gyrogonites are elongated, sub-spherical or oval in shape. There are 9–10 countable convolutions, and the lime spirals cells are flat to concave without ornamentation and inclined from right to left, visible in lateral view. The apical part shows spiral cells of similar width and is flat and slightly convex. The basal part has a central pore.

Table 4.8 Dimensions of gyrogonite specimens of ?*Grambastichara* sp.

	Dimensions (μm)		
Catalogue no.	LPA	LED	ISI
MPL/SK/JML/1754	729	471	154
MPL/SK/JML/1755	582	510	114
MPL/SK/JML/1756	596	520	114
MPL/SK/JML/1757	625	510	122
Mean	588	503	126

Remarks A few well-preserved gyrogonite specimens of ?*Grambastichara* sp. were collected from the Jhilmili intertrappean beds. This species is considerably different from other charophyte species known from the Jhilmili intertrappeans. The recovered species are: *Platychara perlata, P. raoi, P. sahnii, P. compressa, P.* sp. *P. closasi* sp. nov., *Peckichara* cf. *varians, Nemegtichara* cf. *grambasti, Microchara shivarudrappai* sp. nov. and *Chara chhindwaraensis* sp. nov. ?*Grambastichara* sp. differs from *P. perlata* in the height, width and number of volutions. It differs from *P. raoi* in the height, lime spirals and basal part and from *P. sahnii* in width, lime spirals and basal part.

?*Grambastichara* sp. differs from *P. compressa* in the height, lime spirals, basal part and rosettes. It differs from *P. closasi* sp. nov. in the shape, height, width and distinct basal part. It differs from *Peckichara* cf. *varians* by having a different shape, more height, flat lime spirals, lack of rosette and less broad basal part. ?*Grambastichara* sp. differs from *Nemegtichara* cf. *grambasti* in possessing more height and width together with flat lime spirals. Further, it differs from *Microchara shivarudrappai* sp. nov. in overall shape and width and from *Chara chhindwaraensis* sp. nov. by having more height and width, flat lime spirals, a smaller number of convolutions and by a distinct basal part. The genus has been questionably assigned and the scarcity of material does not allow the authors to describe a new species. More and better specimens are needed to confirm the identity of this genus.

Distribution Upper Cretaceous Jhilmili intertrappean beds, Chhindwara District, Madhya Pradesh.

4.2.10 Species Microchara shivarudrappai sp. nov. (Figs. 4.10B–F, 4.11; Table 4.9)

Genus	*Microchara* (Grambast 1959)
Species	***Microchara shivarudrappai*** **sp. nov.**

Holotype MPL/SL/GP/901, moderatley-preserved gyrogonite specimen.

Referred Material More than 20 moderatley preserved gyrogonite specimens.

Horizons, Age and Localities Upper Cretaceous to Lower Palaeocene clayey limestone of unit 3 (JH 17 and JH 19) of the Jhilmili intertrappean beds, Chhindwara District, Madhya Pradesh, India. Hard clayey limestone horizon of the Ghat Parasia intertrappean beds, Chhindwara District, Madhya Pradesh, India.

Etymology Species named after Dr. T.V. Shivarudrappa to honour his contributions to the charophytes of the intertrappean beds, India.

Diagnosis Gyrogonites are medium sized, sub-spherical, ellipsoidal or oval in shape and longer than wide and have a rounded apex and elongated base. Concave lime spirals. Lacks spiral ornamentation and has 7–8 convolutions in lateral view. Possesses a slender periapical depression with rounded basal pore.

Description Gyrogonites medium in size, 574–662 μm length (mean: 604 μm) and 421–518 μm width (mean: 470 μm). Isopolarity index ranges from 117 to 142 (mean: 128). Gyrogonites are sub-spherical, ellipsoidal or oval in shape and are longer than wide. Apex rounded. Lime spirals are concave to flat and without ornamentation, 50–70 μm wide and form 7–8 volutions in lateral view, separated by sharp and thick intercellular ridges. Apical part shows a slender periapical depression and spiral cells of similar width, and it is flat and slightly convex. Base elongated, terminating in small and rounded basal pore, about 30–40 μm in diameter.

Table 4.9 Dimensions of gyrogonite specimens of *Microchara shivarudrappai* sp. nov.

	Dimensions (μm)		
Catalogue no.	LPA	LED	ISI
MPL/SK/JML/901	662	518	127
MPL/SK/GP/902	601	454	132
MPL/SK/JML/903	589	500	117
MPL/SK/JML/904	601	421	142
MPL/SK/GP/905	574	447	128
MPL/SK/JML/906	599	490	124
Mean	604	470	128

Remarks *Microchara shivarudrappai* sp. nov. differs from *Microchara* sp. from the Upper Cretaceous intertrappean beds of Takli, Nagpur (Bhatia and Mannikeri 1976), Asifabad, Andhra Pradesh (Bhatia et al. 1990b), Chandarki (Gurmatkal), Karnataka (Srinivasan et al. 1992, 1994), the Lameta Formation of Jabalpur, Madhya Pradesh (Khosla 2014), and the infratrappean beds of Pisdura, Maharashtra (Khosla et al. 2015, 2016), and from *Microchara sausari* reported from the intertrappean beds of Takli and Asifabad (Bhatia et al. 1990a, b) in shape, size and number of convolutions. With regard to size, *Microchara* sp. from Chandarki (Karnataka) is the largest, being on average 645–390 μm in height, followed by *Microchara shivarudrappai* sp. nov. from Jhilmili at LPA = 574–662 μm in height and 421–518 μm in width, followed by the same genus from Jabalpur, which is 420–716 μm in height and 320–577 μm in width, whilst the smallest is *Microchara* sp. from Takli (Nagpur), which is 475–500 μm in height and 360–400 μm in width.

Microchara shivarudrappai sp. nov. closely resembles in most characteristics to four species, *Microchara cristata* (Grambast 1971), *M. gobica* (Karczewska and Ziembinska-Tworzydło 1969), *M. leiocarpa* (Grambast 1971) and *M. prolix* (Wang et al. 1985) known from the Cretaceous-Palaeogene in the Songliao and Pingyi basins of China, but varies in having an ellipsoidal-ovoidal shape, base elongated or more projected and in the lack of surface tubercles.

Indian representatives of the genus differ from Chinese representatives in several ways. With regard to size, *Microchara leiocarpa* (Grambast 1971) from the Songliao basin is the largest and includes medium-sized gyrogonites, variable in size, 453–605 μm high (mean: 529 μm), and 392–546 μm wide (mean: 469 μm), normally ovoid with an isopolarity index that ranges from 101 to 135, 7–9 convolutions and a tapering base; followed by *M. gobica* (Karczewska and Ziembinska-Tworzydło 1969), which includes medium-sized gyrogonites, variable in size, LPA = 443–584 μm; LED = 357–515 μm; subprolate in shape with an isopolarity index of 103–143 μm; 7–9 (often 8–9) convolutions and with a rounded base. The gyrogonites of a third species, *M. prolix* (Wang et al. 1985), known from the Songliao basin, are also of medium size, variable in size, 407–745 μm high (mean: 491 μm) and 279–393 μm wide (mean: 336 μm), prolate ovoidal in shape with an isopolarity index of 120–170, 8–10 convolutions and a tapering base. Whereas *M. prolix* (Wang et al. 1985), which has been recently reported from the Jiaolai Basin (Tian et al. 2021), is also of medium size but much thicker (variable in size, 452–642 μm high (mean: 547 μm) and 318–437 μm wide (mean: 377.5 μm), prolate spheroidal in shape with an isopolarity index of 125–170, 9–10 convolutions and tapering base) than the specimens recorded from the Songliao Basin. Whilst the smallest is *M. cristata* (Grambast 1971), known from the Jiaolai Basin (Tian et al. 2021), and includes small-sized gyrogonites, LPA = 302–393 μm high (mean: 347.5 μm) and 272–330 μm wide (mean: 301 μm); prolate spheroidal and subprolate in shape with an isopolarity index that ranges from 111 to 140, 8–9 convolutions and a pointed base.

Microchara shivarudrappai sp. nov. also differs from *M. nana* (Vicente et al. 2015), known from the Upper Cretaceous (Maastrichtian) and Lower Danian of the Vallcebre Basin of Catalonia (Spain) and the Pingyi Basin of eastern China (Li et al. 2016). *Microchara nana* from Spain is the smallest known species and is characterised by small-sized gyrogonites, variable in size, LPA = 243–323 μm high; LED = 202–257 μm wide; ovoid or subprolate in shape with an isopolarity index of 102–149 (mean: 120); 6–8 volutions and with a rounded and rarely tapering base. In the Pingyi Basin of eastern China, *M.* cf. *nana* was described by Li et al. (2016), and the gyrogonites are very small in size, LPA = 251–486 μ m high; LED = 209–399 μm; prolate spheroidal or subprolate in shape with an isopolarity index of 103–152; and 7–8 convolutions with rounded or tapering ends.

Thus, morphologically, *Microchara shivarudrappai* sp. nov. resembles five of the above mentioned *Microchara* species known from the Cretaceous-Palaeogene sections of China and Spain. In future, more and better-preserved specimens are needed to confirm the identity of this species.

Distribution *Microchara shivarudrappai* sp. nov. is a new species discovered from the Jhilmili intertrappean beds, Chhindwara area of Madhya Pradesh. However, *Microchara* sp. was widely distributed in the east-central and southern part of peninsular India. It was initially reported in the Upper Cretaceous intertrappean beds of Nagpur (Bhatia and Mannikeri 1976), the infratrappean beds of Jabalpur and Pisdura in Central India (Khosla 2014; Khosla et al. 2015, 2016), the Upper Cretaceous intertrappean beds of Asifabad in Telangana (Bhatia et al. 1990b) and the Upper Cretaceous intertrappean beds of Chandarki in Karnataka (Srinivasan et al. 1992, 1994).

4.2.11 Species Chara chhindwaraensis sp. nov. (Figs. 4.10G–P, 4.12; Table 4.10)

Subfamily	Charoideae (Braun in Migula 1897)
Genus	*Chara* (Vaillant 1719)
Species	***Chara chhindwaraensis* sp. nov.**

Holotype MPL/SK/JML/1001 and MPL/SK/JML/1002, well-preserved gyrogonite specimens.

Referred Material More than 50 moderately preserved gyrogonite specimens.

Horizon, Age and Locality Upper Cretaceous to Lower Palaeocene clayey limestone of unit 3 (JH 17 and JH 19) of the Jhilmili intertrappean beds, Chhindwara District, Madhya Pradesh, India.

Etymology Named after Chhindwara District, Madhya Pradesh, from which it was recovered.

Diagnosis Gyrogonites medium to large in size, elongate, cylindrical, ovoid to prolate or fusiform in shape and with 8–9 countable convolutions. Length is greater than width. Apex rounded and forming an apical rosette, with radial lines. The lime spirals are broader than the peripheral region. The basal pore is rounded or elongated and somewhat tapering and generally shows a pentagonal depression.

Description Gyrogonites medium to large in size, 543–630 μm length (mean: 585 μm) and 331–424 μm width (mean: 331 μm). Isopolarity index ranges from 140 to 158 (mean: 147). Gyrogonites are elongate, cylindrical and ovoid to prolate fusiform in shape. Length is greater than width. Apex rounded and forming an apical rosette, with radial lines. Spiral cells 40–60 μm wide and without ornamentation, with 8–9 convolutions visible in lateral view, flat to concave and inclined from right to left. The lime spirals are broader at the middle part of the gyrogonite than in the peripheral area. The basal pore is rounded or elongated and somewhat tapering and generally shows a pentagonal depression, 30–50 μm in diameter, and the basal plate is unknown.

Table 4.10 Dimensions of gyrognite specimens of *Chara chhindwaraensis* sp. nov.

	Dimensions (μm)		
Catalogue no.	LPA	LED	ISI
MPL/SK/JML/1001	630	424	140
MPL/SK/JML/1002	579	378	153
MPL/SK/JML/C/1158	543	375	145
MPL/SK/JML/1004	593	413	143
MPL/SK/JML/C/1160	561	355	158
MPL/SK/JML/1006	608	419	145
Mean	585	331	147

Remarks *Chara* sp. was first reported in the Upper Cretaceous deposits of the intertrappean beds at Kateru, Rajahmundry of Andhra Pradesh (Rao and Rao 1939). It was later reported in the Upper Cretaceous intertrappean beds of Gurmatkal (Gulbarga District) in Karnataka (Shivarudrappa 1989). Shivarudrappa (1989) recognised five species of *Chara*, which are *C. gurmatkali*, *C. indica*, *C. tubulipora*, *C. hispida* and *C. microcera*. Srinivasan et al. (1992, 1994) described *Chara* sp. from the intertrappean beds of Chandarki and Gurmatkal sections in Karnataka.

In overall gyrogonite outline, shape and number of lime spirals the Gurmatkal specimens resemble *Chara* sp. described by Shivarudrappa (1989) and Srinivasan et al. (1994) from the freshwater Gurmatkal and Chandarki intertrappean beds of Upper Cretaceous age. These specimens of *Chara* sp. (Srinivasan et al. 1994) differ by having slightly larger (LPA = 660–675 μm and LED = 410–435 μm) size than *Chara chhindwaraensis* sp. nov. (LPA = 543–630 μm; LED = 331–424 μm). The present species closely resembles *Chara* sp., reported from the Pingyi Basin of Cretaceous-Palaeogene age in the eastern China (Li et al. 2016). The Indian and Chinese specimens show a similar prolate shape, isopolarity index (105–147) and convolution numbers (8–9). However, the Chinese gyrogonites are smaller in size (LPA = 404–573 μm; LED = 324–413 μm) in contrast to *Chara chhindwaraensis* sp. nov. from Jhilmili.

Overall, Indian members of the genus are similar to *Chara changzhouensis* (Huang and Wang in Wang 1978), which was first recorded from the Upper Palaeocene-Lower Eocene continental deposits of the Yangqi, Zhaizishan and Fangjiahe formations (Jianghan Basin, Wang 1978) and Upper Cretaceous-Palaeocene deposits in Jiangsu (Yang et al. 2005). *Chara changzhouensis* (Huang and Wang 1978 in Z. Wang 1978) was recently described by Li et al. (2019) from the Cretaceous-Palaeogene boundary in China's Songliao Basin. In contrast to *Chara chhindwaraensis* sp. nov., the Chinese gyrogonites are subprolate in shape and larger in size (LPA = 720–798 μm; LED = 613–739 μm, isopolarity index of 106–127, and eight to nine convolutions in lateral view). In future, more and better-preserved specimens are needed to confirm the identity of this species.

Distribution Upper Cretaceous Jhilmili intertrappean beds, Chhindwara District, Madhya Pradesh.

4.3 Ostracods

4.3.1 Species Buntonia whittakerensis sp. nov. (Figs. 4.13A–B; Table 4.11)

Kingdom	Animalia (Linnaeus 1758)
Phylum	Arthropoda (von Siebold 1848)
Subclass	Ostracoda (Latreille 1806)
Order	Podocopida (Müller 1894)
Suborder	Podocopina (Müller 1894)
Superfamily	Cytheroidea (Baird 1850)
Family	Trachyleberididae (Sylvester-Bradley 1948)
Subfamily	Buntoninae (Apostolescu 1961)
Genus	*Buntonia* (Howe 1935)
Type species	*Buntonia shubhutaensis* (Howe 1935)
Species	***Buntonia whittakerensis*** **sp. nov.**

Holotype MPL/SK/JML/2009, a right valve.

Paratype MPL/SK/JML/2070, a left valve.

Horizon, Age and Locality Upper Cretaceous to Lower Palaeocene clayey limestone of unit 3 (Sample JH 21) of the Jhilmili intertrappean beds, Chhindwara District, Madhya Pradesh, India.

Etymology Species named after John E. Whittaker to honour his scientific contributions to the field of ostracod research.

Diagnosis A small, smooth carapace of an elongate to subtriangular species of *Buntonia* with left valve overlapping the right valve, along the entire periphery. Anterior margin is broader and more concave, and posterior margin is sharply pointed.

Description A small carapace elongated, and rectangular to subtriangular shaped in lateral view. The maximum height on the anterior side, the maximum length below the mid-height. The anterior margin is broadly rounded and broader than the posterior margin. Elongated to flat in dorsal view. The posterior margin is less rounded, concave and pointed compared to the anterior margin. The dorsal margin is almost straight and sloping towards the posterior margin. The ventral margin is straight and inclined towards the anterior end. The left valve is larger than the right valve and observable on the entire margins. The carapace has a smooth surface without ornamentation, is sexually dimorphic, and internal structure is not evident.

Remarks Characteristics of *Buntonia whittakerensis* sp. nov. include a distinctly broad anterior margin, a pointed posterior margin, an inclined dorsal margin and a nearly straight ventral margin. *Buntonia* is well-known throughout the African continent, where it is represented by numerous fossil and living species.

Buntonia whittakerensis sp. nov. differs from *Buntonia apatayeriyerii* (Reyment 1963), from the Upper Palaeogene of the Senegal-Mauritanian basin (Sarr 1995), Maastrichtian-Ypresian (Upper Palaeogene) of the Trans-Saharan basin (Carbonnell et al. 1990; Carbonnel and Monciardini 1995; Sarr 2015), Maastrichtian-Palaeogene of the North African basin (Salahi 1966; Bassiouni and Morsi 2000) and Palaeogene of the Gulf of Guinea basin (Reyment 1963; Foster et al. 1983) by its smaller size

Table 4.11 Dimensions of two carapaces of *Buntonia whittakerensis* sp. nov.

	Dimensions (mm)		
Catalogue no.	Length	Width	Height
MPL/SK/JML/2009	0.715	0.250	0.364
MPL/SK/JML/2070	0.648	0.241	0.284

and distinct morphology such as broad anterior and pointed posterior margins with a lack of ornamentation on the valve surface. *Buntonia apatayeriyerii*, on the other hand, has a broader posterior margin and a heavily ornamented (large pores) valve surface. It differs from *Buntonia fortunata* (Apostolescu 1961) from the Maastrichtian-Late Eocene of North Africa (Salahi 1966; Morsi et al. 2011), Upper Palaeogene of the Senegal-Mauritanian basin (Sarr 1995), Palaeogene of the Gulf of Guinea basin (Apostolescu 1961; Foster et al. 1983) and Palaeogene-Ypresian Trans-Saharan basin (Foster et al. 1983; Sarr 2015) by its smaller size, broad anterior and pointed posterior margin and lack of ornamentation. *Buntonia fortunate* has a broader anterior margin, which is less pointed in comparison and heavily ornamented with pores on the valve surface. Further, *Buntonia whittakerensis* sp. nov. differs from *Buntonia tichittensis* (Apostolescu 1961) from the Upper Palaeogene of the Senegal-Mauritian basin (Sarr 1995) and Trans-Saharan basin (Apostolescu 1961; Reyment 1981), Palaeogene of the Gulf of Guinea basin (Reyment 1981) and Upper Palaeogene-Ypresian of the North African basin (Barsotti 1963; Salahi 1966) in having a subrectangular shape, large size and lack of ornamentation with broader anterior and posterior margins. In contrast, *Buntonia tichittensis* has a globular shape and small size and pores are present on the valve surface.

Buntonia whittakerensis sp. nov. differs from *Buntonia attitogonensis* (Apostolescu 1961) from the Palaeogene of Nigeria (Reyment and Reyment 1980; Foster et al. 1983), Palaeogene of Libya (Reyment and Reyment 1980), Early Eocene of Togo (Apostolescu 1961) and Algeria (Grekoff 1969), Eocene of Nigeria (Reyment 1963), and Late Eocene of Sinai (Shahin 2000), in having a rectangular shape, small size and by the absence of ornamentation. *Buntonia attitogonensis* is characterised by possessing an almost equally broad anterior margin and reticulate ornamentation on the valve surface. It differs from *Buntonia livida* Apostolescu (1961), from the Upper Palaeogene of the Senegalese-Mauritanian basin (Sarr 1995), Palaeogene of the Gulf of Guinea basin (Apostolescu 1961; Reyment 1963) and Palaeogene of the Trans-Saharan basin by possessing a broader anterior margin, pointed posterior margin, small size and a smooth surface. In contrast, *Buntonia livida* has a less broad anterior margin and pointed posterior margin with a weakly ornamented valve surface.

Further, it differs from *Buntonia tatteuliensis* (Apostolescu 1961) from the Late Cretaceous to Palaeogene of Libya (El Sogher 1996), Palaeogene of Mali, Nigeria and Libya (Apostolescu 1961; Reyment and Reyment 1980; Reyment 1981; Foster et al. 1983), latest Palaeogene or basal Eocene of Egypt (Bassiouni and Luger 1990) and Middle and Late Eocene of Sinai (Shahin 2000, 2005) by displaying a pointed posterior margin, more broad anterior margin and valve surface with no ornamentation, whereas *Buntonia tatteuliensis* has an elongated to rectangular shape, more broad anterior margin as compared to the posterior margin and a strongly ornamented valve surface. In future, more and better-preserved specimens are needed to confirm the identity of this species.

Distribution The oldest record of the genus *Buntonia* (Howe 1935) is from the Senonian to Palaeogene of Egypt (Sarr 2015) and Eocene sediments of the Shubuta Clay in Eastern Mississippi (Howe and Chambers 1935). It also occurs at the

Cretaceous-Palaeogene transition in many parts of Africa, where it forms three bio-regions: North Africa (Algeria, Libya, Tunisia, Morocco, and Sudan), West Africa (Mali, Togo, Benin, Cote d' Ivore, Senegal, Ghana, Gambia, Nigeria, Cameron, and Mauritania) and South Africa (Namibia and South Africa) (Elewa and Abdelhady 2020). It is also known from Maastrichtian deposits in Jamaica (Puckett et al. 2012) and Madagascar (Benmansour et al. 2016).

4.3.2 Species Neocyprideis raoi (Jain 1978) (Figs. 4.13C–K; Table 4.12)

Superfamily	Cytheroidea (Baird 1850)
Family	Cytherideidae (Sars 1925)
Subfamily	Cytherideinae (Sars 1925)
Genus	*Neocyprideis* (Apostolescu 1956)
Type species	*Cyprideis* (*Neocyprideis*) *durocortoriensis* (Apostolescu 1956)
Species	***Neocyprideis raoi*** (Jain 1978)

1978 *Ovocytheridea raoi*: Jain, pp. 53, Pl. 1, Figs. 7–10.
1995 *Ovocytheridea raoi*: Bhandari, pp. 95–96, Pl. 2, Figs. 1–2.
2002 *Neocyprideis raoi*: Khosla and Nagori, pp. 201–203, Figs. 2.12–2.13.
2009a *Neocyprideis raoi*: Keller et al. Figs. 9.1–9.2.
2009 *Neocyprideis raoi*: Khosla et al. pp. 725, Pl. 1, Figs. 12–15.
2009 *Neocyprideis raoi*: Sharma and Khosla, pp. 201–204, Pl. 1, Figs. N–R, Pl.2, Figs. A–D.
2011a *Neocyprideis raoi*: Khosla et al. pp. 230–231, Pl. 1, Figs. 1–6.
2015 *Neocyprideis raoi*: Khosla: pp. 348, 350, 352, Figs. 5, A–B.

Material A total of 1361 carapaces and some open valve remains.

Horizon, Age and Locality Upper Cretaceous-Lower Palaeocene clayey limestone horizon of unit 3 (JH 16, JH 21–26) of the Jhilmili intertrappean beds, Chhindwara District, Madhya Pradesh, India.

Description Carapace elongate, subrectangular or subovate to somewhat trianguloid in lateral view. The valve has a maximum height slightly anterior to the middle with the greatest length below the mid-height. The left valve is larger than the right and overlapping around all of the margins. Strongly convex to subangulate dorsal margin, almost straight ventral margin and valves curved postero-ventrally and antero-ventrally. The anterior margin is broadly and equally rounded, and there is a narrowly rounded posterior end. The external valve surface is densely covered by normal pore pits or punctate ornamentation. A shallow depression is situated in the antero-dorsal region. Sexually dimorphic and internal structure is not evident.

Table 4.12 Dimensions (in mm) of carapaces of *Neocyprideis raoi* (Jain 1978)

Catalogue no.	Length	Height	Width
MPL/SK/JML/O/5001	0.575	0.357	0.271
MPL/SK/JML/O/5002	0.589	0.342	0.250
MPL/SK/JML/O/5003	0.557	0.372	0.192
MPL/SK/JML/O/5004	0.580	0.338	0.243
MPL/SK/JML/O/2001	0.560	0.362	0.201
MPL/SK/JML/O/5006	0.661	0.362	0.286
MPL/SK/JML/O/5007	0.620	0.351	0.256
MPL/SK/JML/O/5008	0.618	0.359	0.265
MPL/SK/JML/O/5009	0.632	0.397	0.265
MPL/SK/JML/O/5010	0.544	0.323	0.263
MPL/SK/JML/O/5011	0.568	0.314	0.264
MPL/SK/JML/O/5012	0.629	0.403	0.261
MPL/SK/JML/O/5013	0.539	0.305	0.262
MPL/SK/JML/O/5014	0.566	0.309	0.260
MPL/SK/JML/O/5015	0.660	0.377	0.267
MPL/SK/JML/O/5016	0.639	0.385	0.266
MPL/SK/JML/O/5017	0.573	0.314	0.265
MPL/SK/JML/O/5018	0.592	0.364	0.261
MPL/SK/JML/O/5019	0.603	0.343	0.263
MPL/SK/JML/O/5020	0.607	0.380	0.264
MPL/SK/JML/O/5021	0.600	0.319	0.269
MPL/SK/JML/O/5022	0.565	0.317	0.265
MPL/SK/JML/O/5023	0.628	0.424	0.261
MPL/SK/JML/O/5024	0.513	0.372	0.263
MPL/SK/JML/O/5025	0.533	0.320	0.261
MPL/SK/JML/O/5026	0.596	0.357	0.248
MPL/SK/JML/O/5027	0.599	0.336	0.263
MPL/SK/JML/O/5028	0.578	0.353	0.263
MPL/SK/JML/O/5029	0.532	0.342	0.261
MPL/SK/JML/O/5030	0.541	0.336	0.231
MPL/SK/JML/O/5031	0.548	0.352	0.262
MPL/SK/JML/O/5032	0.574	0.314	0.262
MPL/SK/JML/O/5033	0.573	0.275	0.260
MPL/SK/JML/O/5034	0.545	0.357	0.263
MPL/SK/JML/O/5035	0.517	0.367	0.240
MPL/SK/JML/O/5036	0.478	0.361	0.235
MPL/SK/JML/O/5037	0.252	0.325	0.243
MPL/SK/JML/O/5038	0.588	0.324	0.265
MPL/SK/JML/O/5039	0.600	0.330	0.260
MPL/SK/JML/O/5040	0.540	0.309	0.261
MPL/SK/JML/O/5041	0.551	0.298	0.242

(continued)

Table 4.12 (continued)

Catalogue no.	Length	Height	Width
MPL/SK/JML/O/5042	0.609	0.415	0.263
MPL/SK/JML/O/5043	0.529	0.274	0.242
MPL/SK/JML/O/5044	0.568	0.316	0.245
MPL/SK/JML/O/5045	0.567	0.356	0.243
MPL/SK/JML/O/5046	0.523	0.323	0.242
MPL/SK/JML/O/5047	0.548	0.349	0.262
MPL/SK/JML/O/5048	0.593	0.407	0.268
MPL/SK/JML/O/5049	0.522	0.322	0.243
MPL/SK/JML/O/5050	0.524	0.389	0.249

Remarks *Neocyprideis raoi* is one of the most abundant ostracod species in the present collection, accounting for nearly half of the total ostracod sample. The species is very similar in shape, size, finer morphology and pitted ornamentation (punctate) to *Ovocytheridea raoi* Jain (1978), which was described from the intertrappean beds of Duddukuru and Kateru (Rajahmundry) in Andhra Pradesh (Bhandari 1995). Following that, Khosla and Nagori (2002) classified the genus *Ovocytheridea* and the species *Ovocytheridea raoi* as *Neocyprideis* and *Neocyprideis raoi*, respectively, while describing *N. raoi* from the Lower Palaeocene intertrappean beds of Kovur, Rajahmundry (India's east coast). Later, *N. raoi* was found in the Upper Cretaceous-Lower Palaeocene intertrappean beds of Jhilmili, Chhindwara District, Madhya Pradesh (Sharma and Khosla 2009; Keller et al. 2009a, b; Khosla et al. 2009a, b, 2011a, b; Khosla 2015).

Distribution This species was initially described as *Ovocytheridea raoi* from the intertrappean beds of Kateru (Rajahmundry District) and Duddukuru (West Godavari District), Andhra Pradesh (Jain 1978; Bhandari 1995). Later, this species was also reported from the Lower Danian sediments of the Jhilmili intertrappeans, Chhindwara District, Madhya Pradesh (Keller et al. 2009a, b; Sharma and Khosla 2009; Khosla et al. 2011a; Khosla 2015).

4.3.3 Species Limnocythere deccanensis (Khosla et al. 2005) (Figs. 4.13L–P; Table 4.13)

Family	Limnocytheridae (Klie 1938)
Subfamily	Limnocytherinae (Klie 1938)
Genus	*Limnocythere* (Brady 1868)
Type species	*Cythere inopinata* (Baird 1843)
Species	***Limnocythere deccanensis*** (Khosla et al. 2005)

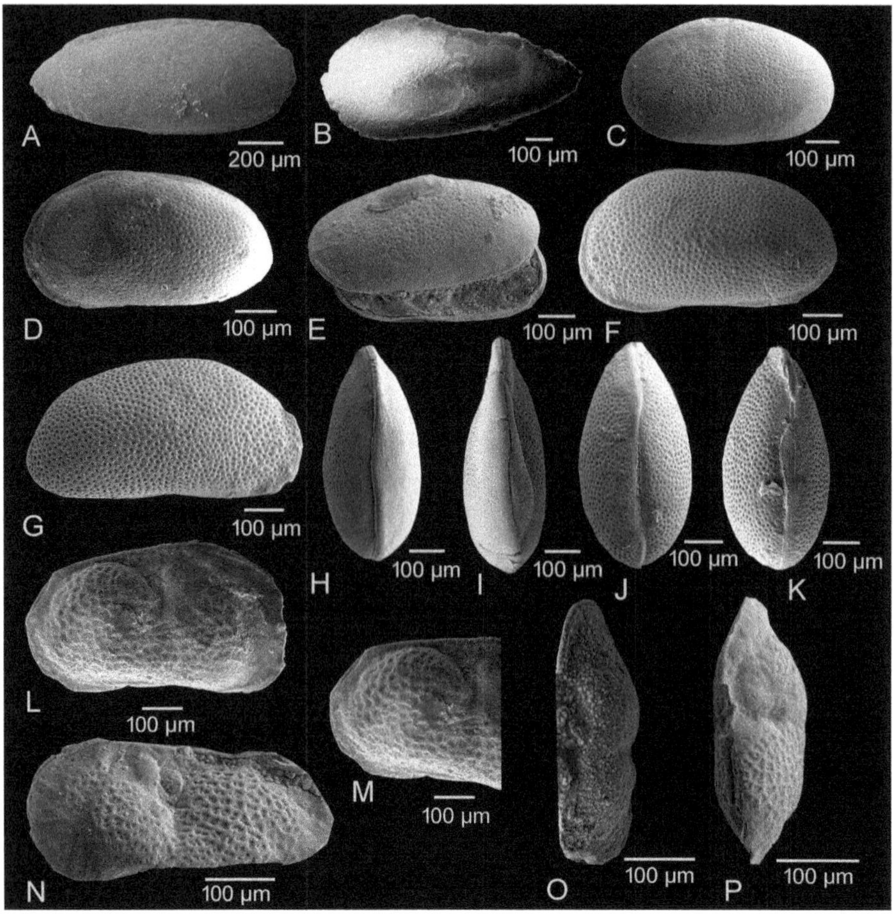

Fig. 4.13 (**A–B**) ***Buntonia whittakerensis*** sp. nov. Carapace, (**A**) Lateral view MPL/SK/JML/O 2009, right valve; (**B**) Lateral view MPL/SK/JML/O/2070, left valve. (**C–K**) ***Neocyprideis raoi*** (Jain 1978). Carapace, left valve; (**C**) Lateral view MPL/SK/JML/O/5001, left valve; (**D**) Lateral view MPL/SK/JML/O/5002, left valve; (**E**) Open valve MPL/SK/JML/O/5003, left valve; (**F**) Lateral view MPL/SK/JML/5004, right valve; (**G**) Lateral view MPL/SK/JML/2001, right valve; (**H**) Dorsal view MPL/SK/JML/O/5006; (**I**) Ventral view MPL/SK/JML/O/5007; (**J**) Dorsal view MPL/SK/JML/O/5008; (**K**) Ventral view MPL/SK/JML/O/5009. (**L–P**) ***Limnocythere deccanensis*** (Khosla et al. 2005). Carapace (**L**) Lateral view MPL/SK/JML/2013, right valve; (**M**) Enlarged view of right valve Fig. L MPL/SK/JML/2013 showing the punctate and reticulate ornamentation; (**N**) Lateral view MPL/SK/JML/O/5051, left valve; (**O**) Dorsal view MPL/SK/JML/O/5052; (**P**) Ventral view MPL/SK/JML/O/5053. (Figure **A**, **G** and **L**, Reproduced from Khosla et al. 2022 with permission from Wiley)

2004 *Limnocythere bhatiai*: Bajpai et al. pp. 150, Pl. 1, Fig. R, Pl. 2, Figs. A–B.
2005 *Limnocythere deccanensis*: Khosla et al. pp. 136, Pl. 1, Figs. 1–2.
2007a *Limnocythere deccanensis*: Khosla et al. 2005, pp. 215, Khosla and Nagori, Pl. 1, Figs. 6–9.
2007b *Limnocythere deccanensis*: Khosla et al. 2005: Khosla and Nagori, pp. 6.

2009a *Limnocythere deccanensis*: Khosla et al. 2005: pp. 51, Keller et al. Figs. 6.9–6.10.
2009a *Limnocythere deccanensis*: Khosla et al., 2005: pp. 725, Khosla et al. Pl. 2, Fig. 12.
2009 *Limnocythere deccanensis*: Khosla et al. 2005: pp. 202, Sharma and Khosla, Pl. 1, Figs. G–J.
2011a *Limnocythere deccanensis*: Khosla et al. 2005: pp. 231, Khosla et al. Pl. 1, Figs. 7–11.
2011b *Limnocythere deccanensis*: Khosla et al. 2005: pp. 225–226, Khosla et al. Pl. 1, Figs. 5–9.
2015 *Limnocythere deccanensis*: Khosla et al. 2005: Khosla, pp. 352, Fig. 5I.
2018 *Limnocythere deccanensis*: Khosla et al. 2005: Rathore, pp. 6.

Material A total of 2280 carapaces and open valves.

Horizons, Age and Localities Upper Cretaceous-Lower Palaeocene clayey limestone horizon from unit 3 (JH 16–26) of the Jhilmili intertrappean beds and reddish chert fossiliferous greenish limestone units of the Upper Cretaceous to ? Lower Palaeocene Ghat Parasia intertrappean beds, Chhindwara District, Madhya Pradesh, India.

Description Carapace elongated, subrectangular to subquadrate in lateral outline. The valve has its maximum height close to the anterior margin and the greatest length at mid-height. In dorsal view, the carapace is fusiform and rather flat sided. Rounded, inflated anterior margin that shows an arcuate depression, and posterior margin is short and pointed; a sulcus is present in the middle of the valve, which divides the valve into two parts; narrows ventrally. Dorsal margin straight, sloping down posteriorly and has compressed ends. Ventral margin is straight from the middle and elevated on the posterior and anterior margins. Left valve larger than right and overlapping along the entire margin. Median vertical sulcus on the valve surface, narrowing ventrally; arcuate anterior depression and shows the antero-ventral rib. The surface has punctate and reticulate ornamentation, some polygonal and trapizohedral, dominating on the posterior margin and less on the anterior and other margins. Sexually dimorphic and internal features are not observable.

Table 4.13 Dimensions of carapaces of *Limnocythere deccanensis* (Khosla et al. 2005)

		Dimensions (mm)	
Catalogue no.	Length	Width	Height
MPL/SK/JML/O/5051	0.422	0.122	0.195
MPL/SK/JML/O/5052	0.434	0.125	0.201
MPL/SK/JML/O/5053	0.430	0.116	0.210
MPL/SK/JML/O/5054	0.516	0.113	0.244
MPL/SK/JML/O/5055	0.508	0.114	0.243
MPL/SK/JML/O/5056	0.496	0.127	0.226
MPL/SK/JML/O/5057	0.453	0.128	0.212
MPL/SK/JML/O/5058	0.513	0.110	0.239
MPL/SK/JML/O/2013	0.486	0.125	0.215
MPL/SK/GP/O/5060	0.509	0.118	0.226

Remarks *Limnocythere bhatiai* was first described by Bajpai et al. (2004) from the Phulsagar intertrappean beds in the Mandla District, Madhya Pradesh. Later, Khosla et al. (2005) discovered that the name *Limnocythere bhatiai* had already been assigned to a species (*Limnocythere bhatiai* Mathur 1972) reported from the Tatrot Formation, Upper Siwalik Subgroup, exposed at Pinjaur, Haryana, so these authors renamed this species *Limnocythere deccanensis*. *Limnocythere* (Brady 1868) now has three representatives: *L. deccanensis*, *L. bajpaii* and *Limnocythere* sp. These species have been found in peninsular India's Upper Cretaceous Lameta Formation and intertrappean beds (Khosla et al. 2005; Khosla and Nagori 2007a). *Limnocythere bajpaii* is distinguished from *L. deccanensis* by a short-inclined rib extending from the antero-ventral to mid-ventral region, a wide and rounded posterior end in dorsal view and size variation. There are significant differences between *L. deccanensis* and *Limnocythere* sp. *Limnocythere* sp. is distinguished from *L. deccanensis* by its arched dorsal margin, curved ventral margin and pointed anterior and posterior margins.

Distribution *Limnocythere deccanensis* was named after the Phulsagar intertrappean locality in Mandla District, Madhya Pradesh (Bajpai et al. 2004). Later, *L. deccanensis* was described from a number of intertrappean localities in peninsular India, including Mohgaon-Haveli and Jhilmili, Chhindwara District, Madhya Pradesh (Khosla et al. 2005, 2011a; Khosla and Nagori 2007a; Keller et al. 2009a; Sharma and Khosla 2009; Khosla 2015; Rathore 2018). *Limnocythere deccanensis* has also been recovered in the Lameta Formation of Jabalpur, Madhya Pradesh (Khosla et al. 2011b).

4.3.4 Species Limnocythere martensi sp. nov. (Figs. 4.14A–C; Table 4.14)

Holotype MPL/SK/JML/O/5061, a left valve.

Paratype MPL/SK/JML/O/5062, a left valve.

Material More than 80 uncatalogued carapaces.

Horizon, Age and Locality Upper Cretaceous to Lower Palaeocene clayey limestone of unit 3 (JH 16–26) of the Jhilmili intertrappean beds, Chhindwara District, Madhya Pradesh, India.

Etymology Named after the Prof. Koen Martens (Belgium) to honour his great scientific contributions to the field of ostracod research.

Diagnosis A small ostracod, elongated, subovate to subtriangular shaped in lateral and dorsal views, the greatest height near the anterior margin, greatest length at mid-height. Anterior margin is broader and concave, and posterior margin is short and less rounded. It has a distinctly arched dorsal margin. Surface shows strongly reticulate ornamentation dominating towards the anterior margin.

Description Carapace is small, subovate, subtriangular and elongated to a flattened shape in lateral and dorsal views. The greatest height and length are in the mid-length and mid-height, respectively. The anterior margin is strongly rounded, flat and arch shaped, while the posterior margin is short, less rounded and pointed compared to the anterior side. The dorsal margin is broadly rounded and arch shaped, has a straight hinge line and is slightly curved in the centre, and the dorsal margin declines from the apex in both directions, posteriorly and anteriorly. The right valve is larger than the left, overlapping the entire margin especially, dorsal and ventral. The surface has strongly reticulate structures, dominating on the anterior margin and less on the posterior side. Sexual dimorphic and internal features are not observable.

Table 4.14 Dimensions of carapaces of *Limnocythere martensi* sp. nov

	Dimensions (mm)		
Catalogue no.	Length	Width	Height
MPL/SK/JML/O/5061	0.340	0.117	0.190
MPL/SK/JML/O/5062	0.345	0.118	0.175
MPL/SK/JML/O/5063	0.339	0.116	0.182
MPL/SK/JML/O/5064	0.341	0.119	0.169
MPL/SK/JML/O/5065	0.344	0.121	0.171
MPL/SK/JML/O/5066	0.346	0.114	0.183
MPL/SK/JML/O/5067	0.342	0.120	0.173
MPL/SK/JML/O/5068	0.345	0.116	0.185
MPL/SK/JML/O/5069	0.338	0.125	0.191

Remarks *Limnocythere* includes species such as *L. deccanensis*, *L. bajpaii*, *Limnocythere* sp. and *Cythere inopinata*. The size of *Limnocythere martensi* sp. nov. differs from that of *L. deccanensis* (L = 0.338–0.346 mm, W = 0.114–0.125 mm and H = 0.169–0.191 mm). *Limnocythere deccanensis*, on the other hand, is larger (L = 0.422–0.516 mm, W = 0.113–0.128 mm and H = 0.195–0.244 mm), with a narrower anterior and pointed posterior margin, a subtriangular shape in lateral view and an arch shaped dorsal margin.

Limnocythere martensi sp. nov. can be distinguished from *L. bajpaii* (Khosla and Nagori 2007a) by its smaller size (L = 0.338–0.346 mm, W = 0.114–0.125 mm and H = 0.169–0.191 mm). *Limnocythere bajpaii*, on the other hand, is larger (L = 0.610 mm, W = 0.190 mm and H = 0.270 mm) and has a broader anterior margin, a less rounded and pointed posterior margin with a depression, an arched (high angle) dorsal margin and a curvy and less straight ventral margin. *Limnocythere bajpaii* also has an almost straight dorsal margin and an inclined ventral margin that

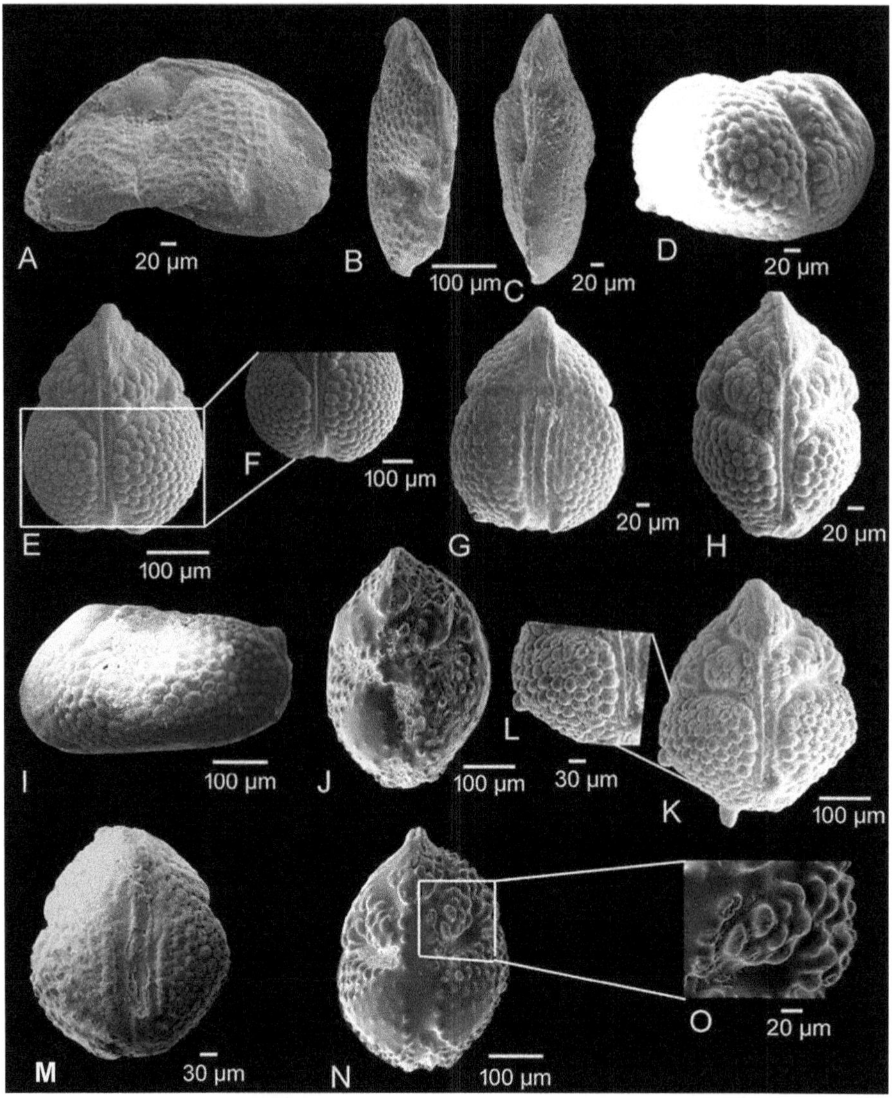

Fig. 4.14 (**A–C**) ***Limnocythere martensi* sp. nov.** Carapace, (**A**) Lateral view MPL/SK/JML/O/5061, left valve; (**B**) Latero-dorsal view MPL/SK/JML/O/5062; (**C**) Ventral view MPL/SK/JML/O/5063. (**D–H**) ***Frambocythere tumiensis anjarensis*** (Bhandari and Colin 1999). Female carapace, (**D**) Lateral view MPL/SK/JML/O/5070, right valve; (**E**) Female carapace in dorsal view MPL/SK/GP/O/2012; (**F**) Enlarged view of Fig. **E** MPL/SK/GP/O/2012 external surface showing the strongly reticulate structures; (**G**) Female carapace in ventral view MPL/SK/JML/O/5072; (**H**) Male carapace in dorsal view MPL/SK/JML/O/5073. (**I–O**) ***Frambocythere tumiensis lakshmiae*** (Whatley and Bajpai 2000a), Carapace, (**I**) Male carapace, lateral view MPL/SK/GW/O/5083, left valve; (**J**) Dorsal view MPL/SK/SW/O/5080; (**K**) Female carapace, ventral view MPL/SK/GW/O/5081; (**L**) Enlarged view of Fig. K MPL/SK/GW/O/5081 showing the large papillate tubercles; (**M**) Female carapace, ventral view of MPL/SK/SW/O/5084; (**N**) Female carapace, dorsal view of MPL/SK/SW/O/5082; (**O**) Enlarged view of MPL/SK/SW/O/5082 showing the large papillate tubercles. (Figures **G**, Reproduced from Khosla et al. 2022 with permission from Wiley, Fig. **H**, Reproduced from Kania et al. 2022 with permission from the Editor of the Himalayan Geology and Khosla et al. 2022 with permission from Wiley)

is curved from the centre, a sulcus close to the posterior side, a weaker sulcus in the middle and strong reticulate ornamentation on the valve surface.

Limnocythere martensi sp. nov. differs from *Limnocythere troelseni* (Krommelbein and Weber 1971) from the Campos Basin, Brazil, in that it is smaller, has a broad anterior and pointed posterior margin, a more rounded and highly arch shaped dorsal margin, a curvy ventral margin and reticulated ornamentation on the entire valve surface. Furthermore, it differs from *Limnocythere* sp. from the Imilchil area of the High Atlas Mountains of Morocco by its small size, broad and more rounded anterior margin, less rounded and pointed posterior margin, subtriangular shape, more arch shaped dorsal margin, curvy ventral margin from the centre and ornamented valve surface (Charriere et al. 2009). In future, more and better-preserved specimens are needed to confirm the identity of this species.

4.3.5 Subspecies *Frambocythere tumiensis anjarensis* (Bhandari and Colin 1999) (Figs. 4.14D–H; Table 4.15)

Subfamily	Timiriaseviinae (Mandelstam 1960)
Genus	*Frambocythere* (Colin 1980)
Species	*Frambocythere tumiensis* (Helmdach 1978)
Subspecies	***Frambocythere tumiensis anjarensis*** (Bhandari and Colin 1999)

1984 *Metacypris strangulata*: Jones 1860: Bhatia and Rana, Pl. 2, Figs. 6–7.
1986 *Metacypris strangulata*: Jones 1860: Prasad, pp. 72.
1990b *Frambocythere tumiensis anjarensis*: Helmdach 1978: Bhatia et al. Pl. 1, Figs. 1–3.
1996 *Frambocythere tumiensis anjarensis*: Helmdach 1978: Bhatia et al. pp. 229.
1999 *Frambocythere tumiensis anjarensis*: Bhandari and Colin, pp. 12–13, Pl. 1. Figs. 1–10.
2002a *Frambocythere tumiensis anjarensis*: Bhandari and Colin 1999: Whatley et al. pp. 166–168, Pl. 1, Figs. 8–9.
2003a *Frambocythere* sp. cf. *F. tumiensis anjarensis*: Bhandari and Colin 1999: Whatley et al. Pl. 1, Figs. 10–11.
2005 *Frambocythere tumiensis anjarensis*: Bhandari and Colin 1999: Khosla et al. pp. 137, Pl.1, Figs. 3–4.
2005 *Frambocythere tumiensis anjarensis*: Bhandari and Colin 1999: Khosla and Nagori, pp. 574, Pl.1, Fig. 4.
2007a *Frambocythere tumiensis anjarensis*: Bhandari and Colin 1999: Khosla and Nagori, pp. 215, Pl. 1, Figs. 10–12.
2007b *Frambocythere tumiensis anjarensis*: Bhandari and Colin 1999: Khosla and Nagori, pp. 6, Pl. 1, Figs. 4–7.
2009a *Frambocythere tumiensis anjarensis*: Bhandari and Colin 1999: Khosla et al. pp. 725, Pl. 2, Fig. 8.

2009 *Frambocythere tumiensis anjarensis*: Bhandari and Colin 1999: Sharma and Khosla, pp. 202, Pl. 1, Figs. D–E.
2010 *Frambocythere tumiensis anjarensis*: Bhandari and Colin 1999: Khosla et al. pp. 118.
2011b *Frambocythere tumiensis anjarensis*: Bhandari and Colin 1999: Khosla et al. pp. 226, 228–229, Pl.1, Figs. 10–11.
2015 *Frambocythere tumiensis anjarensis*: Bhandari and Colin 1999: Khosla, pp. 352, Figs. 5, J–K.
2017 *Frambocythere tumiensis anjarensis*: Bhandari and Colin 1999: Rathore et al. pp. 225, Figs. 4.7–4.10.
2018 *Frambocythere tumiensis anjarensis* Bhandari and Colin 1999: Rathore, pp. 66.
2019 *Frambocythere tumiensis anjarensis*: Bhandari and Colin 1999: Kapur et al. pp. 1152, Figs. 3 K–M.
2019 *Frambocythere tumiensis anjarensis*: Bhandari and Colin 1999: Kshetrimayum and Parmar, pp. 15, Figs. 2d–f.
2021 *Frambocythere tumiensis anjarensis*: Bhandari and Colin 1999: Kshetrimayum et al. pp. 4–5, 6, Figs. 3A–E.

Material More than 230 carapaces and open valves.

Horizons, Age and Localities Upper Cretaceous to Lower Palaeocene clayey limestone of unit 3 (JH 16, JH 25, JH 26) of the Jhilmili intertrappean beds and from Upper Cretaceous to ?Lower Palaeocene hard clayey limestone, as well as reddish chert units of the Ghat Parasia intertrappean beds Chhindwara District, Madhya Pradesh, India.

Description Carapace very small, inflated, subtriangular to subovate shaped in dorsal view and cylindrical in lateral view. It has its greatest height on the posterior side and maximum length at mid-height. Both extremities are rounded, but the anterior margin is more rounded as well. The female carapace differs from the male in that it has a sturdily exaggerated posterior part, whereas males have a slender end with the greatest width median. The carapace is heart shaped in dorsal view. The dorsal and ventral margins are straight to convex. The left valve is larger than the right valve, and overlapping is strongly visible along the antero-dorsal and postero-dorsal margins. The external surface is ornamented with papillate tubercles, tubercles on the dorsal part of the antero-marginal area, much more distinct papillae paralleling the dorsal margin's posterior part and a marked median sulcus.

Remarks Bhandari and Colin (1999) were the first to describe *Frambocythere tumiensis anjarensis* as a subspecies of *Frambocythere tumiensis* (Helmdach 1978), while reporting the ostracod fauna of Anjar, Kachchh District, Gujarat. The current material of *Frambocythere tumiensis anjarensis* differs from *Frambocythere tumiensis lakshmiae* (Whatley and Bajpai 2000a) from the Upper Cretaceous intertrappean beds of Lakshmipur (Gujarat) and Mohagaon-Kalan (Madhya Pradesh) by having distinctive valve overlapping, a larger left valve than a right valve, a less rounded anterior margin and large tubercles on the antero-dorsal margin (Whatley et al. 2002a).

Table 4.15 Dimensions of carapaces of *Frambocythere tumiensis anjarensis* (Bhandari and Colin 1999)

	Dimensions (mm)		
Catalogue no.	Length	Width	Height
MPL/SK/JML/O/5070	0.326	0.265	0.188
MPL/SK/JML/O/5071	0.305	0.267	0.210
MPL/SK/JML/2012	0.309	0.230	0.172
MPL/SK/JML/O/5073	0.324	0.261	0.184
MPL/SK/JML/O/5074	0.316	0.272	0.213
MPL/SK/JML/O/5075	0.308	0.228	0.176
MPL/SK/JML/O/5076	0.319	0.276	0.217
MPL/SK/GP/O/5077	0.329	0.265	0.181
MPL/SK/GP/O/2012	0.311	0.239	0.183
MPL/SK/GP/O/5079	0.329	0.270	0.109

Distribution The subspecies *Frambocythere tumiensis anjarensis* was originally reported from the Upper Cretaceous intertrappean beds exposed at Anjar, Kachchh District, Gujarat (Bhandari and Colin 1999). Later, it was reported from various intertrappean localities in peninsular India such as Chandarki and Yanagundi, Gulbarga District, Karnataka (Whatley et al. 2002a), Mamoni, Kota District, Rajasthan (Whatley et al. 2003a), Jhilmili and Mohgaon-Haveli, Chhindwara District, Madhya Pradesh (Keller et al. 2009a; Khosla 2015; Khosla et al. 2011a, 2009a, b; Khosla and Nagori 2007a; Sharma and Khosla 2009), and Takli, Nagpur District, Maharashtra (Khosla and Nagori 2007b). Very recently, it has been recorded from intertrappean beds of Khar, Manawar, Uthawali and Gujri, (Khargaon and Dhar Districts), Madhya Pradesh (Rathore et al. 2017; Kapur et al. 2019; Kshetrimayum and Parmar 2019; Kshetrimayum et al. 2021) and Pinjaurni, Chandrapur District, Maharashtra (Rathore 2018). It has also been reported from the Lameta Formation of Jabalpur, Madhya Pradesh (Khosla et al. 2011b).

4.3.6 Subspecies *Frambocythere tumiensis lakshmiae* (Whatley and Bajpai 2000a) (Figs. 4.14I–O; Table 4.16)

2000a *Frambocythere tumiensis lakshmiae*: Whatley and Bajpai, pp. 390, Pl. 1, Figs. 6–15.
2002b *Frambocythere tumiensis lakshmiae*: Whatley et al. pp. 107, Pl. 1, Fig. 2.

Material More than 44 carapaces and open valves.

Horizons, Age and Localities Upper Cretaceous (Maastrichtian) black carbonaceous and calcareous shale unit of the intertrappean section exposed in the Shiraj (=Shriwas) well and Upper Cretaceous (Maastrichtian) black calcareous carbonaceous shale unit of the intertrappean section exposed in the Government well, Mohgaon-Kalan village, Chhindwara District, Madhya Pradesh, India.

Description Subtriangular to subovate shaped in dorsal view and cylindrical in lateral view. The greatest height is on the posterior side and greatest length at mid-height. More rounded anterior margin. Large, papillate tubercles are absent on the dorsal portion of the antero-marginal part, and less pronounced papillae are present paralleling the posterior area of the dorsal margin. Left valve larger than right valve, so overlapping is visible on the entire periphery.

Table 4.16 Dimensions of carapaces of *Frambocythere tumiensis lakshmiae* (Whatley and Bajpai 2000a)

	Dimensions (mm)		
Catalogue no.	Length	Width	Height
MPL/SK/SW/O/5080	0.319	0.271	0.219
MPL/SK/GW/O/5081	0.268	0.200	0.178
MPL/SK/SW/O/5082	0.310	0.316	0.160
MPL/SK/GW/O/5083	0.376	0.281	0.221
MPL/SK/GW/O/5084	0.288	0.231	0.183

Remarks This is a junior subspecies of *Frambocythere tumiensis* that is closely related to the subspecies *Frambocythere tumiensis anjarensis* (Bhandari and Colin 1999). It differs from *Frambocythere tumiensis anjarensis* in that it has a rounded anterior margin, no large papillate tubercles on the dorsal portion of the antero-marginal area and less pronounced papillae paralleling the posterior area of the dorsal margin.

Distribution The subspecies *Frambocythere tumiensis lakshmiae* has been reported from Upper Cretaceous intertrappean beds of Lakshmipur, Kachchh District, Gujarat and Mohgaon-Kalan, Chhindwara District, Madhya Pradesh (Whatley and Bajpai 2000a; Whatley et al. 2002b).

4.3.7 Species *Gomphocythere strangulata* (Jones 1860) (Figs. 4.15A–F; Table 4.17)

Genus	*Gomphocythere* (Sars 1924)
Species	***Gomphocythere strangulata*** (Jones 1860)

Table 4.17 Dimensions of carapaces of *Gomphocythere strangulata* (Jones 1860)

Catalogue no.	Dimensions (mm)		
	Length	Width	Height
MPL/SK/JML/O/5085	0.800	0.546	0.470
MPL/SK/JML/O/5086	0.785	0.543	0.468
MPL/SK/JML/O/5087	0.750	0.540	0.451
MPL/SK/JML/O/5088	0.790	0.545	0.466
MPL/SK/JML/O/5089	0.779	0.541	0.465
MPL/SK/JML/O/5090	0.755	0.545	0.468
MPL/SK/SW/O/5091	0.781	0.543	0.460
MPL/SK/SW/O/5092	0.776	0.538	0.449
MPL/SK/GW/O/5093	0.751	0.536	0.441
MPL/SK/GW/O/5094	0.783	0.544	0.467

1860 *Cypris strangulata*: Jones, pp. 187, Pl. 10, Figs. 73a–d.
1984 *Metacypris strangulata*: Jones 1860: Bhatia and Rana, pp. 33, Pl. 2, Figs. 8–9.
1990a *Cytheridella strangulata*: Jones 1860: Bhatia et al. pp. 47, Pl. 3, Figs. 1–2.
1990b *Cytheridella strangulata*: Jones 1860: Bhatia et al. pp. 118, Pl. 1, Figs. 4–5.
1996 *Cytheridella strangulata*: Jones 1860: Bhatia et al. pp. 299, Pl. 3. Figs. 1–2.
1996 *Cytheridella strangulata*: Jones 1860: Udhoji and Mohabey, pp. 413, Pl. 2, Figs. 1–3.
2000a *Limnocythere falsicarinata*: Whatley and Bajpai, pp. 390, Pl. 1, Figs. 1–5.
2002a *Gomphocythere strangulata*: Jones 1860: Whatley et al. pp. 169, Pl. 1, Figs. 10–11.
2005 *Gomphocythere strangulata*: Khosla et al. pp. 137–139, Pl. 1, Figs. 9–10.
2005 *Limnocythere falsicarinata*: Whatley and Bajpai 2000a: Khosla et al. pp. 137, Pl. 1, Figs. 5–6.
2005 *Limnocythere falsicarinata*: Whatley and Bajpai 2000a: Khosla and Nagori, pp. 575, Pl. 1, Fig. 1.
2007b *Gomphocythere falsicarinata*: Whatley and Bajpai 2000a: Khosla and Nagori, pp. 6, Pl. 1, Figs. 8–9.
2009a *Gomphocythere strangulata*: Jones 1860: Khosla et al. pp. 725, Pl. 2, Fig. 10.
2009 *Gomphocythere strangulata*: Jones 1860: Sharma and Khosla, pp. 202, Pl. 1, Figs. A–C.
2011a *Gomphocythere strangulata*: Jones 1860: Khosla et al. pp. 232, Pl. 1, Figs. 16–19.
2011b *Gomphocythere strangulata*: Jones 1860: Khosla et al. pp. 227–228, Pl. 2, Figs. 2–5.
2015 *Gomphocythere strangulata*: Jones 1860: Khosla, Fig. 5 J, pp. 348, 352.
2019 *Gomphocythere strangulata*: Jones 1860: Kshetrimayum and Parmar, pp. 16, Fig. 2c.
2021 *Gomphocythere strangulata*: Jones 1860: Kshetrimayum et al. pp. 5–7, Figs. 3F–J.

Material More than 32 carapaces and open valves.

Horizon, Age and Localities Upper Cretaceous to Lower Palaeocene clayey limestone of unit 3 (JH 25) of the Jhilmili intertrappean beds; Upper Cretaceous (Maastrichtian) black carbonaceous shale unit of the intertrappean section exposed in the Shiraj (=Shriwas) well and Upper Cretaceous (Maastrichtian) black carbonaceous shale unit of the intertrappean section exposed in the Government well, Mohgaon-Kalan village, Chhindwara District, Madhya Pradesh, India.

Description Carapace large, elongated, inflated to subtriangular in shape in dorsal view with greatest height in dorsal view, maximum length at mid-height. Anterior margin is pointed and less rounded, posterior margin is well rounded and broad compared to the anterior, and in the middle a sulcus is present. Dorsal margin is straight and inclined on the postero-ventral margin and almost straight on the ventral margin. Right valve is slightly larger than left valve and overlaps the entire margins. Two sulci are visible that meeting at the hinge line. The surface is strongly ornamented with polygonal to hexagonal reticulate structures with 4–5 puncta and a narrow ventro-lateral rib. The main median sulcus is deep and sinuous, with a smaller sulcus anteriorly; two tubercles occur dorsally between the two sulci, and a flatter tubercle appears just above the adductor scars. Male and female are easy observable in dorsal view; females are more rounded, and males are less rounded in the lower part. Internal features are not visible.

Remarks *Cypris strangulata* was first described by Jones (1860) from the intertrappean beds of Nagpur, Maharashtra. This species was later renamed *Gomphocythere strangulata* by Bhatia et al. (1990a, 1996). The collected carapaces are identified as *Gomphocythere strangulata* because they have an uneven rare reticulation posterior to the median sulcus and tubercles in the anterior marginal area. *Gomphocythere akalypton* (Whatley et al. 2002a) is a strongly inflated species with a reticulum of smaller fossae and is more acuminate anteriorly. *Gomphocythere dasyderma* (Whatley et al. 2002a) has a carapace that is acuminate at both margins and not strongly inflated in dorsal view, with a stronger median sulcus and a low, circular dorso-ventral mound between both sulci. A soft marginal rim develops anteriorly, as do several parallel ribs ventrally. *Gomphocythere whatleyi* (Khosla and Nagori 2007b), on the other hand, has a strongly arched dorsal margin with smooth valve surfaces and a ventral carina.

Distribution *Gomphocythere strangulata* has been recorded from the Upper Cretaceous Lameta Formation exposed in Jabalpur (Madhya Pradesh) and Dongargaon, Maharashtra (Khosla et al. 2005, 2011b). It has also been reported from numerous intertrappean beds situated at Asifabad, Telangana; Anjar and Lakshmipur, Kachchh District, Gujarat; Mamoni, Kota District, Rajasthan; Takli (Nagpur District), and Pinjaurni (Chandrapur District), Maharashtra and Jhilmili (Chhindwara District), Khar (Khargaon District), Manawar, Uthawali and Gujri (Dhar District), Madhya Pradesh (Bhatia et al. 1990b; Whatley and Bajpai 2000a;

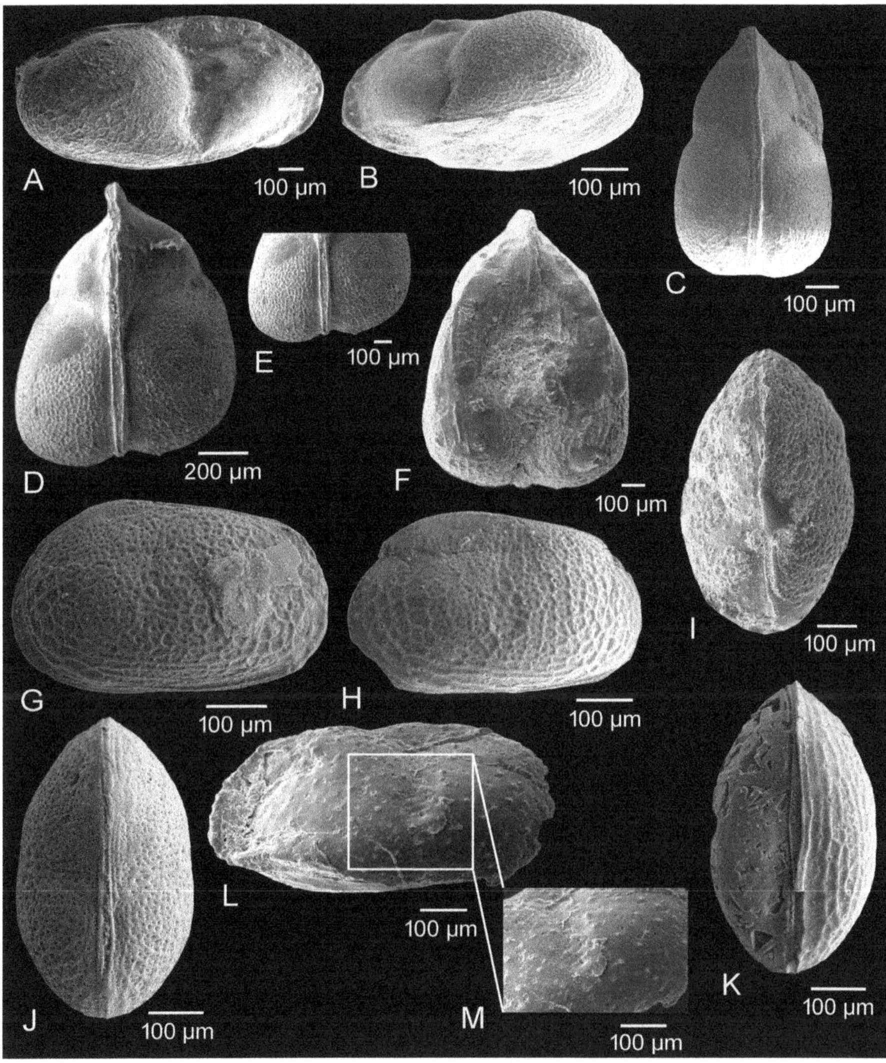

Fig. 4.15 (**A**–**F**) ***Gomphocythere strangulata*** Jones (1860), Carapace, (**A**) Lateral view MPL/SK/JML/O/5086, right valve; (**B**) Latero-ventral view MPL/SK/JML/O/5085, left valve; (**C**) Dorsal view MPL/SK/JML/O/5087; (**D**) Female carapace, ventral view MPL/SK/JML/O/5088; (**E**) Enlarged view of MPL/SK/JML/O/88 showing the surface, which is strongly ornamented with polygonal to hexagonal reticulate structures; (**F**) Ventral view MPL/SK/SW/O/5092. (**G**–**K**) ***Gomphocythere paucisulcatus*** (Whatley et al. 2002b), Carapace, (**G**) Male carapace, lateral view MPL/SK/JML/O/5095, left valve; (**H**) Male carapace, lateral view MPL/SK/JML/O/5096, right valve; (**I**) Dorsal view MPL/SK/SW/O/5097; (**J**) Female carapace, dorsal view MPL/SK/GP/O/5098; (**K**) Ventral view MPL/SK/GW/O/5099. (**L**–**M**) ***Gomphocythere dasyderma*** (Whatley et al. 2002a), Carapace, (**L**) Lateral view MPL/SK/JML/O/6000, latero-ventral view, (**M**) Enlarged lateral view of MPL/SK/JML/O/6000 showing pitted discontinuous ornamentation

Whatley et al. 2002a; Khosla and Nagori 2005, 2007b; Sharma and Khosla 2009; Khosla et al. 2009a, 2011a, b; Khosla 2015; Kshetrimayum and Parmar 2019; Kshetrimayum et al. 2021).

4.3.8 Species *Gomphocythere paucisulcatus* (Whatley et al. 2002b) (Figs. 4.15G–K; Table 4.18)

Genus	*Gomphocythere* (Sars 1924)
Species	***Gomphocythere paucisulcatus*** (Whatley et al. 2002b)

1999 *Gomphocythere*? sp. 1: Bhandari and Colin, pp. 13, Pl. 1, Figs. 11–13.
2002b *Gomphocythere paucisulcatus*: Whatley et al. pp. 107–109, Pl. 1, Figs. 1–6.
2005 *Gomphocythere paucisulcatus*: Whatley et al. 2002b: Khosla et al., pp. 137, Pl. 1, Figs. 7–8.
2005 *Gomphocythere paucisulcatus*: Whatley et al. 2002b: Khosla and Nagori, pp. 574, Pl. 1, Fig. 3.
2007b *Gomphocythere paucisulcatus*: Whatley et al. 2002b: Khosla and Nagori, pp. 8, Pl. 1, Figs. 10–11,
2008 *Gomphocythere paucisulcatus*: Whatley et al. 2002b: Sharma et al. pp. 178, Pl. 1, Figs. E–G.
2009a *Gomphocythere paucisulcatus*: Whatley et al. 2002b: Khosla et al. pp. 725, Pl. 2, Fig. 9.
2010 *Gomphocythere paucisulcatus*: Whatley et al. 2002b: Khosla et al. pp. 118.
2011a *Gomphocythere paucisulcatus*: Whatley et al. 2002b: Khosla et al. pp. 232, Pl. 1, Fig. 15.
2011b *Gomphocythere paucisulcatus*: Whatley et al. 2002b: Khosla et al. pp. 227, Pl. 1, Figs. 12–13, Pl. 2, Figs. 1a, b.
2015 *Gomphocythere paucisulcatus*: Whatley et al. 2002b: Khosla, pp. 348.
2015 *Gomphocythere paucisulcatus*: Whatley et al. 2002b: Khosla et al. pp. 92, Fig. 3c.
2016 *Gomphocythere paucisulcatus*: Whatley et al. 2002b: Khosla et al. pp. 176, Fig. 2.5.
2018 *Gomphocythere paucisulcatus*: Whatley et al. 2002b: Rathore, pp. 6.
2019 *Gomphocythere paucisulcatus*: Whatley et al. 2002b: Kapur et al. Figs. 3H–J.
2019 *Gomphocythere paucisulcatus*: Whatley et al., 2002b: Kshetrimayum and Parmar, pp. 15–16, Figs. 2a–b.
2021 *Gomphocythere paucisulcatus*: Whatley et al., 2002b: Kshetrimayum et al. pp. 6, 7, 9, Figs. 3K–N.

Material A total of 46 carapaces and open valves.

Horizons, Age and Localities Upper Cretaceous to Lower Palaeocene clayey limestone of unit 3 (JH 25) of the Jhilmili intertrappean beds, Chhindwara District, Madhya Pradesh. Upper Cretaceous (Maastrichtian) black carbonaceous shale unit of the intertrappean section exposed in the Shiraj (=Shriwas) well and Upper Cretaceous (Maastrichtian) black calcareous carbonaceous shale unit of the intertrappean section exposed in the Government well, Mohgaon-Kalan village, Chhindwara District, Madhya Pradesh. Upper Cretaceous to ? Lower Palaeocene fossiliferous black chert, fossiliferous clayey limestone, and reddish chert of the Ghat Parasia intertrappean beds, Chhindwara District, Madhya Pradesh, India.

Description A medium-sized species of the genus *Gomphocythere* with a subovate to subrectangular to cylindrical shape in lateral and ventral views. The carapace is subfusiform and almost consistently acuminate at each margin in dorsal view, with a very slender median sulcus. The greatest height is close to the anterior margin, and the greatest length is at mid-height. The anterior margin is well rounded, and the posterior margin is more evenly rounded, with the apex at or just above mid-height. Dorsal margin long, straight and inclined towards the anterior margins, and the ventral margin shows an oral concavity. Hinge line is visible. Left valve considerably larger than right valve, overlapping along the margins. The external surface is coarsely reticulate, and the large reticulae include 1–4 secondary puncta inside each fossa. Across the valve, there are some small conjunctive pore conuli. The reticulum's concentric muri predominate around the free margins, resulting in a number of parallel ribs paralleling the margins. A thin rib runs parallel to the dorsal margin. An anteriorly developed marginal rim is weak. Sexually dimorphic and internal features are not distinguishable.

Table 4.18 Dimensions of carapaces of *Gomphocythere paucisulcatus* (Whatley et al. 2002b)

	Dimensions (mm)		
Catalogue no.	Length	Width	Height
MPL/SK/JML/O/5095	0.645	0.374	0.371
MPL/SK/JML/O/5096	0.548	0.344	0.357
MPL/SK/SW/O/5097	0.485	0.290	0.245
MPL/SK/GP/O/5098	0.501	0.297	0.280
MPL/SK/GW/O/5099	0.483	0.281	0.252

Remarks The genus *Gomphocythere* is represented by seven species, *G. paucisulcatus* (Whatley et al. 2002b), *G. strangulata* (Jones 1860), *G. gomphomatos* (Whatley and Bajpai 2000a), *G. akalypton* (Whatley et al. 2002a), *G. dasyderma* (Whatley et al. 2002a), *G. whatleyi* (Khosla and Nagori 2007b) and *G. testudo* (Kshetrimayum et al. 2021). The present collection of carapaces displays a subfusiform outline in dorsal view, feebly developed median sulcus, presence of a weak rib paralleling the dorsal margin and the presence of large recticulate (with between 1 and 4 secondary puncta) ornamentation within fossae, which place them under the species *G. paucisulcatus* (Whatley et al. 2002a, b c; Khosla et al. 2011a, b).

Distribution Originally, this species was described as ?*Gomphocythere* sp. from the intertrappean bed of Anjar, Kachchh District, Gujarat (Bhandari and Colin 1999). Subsequently, (Whatley et al. 2002b) renamed this species as *Gomphocythere paucisulcatus* while reporting it from the Mohgaon-Kalan intertrappean beds, Chhindwara District, Madhya Pradesh. This species has been collected from a number of intertrappean localities in peninsular India namely, Anjar (Kachchh District), Gujarat (Khosla and Nagori 2005), Takli, Maharashtra (Khosla and Nagori 2007b), Papro (Lalitpur District), Uttar Pradesh (Sharma et al. 2008) and Jhilmili (Chhindwara District), Madhya Pradesh (Khosla et al. 2009a, 2011a; Khosla 2015). It is also known from the Lameta Formation of the Jabalpur, Dongargaon and Pisdura areas (Khosla et al. 2005, 2010, 2011b, 2015, 2016).

4.3.9 Species *Gomphocythere dasyderma* (Whatley et al. 2002a) (Figs. 4.15L–M)

2002a *Gomphocythere dasyderma*: Whatley et al. 2002a, pp. 171–172, Pl. 2, Figs. 6–20.
2003a *Gomphocythere dasyderma*: Whatley et al. 2002a: Whatley et al., Pl. 1, Figs. 14, 15.

Material One moderately preserved carapace.

Horizon, Age and Locality Upper Cretaceous to Lower Palaeocene clayey limestone of unit 3 (JH 25) of the Jhilmili intertrappean beds, Chhindwara District, Madhya Pradesh, India.

Description Carapace subrectangular, subovate to cylindrical in lateral and ventral views with greatest height close to the posterior margin and maximum length at mid-height. Carapace has acuminate anterior and posterior margins in dorsal views, is feebly inflated and contains a well-developed median suclus. The anterior margin is broadly arched, and the posterior margin is more rounded and flatter than the anterior margin. Dorsal margin is straight, and a depression occurs in the centre. Ventral margin nearly straight, inclined on the posterior side and elevated on the anterior side. Hinge line is clearly visible in lateral view. Left valve larger than right valve, overlapping visible along the entire margin. The surface bears pitted discontinuous reticulo-tuberculate ornamentation. Sexually dimorphic and internal features are not observable. MPL/SK/JML/O/6000 measures length 0.726 mm, height 0.379 mm and width 0.421 mm.

Remarks The present collected carapace of *Gomphocythere dasyderma* (Whatley et al. 2002a) differs considerably from the contemporary species of the genus like *G. paucisulcatus* (Whatley et al. 2002b), *G. strangulata* (Jones 1860), *G. gompho-*

matos (Whatley and Bajpai 2000a), *G. akalypton* (Whatley et al. 2002a), *G. whatleyi* (Khosla and Nagori 2007b) and *G. testudo* (Kshetrimayum et al. 2021). It yields numerous typical characters such as acuminate anterior and posterior margins of the carapace in dorsal view, which is feebly inflated, contains a well-developed median sulcus and surface bears pitted discontinuous ornamentation. These features justify assignment to *Gomphocythere dasyderma.*

Distribution This species, *Gomphocythere dasyderma* (Whatley et al. 2002a), has previously been reported from the intertrappean beds of Chandarki and Yanagundi, (Gulbarga District), Karnataka (Whatley et al. 2002a) and Mamoni (Kota District), Rajasthan (Whatley et al. 2003a).

4.3.10 Species Gomphocythere sp. 1 (Figs. 4.16A–C; Table 4.19)

Material A total of five well-preserved carapaces.

Horizon, Age and Locality Upper Cretaceous (Maastrichtian) black calcareous carbonaceous shale unit of intertrappean section exposed in the Government well, Mohgaon-Kalan, Chhindwara District, Madhya Pradesh, India.

Description The carapace is medium in size. It is elongate, subrectangular, elliptical or oval shaped in dorsal and lateral view. The greatest height is near the middle. Anterior and posterior margins are evenly rounded, and the apex is at or just above mid-height. Dorsal margin is almost straight and inclined towards the anterior margin. Ventral margin is almost straight and arched towards the centre. Left valve larger than the right. Surface ornamentation varies between specimens; some parts have delicate reticulation, while others have secondary punctuation. Sexual dimorphism and internal structures are not visible.

Table 4.19 Dimensions of carapaces of *Gomphocythere* sp. 1

	Dimensions (mm)		
Catalogue no.	Length	Width	Height
MPL/SK/GW/O/5534	0.691	0.255	0.243
MPL/SK/GW/O/5535	0.496	0.243	0.325
MPL/SK/GW/O/5536	0.457	0.217	0.319

Remarks The numerous distinguishing characteristics of carapaces recovered from the Government well, such as the reversed lophodont hinge, the presence of a terminal bar in the left valve and the heavily ornamented valve surface, are useful

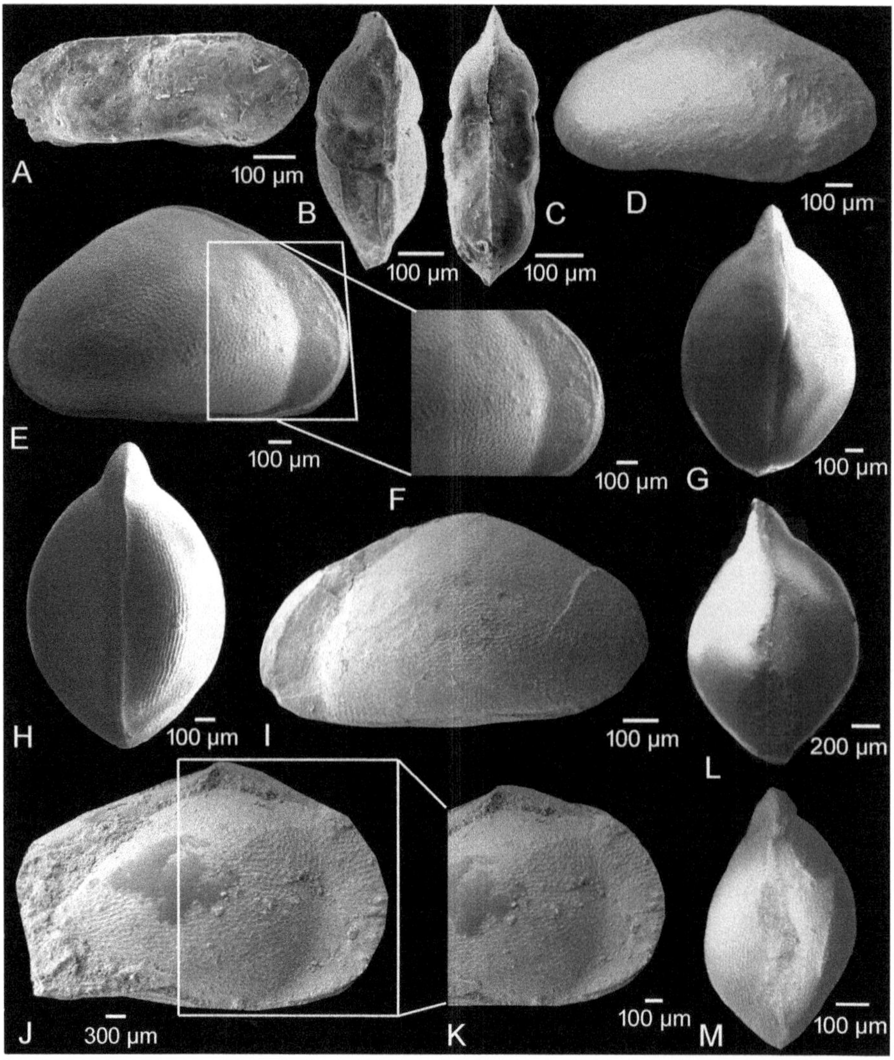

Fig. 4.16 (**A**–**C**) ***Gomphocythere* sp. 1.** Carapace, (**A**) Lateral view MPL/SK/GW/O/5534, right valve; (**B**) Dorsal view MPL/SK/GW/O/5535; (**C**) Ventral view MPL/SK/GW/O/5536. (**D**–**H**) ***Paracypretta subglobosa*** (Sowerby 1840), Carapace, right valve; (**D**) Lateral view MPL/SK/JML/O/5150, right valve; (**E**) Lateral view MPL/SK/JML/O/5102, right valve; (**F**) Enlarged view of MPL/SK/JML/O/5102 surface bearing fine papillate ornamentation, which is neither parallel nor concentrically oriented with respect to the margin; (**G**) Dorsal view MPL/SK/JML/O/5103; (**H**) Ventral view MPL/SK/JML/O/5104. (**I**–**M**) ***Paracypretta jonesi*** (Bhatia and Rana 1984), Carapace, (**I**) Lateral view MPL/SK/JML/O/5152, left valve; (**J**) Lateral view MPL/SK/JML/O/5153, right valve; (**K**) Enlarged view of **J** MPL/SK/JML/O/5153 valve surface bears fine to coarse, very dense punctae, and the fine rib permits longitudinal orientation; (**L**) Dorsal view MPL/SK/JML/O/5154; (**M**) Ventral view MPL/SK/JML/O/5155

for identifying *Gomphocythere* at the genus level (Whatley et al. 2002a, b). However, the present material bears no resemblance to any known species of the genus *Gomphocythere,* including *G. paucisulcatus* (Whatley et al. 2002b), *G. strangulata* (Jones 1860), *G. gomphomatos* (Whatley and Bajpai 2000a), *G. akalypton* (Whatley et al. 2002a) and *G. whatleyi* (Khosla and Nagori 2007b). As a result, the current material is identified as *Gomphocythere* sp. 1.

Distribution Prior to this, a ?*Gomphocythere* sp. 1 was known from the intertrappean beds of Anjar (Kachchh District), Gujarat (Bhandari and Colin 1999).

4.3.11 Species Paracypretta subglobosa (Sowerby 1840) (Figs. 4.16D–H; Table 4.20)

Superfamily	Cypridoidea (Baird 1845)
Family	Cyprididae (Baird 1845)
Subfamily	Cypridinae (Baird 1845)
Genus	*Paracypretta* Sars (1924)
Type species	*Paracypretta ampullaceal* (Sars 1924)
Species	***Paracypretta subglobosa*** (Sowerby 1840)

1840 *Cypris subglobosa*: Sowerby in Malcolmson, Pl. 47, Fig. 3.
1859 *Cypria subglobosa*: Sowerby in Malcolmson 1840: Baird, pp. 232, Pl. 63, Fig. 2c.
1976 *Cypris subglobosa*: Sowerby in Malcolmson 1840: Neale, pp. 125, Pl. 3. Figs. 128–132.
1886 *Cypria subglobosa*: Sowerby in Malcolmson, 1840: Brady, pp. 300, Pl. 38, Figs. 24–27a.
1990b *Altanicypris sczcechurae*: Stankevitch 1974: Bhatia et al. pp. 118, Pl. 1, Figs. 9, 10.
2003a *Paracypretta subglobosa*: Sowerby in Malcolmson 1840: Whatley et al. pp. 79–80, Pl.1, Figs. 1, 2.
2003c *Paracypretta subglobosa*: Sowerby in Malcolmson 1840: Whatley et al. pp. 1287–1290, Figs. 1–19; Figs. 2A–C.
2009 *Paracypretta subglobosa*: Sowerby in Malcolmson 1840: Sharma and Khosla, pp. 204, Pl. 2, Figs. G–H.
2015 *Paracypretta subglobosa*: Sowerby in Malcolmson 1840: Khosla, pp. 348.

Material More than 125 carapaces and open valves.

Horizon, Age and Locality Upper Cretaceous to Lower Palaeocene clayey limestone of unit 3 (JH17, 19, 20, 24, 25) of the Jhilmili intertrappean beds, Chhindwara District, Madhya Pradesh, India.

Description Carapace large, rounded, subovate to subtriangular in lateral view, fusiform in ventral and dorsal views, with a strongly compressed anterior margin and rounded posteriorly. The dorsal margin is strongly arched, and there is a flattened ventral margin. Convex lateral surface and a shoulder-like break of slope around one-sixth from the anterior margin (Whatley et al. 2003a). Ventral margin is flattened in lateral view and has a long antero-dorsal slope. Valves are subequal, left valve larger than the right one. The lateral surface bears fine papillate ornamentation, which is neither parallel nor concentrically oriented with respect to the margin. Sexually dimorphic and internal features are not observable.

Table 4.20 Dimensions of carapaces of *Paracypretta subglobosa* (Sowerby 1840)

	Dimensions (mm)		
Catalogue no.	Length	Width	Height
MPL/SK/JML/O/5102	1.350	0.930	0.845
MPL/SK/JML/O/5103	1.470	0.968	0.921
MPL/SK/JML/O/5104	1.412	0.925	0.887
MPL/SK/JML/O/5105	1.345	0.929	0.864
MPL/SK/JML/O/5106	1.440	0.945	0.863
MPL/SK/JML/O/5107	1.061	0.892	0.735
MPL/SK/JML/O/5108	1.373	0.875	0.900
MPL/SK/JML/O/5109	1.383	0.916	0.926
MPL/SK/JML/O/5110	1.268	0.902	0.869
MPL/SK/JML/O/5111	1.502	0.980	0.862
MPL/SK/JML/O/5112	1.384	0.902	0.835
MPL/SK/JML/O/5113	1.515	0.970	0.936
MPL/SK/JML/O/5114	1.383	0.896	0.825
MPL/SK/JML/O/5115	1.396	0.904	0.933
MPL/SK/JML/O/5116	1.402	0.905	0.862
MPL/SK/JML/O/5117	1.419	1.005	0.912
MPL/SK/JML/O/5118	1.383	0.961	0.946
MPL/SK/JML/O/5119	1.268	0.941	0.835
MPL/SK/JML/O/5120	1.470	0.912	0.965
MPL/SK/JML/O/5121	1.599	1.001	0.967
MPL/SK/JML/O/5122	1.383	0.929	0.800
MPL/SK/JML/O/5123	1.405	0.911	0.856
MPL/SK/JML/O/5124	1.449	0.967	0.943
MPL/SK/JML/O/5125	1.415	0.934	0.906
MPL/SK/JML/O/5126	1.381	0.951	0.942
MPL/SK/JML/O/5127	1.380	0.825	0.941
MPL/SK/JML/O/5128	1.410	0.874	0.901
MPL/SK/JML/O/5129	1.382	0.849	0.938
MPL/SK/JML/O/5130	1.413	0.788	0.836
MPL/SK/JML/O/5131	1.393	0.880	0.893

(continued)

Table 4.20 (continued)

	Dimensions (mm)		
Catalogue no.	Length	Width	Height
MPL/SK/JML/O/5132	1.370	0.952	0.955
MPL/SK/JML/O/133	1.376	0.900	0.935
MPL/SK/JML/O/5134	1.402	0.893	0.903
MPL/SK/JML/O/5135	1.378	0.924	0.932
MPL/SK/JML/O/5136	1.591	0.950	1.017
MPL/SK/JML/O/5137	1.477	0.796	0.906
MPL/SK/JML/O/5138	1.348	0.917	0.932
MPL/SK/JML/O/5139	1.393	0.883	0.901
MPL/SK/JML/O/5140	1.472	0.973	0.904
MPL/SK/JML/O/5141	1.522	0.881	1.001
MPL/SK/JML/O/5142	1.378	0.866	0.935
MPL/SK/JML/O/5143	1.472	0.905	0.937
MPL/SK/JML/O/5144	1.396	0.922	0.920
MPL/SK/JML/O/5145	1.374	0.870	0.935
MPL/SK/JML/O/5146	1.477	0.837	0.909
MPL/SK/JML/O/5147	1.522	0.858	0.935
MPL/SK/JML/O/5148	1.379	0.919	0.933
MPL/SK/JML/O/5149	1.472	0.910	0.939
MPL/SK/JML/O/5150	1.393	0.876	0.926
MPL/SK/JML/O/5151	1.369	0.960	0.938

Remarks *Paracypretta jonesi* (Bhatia and Rana 1984), *P. subglobosa* (Sowerby 1840), *P. verruculosa* (Whatley et al. 2002a) and *P. elizabethae* (Whatley et al. 2003c) have all been described from various localities of the infra- and intertrappean beds in peninsular India. The overall morphology of the carapaces collected clearly indicates that they are of the species *Paracypretta subglobosa* (Sowerby 1840). However, the present *P. subglobosa* material differs from *P. elizabethae* and *P. jonesi* in that it has fine papillae that are neither parallel nor concentrically oriented with respect to the margin. *Paracypretta verruculosa* is distinguished from Jhilmili *P. subglobosa* by its distinct outline, unusual hinge and clearly oriented fine papillae.

Distribution This species was originally reported as *Cypris subglobosa* from the intertrappean beds of the Sichel hills, Telangana (Sowerby in Malcolmson 1840). It has been reported from the intertrappean beds of Nagpur, Maharashtra (Jones 1860), Mamoni (Kota District), Rajasthan (Bhatia et al. 1990b; Whatley et al. 2003c) and Jhilmili (Chhindwara District), Madhya Pradesh (Sharma and Khosla 2009).

4.3.12 Species *Paracypretta jonesi* (Bhatia and Rana 1984) (Figs. 4.16I–M; Table 4.21)

Genus	*Paracypretta* (Sars 1924)
Species	***Paracypretta jonesi*** (Bhatia and Rana (1984)

1984 *Paracypretta jonesi*: Bhatia and Rana, pp. 30–33, Pl. 2, Figs. 1–3.
1994 *Altanicypris* sp.: Sahni and Khosla, pp. 458, Figs. 2n–p.
1996 *Paracypretta jonesi* Bhatia and Rana 1984: Udhoji and Mohabey, pp. 413, Pl. 2. Figs. 4–6.
1996 *Leiria jonesi*: Bhatia et al. pp. 304, Pl. 3, Fig. 7.
2000 *Altanicypris bhatiai* Bhatia and Rana 1984: Khosla and Sahni, pp. 58–59, Pl.1, Figs. a–g.
2000b *Paracypretta bhatiai*: Khosla and Sahni, 2000: Whatley and Bajpai, pp. 174–176, Pl. 1, Figs. 1–3.
2001 *Paracypretta jonesi*: Bhatia and Rana 1984: Bajpai and Whatley, pp. 95–96, Pl. 1, Figs. 2, 4.
2002c *Paracypretta jonesi*: Bhatia and Rana 1984: Whatley et al. pp. 166–168, Pl. 1, Figs. 8–9.
2003c *Paracypretta jonesi*: Bhatia and Rana 1984: Whatley et al. pp. 1293–1294, Pl. 2, Figs. 14, 17.
2005 *Paracypretta anjarensis*: Khosla and Nagori, pp. 576–578, Pl. 1, Figs. 13–16.
2005 *Paracypretta anjarensis*: Khosla and Nagori 2005: Khosla et al. pp. 139, Pl. 1, Figs. 11–12.
2007a *Paracypretta jonesi*: Bhatia and Rana, 1984: Khosla and Nagori, pp. 215–217, Pl. 1, Figs. 13–16; Pl. 2, Figs. 1–3.
2007b *Paracypretta jonesi*: Bhatia and Rana 1984: Khosla and Nagori, pp. 8, Pl. 1, Figs. 15–16.
2009a *Paracypretta jonesi*: Bhatia and Rana 1984: Khosla et al. pp. 725, Pl. 2, Fig. 13.
2009 *Paracypretta jonesi*: Bhatia and Rana 1984: Sharma and Khosla, pp. 204, Pl. 2, Figs. I–N.
2010 *Paracypretta jonesi*: Bhatia and Rana 1984: Khosla et al. pp. 118, Figs. 3a–c.
2011a *Paracypretta jonesi*: Bhatia and Rana 1984: Khosla et al. pp. 233–234, Pl. 2, Figs. 3–4.
2011b *Paracypretta jonesi*: Bhatia and Rana 1984: Khosla et al. pp. 233–235, Pl. 3, Figs. 1–5.
2015 *Paracypretta jonesi*: Bhatia and Rana 1984: Khosla, pp. 348, 352, Fig. 5d.
2017 *Paracypretta jonesi*: Bhatia and Rana 1984: Rathore et al. pp. 221, Pl. 1, Figs. 3.1–3.3.
2019 *Paracypretta jonesi*: Bhatia and Rana 1984: Kshetrimayum and Parmar, pp. 15, Figs. 2g–h.
2021 *Paracypretta jonesi*: Bhatia and Rana 1984: Kshetrimayum et al. pp. 11, 13, Figs. 4N–R.

Material More than 45 carapaces and open valves.

Horizon, Age and Locality Upper Cretaceous to Lower Palaeocene clayey limestone of unit 3 (JH 16–17, 19, 23–26) of the Jhilmili intertrappean beds, Chhindwara District, Madhya Pradesh, India.

Description Carapace is very large, subovate and subtriangular in lateral view and inflated to sub-umbonate in dorsal view with nearly one-sixth of the length of the anterior compressed. The greatest height and width are nearly equal and cover about two-thirds of the length. Both the anterior and posterior margins are generally rounded and have a posterior margin that is narrower and more rounded than the anterior. Symmetrically convex and umbonate dorsal margin, which is subangulate in the middle, and a straight ventral margin that is slightly concave in the middle. The left valve is larger than the right valve, which is overlapping the entire margin. Carapaces are compressed and lip-like, with nearly equal maximum height and width and about two-thirds of the length. The valve surface bears fine to coarse, very dense punctae and the fine riblets oriented in longitudinal fashion and parallel to the lower half of the ventral margin.

Table 4.21 Dimensions of carapaces of *Paracypretta jonesi* (Bhatia and Rana 1984)

	Dimensions (mm)		
Catalogue no.	Length	Width	Height
MPL/SK/JML/O/5152	1.655	0.841	0.886
MPL/SK/JML/O/5153	1.451	0.918	0.982
MPL/SK/JML/O/5154	1.410	0.861	0.881
MPL/SK/JML/O/5155	1.387	0.822	0.907
MPL/SK/JML/O/5156	1.412	0.850	0.887
MPL/SK/JML/O/5157	1.434	0.865	0.882
MPL/SK/JML/O/5158	1.307	0.776	0.812
MPL/SK/JML/O/5159	1.345	0.823	0.829
MPL/SK/JML/O/5160	1.411	0.805	0.932
MPL/SK/JML/O/5161	1.274	0.946	0.800
MPL/SK/JML/O/5162	1.476	0.853	1.09
MPL/SK/JML/O/5163	1.262	0.765	0.970
MPL/SK/JML/O/5164	1.505	0.747	1.117
MPL/SK/JML/O/5165	1.088	0.799	0.807
MPL/SK/JML/O/5166	1.382	0.816	0.866
MPL/SK/JML/O/5167	1.314	0.863	0.898
MPL/SK/JML/O/5168	1.228	0.827	0.813
MPL/SK/JML/O/5169	1.310	0.930	0.825
MPL/SK/JML/O/5170	1.498	0.917	0.924
MPL/SK/JML/O/5171	1.650	0.755	0.880
MPL/SK/JML/O/5172	1.305	0.775	0.810
MPL/SK/JML/O/5173	1.281	0.848	0.921
MPL/SK/JML/O/5174	1.441	0.912	0.840
MPL/SK/JML/O/5175	1.328	0.889	0.907

(continued)

Table 4.21 (continued)

Catalogue no.	Dimensions (mm)		
	Length	Width	Height
MPL/SK/JML/O/5176	1.420	0.965	0.845
MPL/SK/JML/O/5177	1.259	0.824	0.912
MPL/SK/JML/O/5178	1.339	0.845	0.902
MPL/SK/JML/O/5179	1.416	0.770	0.866
MPL/SK/JML/O/5180	1.554	0.899	1.001
MPL/SK/JML/O/5181	1.410	0.811	0.865

Remarks *Paracyptetta verruculosa* (Whatley et al. 2002a), *P. elizabethae* (Whatley et al. 2003c) *P. subglobosa* (Sowerby, in Malcolmson 1840), and *Paracypretta jonesi* (Bhatia and Rana, 1984) have all been reported from diverse places in the infra and intertrappean beds in peninsular India. According to Bhatia and Rana (1984), the general morphology of the carapaces recovered clearly identifies them as belonging to the species *Paracypretta jonesi*. Current *P. jonesi* specimens have fine to coarse papillae that run longitudinally parallel to the lower half of the ventral margin, which sets them apart from *P. elizabethae* and P. verruculosa. When compared to *P. jonesi, Paracypretta verruculosa* differs in that it has a distinctive contour, a peculiar hinge, and fine papillae that are clearly orientated.

Distribution Sars (1924) was the first to describe the genus *Paracypretta*, from South Africa. One of the species, *P. jonesi* (Bhatia and Rana 1984), was originally described from the intertrappean bed of Gitti Khadan (Nagpur District), Maharashtra (Bhatia and Rana 1984). Now, it is known from the outcrops of the Lameta Formation exposed at Jabalpur (Madhya Pradesh) and in the Nand-Dongargaon area (Chandrapur District), Maharashtra and also from numerous intertrappean sites, Anjar and Kora (Kachchh District), Gujarat; Chandarki (Gulbarga District), Karnataka, and Mohgaon-Haveli and Jhilmili (Chhindwara District), Khar (Khargone District), Utawali and Gujri (Dhar District), Madhya Pradesh (Sahni and Khosla 1994; Khosla and Sahni 2000; Whatley and Bajpai 2000b; Bajpai and Whatley 2001; Khosla and Nagori 2005, 2007a; Khosla et al. 2009a, 2011a; Sharma and Khosla 2009; Khosla 2015; Rathore et al. 2017; Kshetrimayum et al. 2021).

4.3.13 Species Paracypretta verruculosa (Whatley et al. 2002a) (Figs. 4.17A–E; Table 4.22)

2002a ?*Eucypris verruculosa*: Whatley et al. pp. 177, Pl. 4, Figs. 8, 9, 12–19.

2003b ?*Eucypris verruculosa*: Whatley et al. 2002a: Whatley et al. pp. 81–82, Figs. 2J–K.

2007b ?*Eucypris verruculosa*: Whatley et al. 2002a: Khosla and Nagori, pp. 12, Pl. 3, figs. 1–2.
2009a ?*Eucypris verruculosa*: Whatley et al. 2002a: Khosla et al. pp. 725, Pl. 2, Fig. 7.
2009 *Paracypretta* sp. : Sharma and Khosla, pp. 204, Pl. 2, Figs. O–R.
2011a *Paracypretta verruculosa*: Whatley et al. 2002a: Khosla et al. pp. 234, Pl. 2, Figs. 5–10.
2011b *Paracypretta verruculosa*: Whatley et al. 2002a: Khosla et al. pp. 233–234, Pl. 3, Figs. 6–7.
2015 *Paracypretta verruculosa* Whatley et al. 2002a: Khosla, pp. 348.

Material More than 31 carapaces and open valves.

Horizon, Age and Locality Upper Cretaceous to Lower Palaeocene clayey limestone of unit 3 (JH 17, 20, 24 and 25) of the Jhilmili intertrappean beds, Chhindwara District, Madhya Pradesh, India.

Description Carapace large and elongate to subtriangular in lateral view. Carapace fusiform when viewed dorsally. The greatest height occurs at the anterior cardinal angle. Both the ends are compressed, and the anterior end is more compressed than the posterior end. The anterior margin is broad and indirectly rounded, while the posterior margin is straight in the upper part and ventrally rounded in the lower part. The dorsal margin is straight, and the ventral margin is concave. The left valve is marginally larger than the right valve. The external surface of the carapace has minute reticulation oriented in striations parallel to the ventral margin in the lower half, as well as scattered papillae.

Table 4.22 Dimensions of carapaces of *Paracypretta verruculosa* (Whatley et al. 2002a)

	Dimensions (mm)		
Catalogue no.	Length	Width	Height
MPL/SK/JML/O/5182	0.576	0.421	0.326
MPL/SK/JML/O/5183	0.520	0.411	0.336
MPL/SK/JML/O/5184	0.493	0.409	0.329
MPL/SK/JML/O/5185	1.123	0.515	0.522
MPL/SK/JML/O/5186	0.770	0.422	0.470
MPL/SK/JML/O/5187	0.768	0.420	0.471
MPL/SK/JML/O/5189	0.771	0.430	0.419
MPL/SK/JML/O/5190	0.656	0.419	0.430
MPL/SK/JML/O/5191	0.705	0.428	0.469
MPL/SK/JML/O/5192	0.570	0.419	0.439

Remarks The carapaces of *Paracypretta verruculosa* collected from Jhilmili show an unusual hinge, distinctive shape and positive features in the larger valve and negative features in the smaller that are characters that differentiate them from the other species of *Paracypretta*, *P. jonesi* (Bhatia and Rana 1984), *P. subglobosa* (Sowerby, in Malcolmson 1840) and *P. elizabethae* (Whatley et al. 2003c) known from several localities of infra- and intertrappean beds of peninsular India. Additionally, they possess a typical *Paracypretta*-like valve surface ornamentation, on which basis they are placed under the genus *Paracypretta*.

Distribution Initially, *Paracypretta verruculosa* was reported as *Eucypris verruculosa* from the Chandarki intertrappean beds in the Gulbarga District of Karnataka (Whatley et al. 2002a). Whatley et al. (2003b) later described the material consisting of *Paracypretta verruculosa* recorded from the intertrappean collection of Sichel hill (Telangana) housed in the Natural History Museum, London. This species is currently known from the Lameta Formation of Jabalpur, Madhya Pradesh (Khosla et al. 2011b), and the intertrappeans beds of Takli, Maharashtra (Khosla and Nagori 2007b), and Jhilmili, Madhya Pradesh (Khosla et al. 2009a; Sharma and Khosla 2009; Khosla 2015).

4.3.14 Species Strandesia jhilmiliensis (Khosla et al. 2011a) (Fig. 4.17F)

Genus	*Starndesia* (Stuhlmann 1888)
Type species	*Cypris* (*Strandesia*) *mercatorum* (Vávra 1895)
Species	***Strandesia jhilmiliensis*** (Khosla et al. 2011a)

2011a *Strandesia jhilmiliensis*: Khosla et al. pp. 235, Pl. 2, Figs. 12–13; Pl. 3, Fig. 1.

Material A single carapace.

Horizon, Age and Locality Upper Cretaceous to Lower Palaeocene clayey limestone of unit 3 (JH 25) of the Jhilmili intertrappean beds, Chhindwara District, Madhya Pradesh, India.

Description Carapace of medium size, subtriangular to elongate in lateral view. The maximum height is close to the mid-length, and the maximum length is close to the ventral margin. Anterior margin is broad and rounded, whereas posterior margin is slightly rounded. Dorsal margin is broad and arched, subangulate and posterodorsally inclined straight down posteriorly. Ventral margin is almost straight

Fig. 4.17 (**A–E**) ***Paracypretta verruculosa*** (Whatley et al. 2002a), Carapace, (**A**) Lateral view MPL/SK/JML/O/5182, right valve; (**B**) Lateral view MPL/SK/JML/O/5183, right valve; (**C**) Lateral view MPL/SK/JML/O/5184, left valve; (**D**) Ventral view MPL/SK/JML/O/5185; (**E**) Dorsal view MPL/SK/JML/O/5186. **F.** ***Strandesia jhilmiliensis*** (Khosla et al. 2011a), Carapace, (**F**) Lateral view MPL/SK/JML/O/5519, left valve; (**G–J**) ***Stenocypris cylindrica*** (Sowerby in Malcolmson 1840), Carapace, (**G**) Lateral view MPL/SK/JML/O/7301, right valve; (**H**) Lateral view MPL/SK/JML/O/7302, left valve; (**I**) Lateral view MPL/SK/JML/O/7303, left valve; (**J**) Lateral view MPL/SK/JML/O/7304, left valve; (**K–P**) ***Periosocypris megistus*** (Whatley et al. 2012), Carapace, (**K**) Lateral view MPL/SK/GP/O/5196, left valve; (**L**) Lateral view MPL/SK/JML/O/7193, left valve; (**M**) Lateral view MPL/SK/JML/O/5194, left valve; (**N**) Lateral view MPL/SK/JML/O/5195, left valve; (**O**) Dorsal view MPL/SK/GP/O/5196; (**P**) Ventral view MPL/SK/GP/O/5196. (Figure **K**, **O**, and **P**, Reproduced from Kania et al. 2022 with permission from the Editor of the Himalayan Geology)

and a little concave anteriorly. Left valve larger than right valve, so there is overlap around the entire periphery, especially on dorsal and ventral margins. The surface of the valve is smooth. Sexually dimorphic and internal structures are not observable. The carapace, MPL/SK/JML/O/5519, measures 0.588 mm in length, 0.212 mm in width and 0.368 mm in height.

Remarks Morphologically, MPL/SK/JML/O/5519 bears a close resemblance to the material of *Strandesia jhilmiliensis* described by Khosla et al. (2011a), so the carapace described and illustrated here is identified as belonging to the species *S. jhilmiliensis*. It differs from *Strandesia indica*, which was described from a pond at Ernakulam, Travancore, and several other localities in India by Hartmann (1964), by having an asymmetrically arched dorsal margin and lacks sub-angulation anterior to the middle.

Distribution This species is only known from the Jhilmili intertrappean beds, Chhindwara District, Madhya Pradesh.

4.3.15 Species Stenocypris cylindrica (Sowerby in Malcolmson 1840) (Figs. 4.17G–J; Table 4.23)

Subfamily	Herpetocypridinae (Kaufmann 1900)
Genus	*Stenocypris* (Sars 1889)
Species	***Stenocypris cylindrica*** (Sowerby in Malcolmson 1840)

1840 *Cypris cylindrica*: Sowerby in Malcolmson 1840, Pl. 67, Fig. 2.
1988 *Mongolianella* sp.: Mathur and Verma, pp. 172, Pl. 1, Figs. 5a, b.
1990b *Mongolianella* sp. Bhatia et al. pp. 118, Pl. 1, Fig. 6.
1990b *Candoniella altanica*: Stankevitch in Stankevitch and Sochava: Bhatia et al. pp. 118, Pl. 1, Fig. 11.
1994 *Mongolianella palmosa*: Mandelstam: Sahni and Khosla, pp. 458, Figs. 2q–r.
1999 ?*Moenocypris* sp. : Bhandari, pp. 8, Pl. 2, Fig. 11.
2000 *Mongolianella palmosa*: Mandelstam: Khosla and Sahni, pp. 59, Figs. 3k, l; 4 a–e.
2000a *Mongolianella cylindrica*: Sowerby in Malcolmson 1840: Whatley and Bajpai, pp. 403–404, Pl. 6, Figs. 1–8.
2001 *Mongolianella cylindrica*: Sowerby in Malcolmson 1840: Bajpai and Whatley, pp. 103–104, Pl. 3, Figs. 6–9.
2002a *Mongolianella cylindrica*: Sowerby in Malcolmson 1840: Whatley et al. pp. 176–177, Pl. 4, Fig. 10.

2003a *Mongolianella cylindrica*: Sowerby in Malcolmson 1840: Whatley et al. pp. 80–83, Text Fig. 1, Fig.1–4.
2003b *Mongolianella cylindrica*: Sowerby in Malcolmson 1840: Whatley et al. pp. 82–83, Figs. 2l–O, Q, R; 3A–L.
2004 *Mongolianella cylindrica*: Sowerby in Malcolmson 1840: Bajpai et al. pp. 154–155, Pl. 2, Figs. n–o.
2005 *Mongolianella cylindrica*: Sowerby in Malcolmson 1840: Khosla and Nagori, pp. 576. Pl. 1, Fig. 9.
2007b *Mongolianella cylindrica*: Sowerby in Malcolmson 1840: Khosla and Nagori, pp. 9, Pl. 2, Figs. 10–11.
2008 *Mongolianella cylindrica*: Sowerby in Malcolmson 1840: Sharma et al. pp. 180, Pl. 2, Figs. I–K.
2009 *Mongolianella cylindrica*: Sowerby in Malcolmson 1840: Sharma and Khosla, pp. 204, Pl. 3, Fig. K.
2009a *Stenocypris cylindrica*: Sowerby in Malcolmson 1840: Khosla et al. pp. 725, Pl. 2, Fig. 14.
2009b *Stenocypris cylindrica*: Sowerby in Malcolmson 1840: Khosla et al. pp. 580–583, Pl. 1, Figs. 1–7; Fig. 3a.
2010 *Stenocypris cylindrica*: Sowerby in Malcolmson 1840: Khosla et al. pp. 118, Figs. 3d–e.
2011a *Stenocypris cylindrica*: Sowerby in Malcolmson 1840: Khosla et al. pp. 234–235, Pl. 2, Fig. 11.
2011b *Stenocypris cylindrica*: Sowerby in Malcolmson 1840: Khosla et al. pp. 234–237, Pl. 3, Figs. 10–14.
2017 *Stenocypris cylindrica*: Sowerby in Malcolmson 1840: Rathore et al. pp. 221, Figs. 3–4.
2021 *Stenocypris cylindrica*: Sowerby in Malcolmson 1840: Kshetrimayum et al. pp. 12. 15–16, Figs. 5P–R.

Material A total of eight carapaces.

Horizon, Age and Locality Upper Cretaceous to Lower Palaeocene clayey limestone of unit 3 (JH 25) of the Jhilmili intertrappean beds, Chhindwara District, Madhya Pradesh, India.

Description Carapace elongate, inflated, subrectangular to subcylindrical in lateral view and fusiform in dorsal view. It shows variation in the degree of dorsal margin curvature from gentle to protuberant, the length/height ratio and the projecting overlap along the ventral margin. The greatest height is in the middle and greatest width towards the anterior margin. Rounded and laterally compressed anterior margin, convex to straight dorsal margin. Left valve larger than right valve, ventral margin concave in the middle. Valve surface is smooth. Sexual dimorphism is visible, and female carapaces are smaller than the male. The internal features are not observable.

Table 4.23 Dimensions of carapaces of *Stenocypris cylindrica* (Sowerby in Malcolmson 1840)

	Dimensions (mm)		
Catalogue no.	Length	Width	Height
MPL/SK/JML/O/7301	1.763	0.728	1.113
MPL/SK/JML/O/7302	1.752	0.730	1.037
MPL/SK/JML/O/7303	1.758	0.725	1.04
MPL/SK/JML/O/7304	1.765	0.729	1.045

Remarks Originally named *Cypris cylindrica* (Sowerby in Malcolmson 1840), this species was later renamed *Mongolianella cylindrica* (Whatley and Bajpai 2000a). Khosla et al. (2009b) studied carapaces recovered from the Lakshmipur intertrappean beds of Gujarat and assigned this species to the genus *Stenocypris* (Sars 1889) based on the internal and external morphologies of the carapaces. As a result, all *Mongolianella* species known from the infra- and intertrappean beds of peninsular India, such as *M. hislopi*, *M. khamariensis*, and *M. subarcuata*, are here treated as species of *Stenocypris*.

The arched dorsal margin and tapering posterior end differentiate *Mongolianella hislopi* (Jones 1860) from the Jhilmili carapaces of *Stenocypris cylindrica*. *Mongolianella khamariensis* (Galeeva 1955) differs from *S. cylindrica* by possessing a convex postero-dorsal margin and an apex of the posterior margin situated towards the ventral margin. *Stenocypris cylindrica* is larger in size than *M. subarcuata* (Whatley et al. 2003a). Further, *M. subarcuata* possess distinct morphological details such as a medianly concave ventral margin, sub-arcuate shape and valve surface that bears fine punctae, which also differentiate it from *S. cylindrica*.

Distribution This species is known from infra- and intertrappean localities in peninsular India. It is known from Upper Cretaceous outcrops of the Lameta Formation exposed at the Chui hill and Bara Simla hill sections in Jabalpur (Madhya Pradesh) and Pisdura hill, Maharashtra (Sahni and Khosla 1994; Khosla and Sahni 2000; Khosla et al. 2010, 2011b). It is also known from Upper Cretaceous intertrappean beds at Sichal hills, Telangana (Sowerby in Malcolmson 1840), Mamoni (Kota District), Rajasthan (Bhatia et al. 1990b; Whatley et al. 2003a), Lakshmipur, Kora and Anjar (Kachhch District), Gujarat (Whatley and Bajpai 2000a; Bajpai and Whatley 2001; Khosla and Nagori 2005; Khosla et al. 2009b), Chandarki and Yanagundi (Gulbarga District), Karnataka (Whatley et al. 2002a), Phulsagar, Khar and Jhilmili, Madhya Pradesh (Bajpai et al. 2004; Sharma and Khosla 2009; Khosla et al. 2009a, 2011a; Rathore et al. 2017) and Takli (Nagpur District), Maharashtra (Khosla and Nagori 2007b). It has also been reported from the Eocene Cambay shales, Gujarat (Bhandari 1998).

4.3.16 Species *Periosocypris megistus* (Whatley et al. 2012) (Figs. 4.17K–P; Table 4.24)

Subfamily	*Cypridinae* (Baird 1845)
Genus	*Periosocypris* (Whatley et al. 2012)
Species	***Periosocypris megistus*** (Whatley et al. 2012)

1983 *Mongolocypris* cf. *M. gigantea*: Ye, Gou and Cao, pp. 392, Pl. 4, Fig. 3.
1994 *Mongolocypris* cf. *M. gigantea*: Ye: Sahni and Khosla, pp. 456–458, Figs. 2a–d.
2000 *Mongolocypris* cf. *M. gigantea*: Ye: Khosla and Sahni, pp. 61–62. Pl. 2, Figs. J–L; Pl. 3, Figs. A–H.
2005 *Mongolocypris* sp.: Khosla et al. pp. 143, Pl. 2, Figs. 11–12, Pl. 3, Fig. 1.
2009a *Heterocypris* sp.: Khosla et al. pp. 725, Pl. 2, Fig. 11.
2011a *Heterocypris* sp.: Khosla et al. pp. 233, Pl. 2, Fig. 2.
2011b *Mongolocypris* cf. *M. gigantea*: Khosla et al. pp. 230–231, Pl. 2, Figs. 6–10.
2012 *Periosocypris megistus*: Whatley et al. pp. 113–116, Pl. 1, Figs. 1–10; Figs. 1a–c.
2021 *Periosocypris megistus*: Khosla et al. 2012: Kshetrimayum et al. pp. 12, 13, Figs. 5A–D.

Material More than 20 carapaces and open valves.

Horizons, Age and Localities Upper Cretaceous to Lower Palaeocene clayey limestone of unit 3 (JH 25) of the Jhilmili intertrappean beds, Chhindwara District, Madhya Pradesh. Upper Cretaceous to ?Lower Palaeocene hard clayey limestone unit of the Ghat Parasia intertrappean beds, Chhindwara District, Madhya Pradesh, India.

Description Large carapace, thick-shelled, elongate, subrectangular to subovate in side view; dorsal margin is extremely arched; apex at about mid-length. The postero-dorsal angle is acute. The anterior margin is broadly but unevenly rounded, with an apex that is lower than mid-height. The ventral margin is almost perfectly straight. The dorsal margin's antero-ventral slope is short and firmly convex, whereas the anterodorsal slope is elongate and gently curved. The posterior margin is straight and sloping down in the upper part and well-bowed in the lower part. The greatest height occurs slightly posterior to the median, and the greatest width occurs posterior to the middle or one-third of the carapace. A gentle slope leading to a sub-rounded anterior and a smaller slope leading to the posterior has been observed. In dorsal view, biconvex, subhastate with laterally compressed posterior end; left valve larger than right and overlapping along ventral and posterior margins apart from dorsal and postero-dorsal margins. The valve surface is smooth, and no internal features are known.

Table 4.24 Dimensions of carapaces of *Periosocypris megistus* (Whatley et al. 2012)

	Dimensions (mm)		
Catalogue no.	Length	Width	Height
MPL/SK/JML/O/7193	2.589	1.294	1.737
MPL/SK/JML/O/5194	2.474	1.112	1.662
MPL/SK/JML/O/5195	1.716	0.830	1.167
MPL/SK/GP/O/5196	1.699	0.629	1.215

Remarks The large size, together with the other morphological features described above, aid in the identification of *Periosocypris megistus* material recovered from Jhilmili and Ghat Parasia. It is also found that the morphological characteristics of *Periosocypris megistus* carapaces recovered from Jhilmili and Ghat Parasia are strikingly similar to those described from the Lameta Formation of Jabalpur and Dongargaon. Thus, the Jhilmili and Ghat Parasia material is assigned to *Periosocypris megistus*.

Distribution Initially, this species was described as *Mongolocypris* cf. *M. gigantea* from the Upper Cretaceous Lameta Formation of the Bara Simla and the Chui hill areas, Jabalpur, Madhya Pradesh and Dongargaon hill, Chandrapur District, Maharashtra (Sahni and Khosla 1994; Khosla and Sahni 2000; Khosla et al. 2011b; Whatley et al. 2012). Later, it was reported from the intertrappean beds of Jhilmili and Gujri, Madhya Pradesh (Khosla et al. 2009a, 2011a; Kshetrimayum et al. 2021).

4.3.17 *Species Zonocypris spirula (Whatley and Bajpai 2000a) (Figs. 4.18A–E; Table 4.25)*

Subfamily	Cypridopsinae (Kaufmann 1900)
Genus	*Zonocypris* (Müller 1898)
Type species	*Zonocypris madagascarensis* (Müller 1898)
Species	***Zonocypris spirula*** (Whatley and Bajpai 2000a)

2000a *Zonocypris spirula*: Whatley and Bajpai, pp. 396–397, Pl. 3, Figs. 1–7, 9.
2002a *Zonocypris spirula*: Whatley and Bajpai 2000a: Whatley et al. pp. 173, Pl. 3, Figs. 6–7.
2002c *Zonocypris spirula*: Whatley and Bajpai 2000a: Whatley et al., pp. 168, Pl. 1, Figs. 11–12.
2005 *Zonocypris spirula*: Whatley and Bajpai 2000a: Khosla et al., pp. 139, Pl. 1, Figs. 13–14.

2005 *Zonocypris spirula*: Whatley and Bajpai 2000a: Khosla and Nagori, pp. 576, Pl. 1, Fig. 19.
2007a *Zonocypris spirula*: Whatley and Bajpai 2000a: Khosla and Nagori, pp. 217, Pl. 2, Figs. 4–5.
2007b *Zonocypris spirula*: Whatley and Bajpai 2000a: Khosla and Nagori, pp. 9. Pl. 2, Figs. 5–6.
2008 *Zonocypris spirula*: Whatley and Bajpai 2000a: Sharma et al. pp. 182, Pl. 2, Figs. O–P.
2009a *Zonocypris spirula*: Whatley and Bajpai 2000a: Keller et al. Figs. 9.3–9.4.
2009a *Zonocypris spirula*: Whatley and Bajpai 2000a: Khosla et al. pp. 725, Pl. 2, Fig. 16.
2009 *Zonocypris spirula*: Whatley and Bajpai 2000a: Sharma and Khosla, pp. 204, Pl. 3, Figs. C–E.
2011a *Zonocypris spirula*: Whatley and Bajpai 2000a: Khosla et al. pp. 235–236, Pl. 3, Figs. 2–3.
2011b *Zonocypris spirula*: Whatley and Bajpai 2000a: Khosla et al. pp. 237–238, Pl. 4, Figs. 6–7.
2015 *Zonocypris spirula*: Whatley and Bajpai 2000a: Khosla, pp. 348, 352, Fig. 5E.
2018 *Zonocypris spirula*: Whatley and Bajpai 2000a: Rathore, pp. 66.

Material More than 88 carapaces and open valves.

Horizon, Age and Locality Upper Cretaceous to Lower Palaeocene clayey limestone of unit 3 (JH 25 and 26) of the Jhilmili intertrappean beds, Chhindwara District, Madhya Pradesh, India.

Description Carapace small, elongate, subrectangular, strongly tumid and fusiform in dorsal and lateral views. The greatest height close to anterior end, the greatest length below the mid-height. Anterior margin is flat and rounded as compared to the posterior margin, which is less rounded, pointed and convex. Strongly tumid and fusiform dorsally. Dorsal margin is broad and well-rounded and convex, declined from the maximum height to posterior margin and gently inclined from the anterior side to the apex. Ventral margin is almost flat and slightly convex from the middle. Hinge line is straight, clearly observable in dorsal view, and the ventral view is strongly elevated. Right valve greater than the left valve and overlapping the entire periphery, especially the dorsal and ventral margins. The valve surface is strongly ornamented with lines that appear parallel in ventral and dorsal views; spirally coiled begins from the middle. It has a distinctive ornamentation consisting of a single rib and a single helix, a spirally coiled mid-valve surface, outer coils more or less angular and inner coils that are circular (Sharma and Khosla 2009). Sexually dimorphic and internal structures are not observable.

Table 4.25 Dimensions of carapaces of *Zonocypris spirula* (Whatley and Bajpai 2000a)

Catalogue no.	Dimensions (mm)		
	Length	Width	Height
MPL/SK/JML/O/5197	0.329	0.248	0.223
MPL/SK/JML/O/5198	0.321	0.255	0.220
MPL/SK/JML/O/5199	0.316	0.242	0.213
MPL/SK/JML/O/2010	0.205	0.239	0.187
MPL/SK/JML/O/5201	0.325	0.243	0.216
MPL/SK/JML/O/5202	0.299	0.256	0.197
MPL/SK/JML/O/5203	0.333	0.259	0.218
MPL/SK/JML/O/5204	0.319	0.247	0.223
MPL/SK/JML/O/5205	0.299	0.252	0.226
MPL/SK/JML/O/5206	0.310	0.261	0.229

Remarks Five species of the genus *Zonocypris* are *Z. spirula* (Whatley and Bajpai 2000a), *Z. viriensis* (Khosla and Nagori 2005), *Z. gujaratensis* (Bhandari and Colin 1999), *Z. pseudospirula* (Khosla et al. 2010) and *Zonocypris* sp. have been reported from the infra- and intertrappean localities in peninsular India. The distinct valve surface ornamentation – strongly tumid and fusiform dorsally – are the characters that differentiate the Jhilmili carapaces of *Z. spirula* from other species of *Zonocypris*.

Distribution Originally, *Zonocypris spirula* was reported from the intertrappean locality of Lakshmipur (Kachchh District), Gujarat (Whatley and Bajpai 2000a). Subsequently, this species has been reported from a number of intertrappean localities in peninsular India such as Yanagundi (Gulbarga District), Karnataka (Whatley et al. 2002a), Jhilmili and Mohgaon-Haveli (Chhindwara District), Khar (Khargaon District), Madhya Pradesh (Khosla and Nagori 2007a; Keller et al. 2009a; Khosla et al. 2009a, 2011a; Sharma and Khosla 2009; Khosla 2015), Takli (Nagpur District) and Pinjaurni (Chandrapur District), Maharashtra (Khosla and Nagori 2007b; Rathore et al. 2017). This species was also reported from the Lameta Formation of the Jabalpur and Dongargaon areas (Khosla et al. 2005, 2011b) and Papro (Lalitpur District), Uttar Pradesh (Sharma et al. 2008).

4.3.18 Species Zonocypris viriensis (Khosla and Nagori 2005) (Figs. 4.18F–H; Table 4.26)

Genus	*Zonocypris* (Müller 1898)
Species	***Zonocypris viriensis*** (Khosla and Nagori 2005)

2000a *Zonocypris viriensis*: Whatley and Bajpai, pp. 395–397, Pl. 3, Figs. 7, 9.
2005 *Zonocypris viriensis*: Khosla and Nagori, pp. 578, Pl. 1, Figs. 20–22.
2009a *Zonocypris viriensis* Khosla and Nagori 2005: Keller et al. Figs. 6.5–6.6.
2009a *Zonocypris viriensis*: Khosla and Nagori 2005: Khosla et al. pp. 729, Pl. 2, Fig. 17.
2009 *Zonocypris viriensis*: Khosla and Nagori 2005: Sharma and Khosla, pp. 202–203, Pl. 3, Fig. F–J.
2011a *Zonocypris viriensis*: Khosla and Nagori 2005: Khosla et al. pp. 235–236, 240, 241, Pl. 3, Figs. 4–6.
2015 *Zonocypris viriensis*: Khosla and Nagori 2005: Khosla, pp. 348.

Material A total of 209 carapaces and open valves.

Horizon, Age and Locality Upper Cretaceous to Lower Palaeocene clayey limestone of unit 3 (JH 16–19, 21–23, 25) of the Jhilmili intertrappean beds, Chhindwara District, Madhya Pradesh, India.

Description Carapaces subovate and subtriangular shaped in lateral view; the maximum height is three-quarters of the length and a little anterior to the middle. Anterior margin is more convex, whereas posterior margin is less rounded to slightly pointed. The ventral margin is sinuate, and the dorsal margin is well rounded, flattened and arch shaped. The apex is in the middle, slightly inclined from the posterior to anterior margin and gently inclined to the posterior margin. Right valve overlaps the left valve on the entire periphery. The valve surface bears a distinctive ornamentation consisting of fine striations concentrically oriented in the marginal region and irregularly present in the middle (Sharma and Khosla 2009; Khosla et al. 2011a). Sexually dimorphic and internal structures are not distinguishable.

Table 4.26 Dimensions of carapaces of *Zonocypris viriensis* (Khosla and Nagori 2005)

	Dimensions (mm)		
Catalogue no.	Length	Width	Height
MPL/SK/JML/O/2011	0.373	0.315	0.277
MPL/SK/JML/O/5208	0.331	0.271	0.261
MPL/SK/JML/O/5209	0.335	0.245	0.248
MPL/SK/JML/O/5210	0.308	0.256	0.229
MPL/SK/JML/O/5211	0.328	0.266	0.253
MPL/SK/JML/O/5212	0.365	0.283	0.255
MPL/SK/JML/O/5213	0.346	0.263	0.253
MPL/SK/JML/O/5214	0.310	0.259	0.230
MPL/SK/JML/O/5215	0.327	0.263	0.251
MPL/SK/JML/O/5216	0.336	0.241	0.240
MPL/SK/JML/O/5224	0.315	0.219	0.240
MPL/SK/JML/O/5225	0.263	0.218	0.237
MPL/SK/JML/O/5226	0.299	0.203	0.234
MPL/SK/JML/O/5227	0.270	0.230	0.263

Fig. 4.18 (**A**–**E**) ***Zonocypris spirula*** (Whatley and Bajpai, 2000a), female carapace; (**A**) Lateral view MPL/SK/JML/O/5197; right valve; (**B**) Lateral view MPL/SK/JML/O/5198; right valve; (**C**) Lateral view MPL/SK/JML/O/5199, left valve; (**D**) Dorsal view MPL/SK/JML/O/2010; (**E**) Ventral view MPL/SK/JML/O/5201. (**F**–**H**) ***Zonocypris viriensis*** (Khosla and Nagori 2005), Carapace,

Remarks The distinctive valve surface ornamentation, consisting of fine striations concentrically oriented in the peripheral region and irregularly present in the middle help to identify the Jhilmili material as *Zonocypris viriensis* Khosla and Nagori (2005) and also to differentiate *Zonocypris viriensis* from the other five species of the genus, *Z. spirula* (Whatley and Bajpai 2000a), *Z. gujaratensis* (Bhandari and Colin 1999), *Z. pseudospirula* (Khosla et al. 2010), *Z. labyrinthicos* (Whatley et al. 2002b) and *Zonocypris* sp., known from the infra- and intertrappean beds.

Distribution *Zonocypris viriensis* has been described from the intertrappean beds of Lakshmipur and Anjar (Kachchh District), Gujarat, and Jhilmili (Chhindwara District), Madhya Pradesh (Whatley and Bajpai 2000a; Khosla and Nagori 2005; Keller et al. 2009a; Khosla et al. 2009a, 2011a; Sharma and Khosla 2009; Khosla 2015).

4.3.19 Species Zonocypris labyrinthicos (Whatley et al. 2002b) (Fig. 4.18I; Table 4.27)

2002b *Zonocypris labyrinthicos*: Whatley et al. pp. 110–111, Pl. 2, Figs. 3–9.

Material More than 20 carapaces and open valves.

Horizon, Age and Locality Upper Cretaceous (Maastrichtian) greenish laminated shale and black carbonaceous shale unit of the intertrappean section exposed in the Shiraj (=Shriwas) well, Mohgaon-Kalan village, Chhindwara District, Madhya Pradesh, India.

Description Carapace small, elongate, subovate to subtriangular in lateral view and fusiform in dorsal view. The greatest height is at mid-height, the greatest width is at the back of mid-length, and the greatest height is at the front of mid-length (Whatley et al. 2002b). The anterior margin is more rounded compared to the posterior margin. The posterior margin is flat with a small arch and convex. Dorsal margin is arch shaped and strongly convex. Ventral margin is almost straight in lateral view with a

Fig. 4.18 (continued) (**F**) Lateral view MPL/SK/JML/O/2011, left valve; (**G**) Latero-ventral view MPL/SK/JML/O/5208, right valve; (**H**) Dorsal view MPL/SK/JML/O/5209. (**I**) ***Zonocypris labyrinthicos*** (Whatley et al. 2002b), Carapace (**I**) Ventral view MPL/SK/SW/O/5228. (**J**–**K**) ***Zonocypris gujaratensis*** (Bhandari and Colin 1999), Carapace, (**J**) Lateral view MPL/SK/SW/O/7355, right valve; (**K**) Enlarged view **J** MPL/SK/SW/O/7355 showing strong spiral ribs extending over several whorls and at intervals dividing into secondary ribs. (**L**–**O**) ***Zonocypris penchi* sp. nov.** Carapace, (**L**) Lateral view MPL/SK/JML/O/6217; left valve; (**M**) Lateral view MPL/SK/JML/O/6218, right valve; (**N**) Dorsal view MPL/SK/JML/O/6219; (**O**) Ventral view MPL/SK/JML/O/6220. (Figure **F**, and **G**, Reproduced from Khosla et al. 2022 with permission from Wiley)

small concavity. Hinge line is clearly visible in dorsal view. Right valve is greater than the left valve and overlaps around the entire margins. The valve surface is strongly ornamented with concentrically arranged spiral ribs visible in the ventral, dorsal and lateral views. Sexually dimorphic and internal structures are not observable.

Table 4.27 Dimensions of carapaces of *Zonocypris labyrinthicos* (Whatley et al. 2002b)

	Dimensions (mm)		
Catalogue no.	Length	Width	Height
MPL/SK/SW/O/5228	0.400	0.285	0.201
MPL/SK/SW/O/5229	0.348	0.269	0.212
MPL/SK/SW/O/5230	0.365	0.282	0.202

Remarks The carapaces of *Zonocypris labyrinthicos* (Whatley et al. 2002b) recovered from Shiraj (=Shriwas) well are similar in size and morphology to those described by Whatley et al. (2002b) from the Mohagaon-Kalan intertrappean beds. However, they are much smaller in size and have a distinct ornamentation pattern that distinguishes *Zonocypris labyrinthicos* from the other five species of the genus *Zonocypris*, *Z. spirula* (Whatley and Bajpai 2000a), *Z. gujaratensis* (Bhandari and Colin 1999), *Z. pseudospirula* (Khosla et al. 2010), *Z. viriensis* (Khosla and Nagori 2005) and *Zonocypris* sp., known from the infra- and intertrappean beds of peninsular India.

Distribution *Zonocypris labyrinthicos* has been reported from the intertrappean beds of Anjar and Lakshmipur (Kachchh District), Gujarat, and Mohagaon-Kalan (Chhindwara District), Madhya Pradesh (Bhandari and Colin 1999; Whatley et al. 2002b).

4.3.20 Species Zonocypris gujaratensis (Bhandari and Colin 1999) (Figs. 4.18J–K)

1999 *Zonocypris gujaratensis*: Bhandari and Colin 1999: Bhandari and Colin: pp. 17, Pl. 2, Figs. 1–4.

2005 *Zonocypris gujaratensis*: Bhandari and Colin 1999: Khosla and Nagori, pp. 576, Pl. 1, Fig. 18.

2007a *Zonocypris gujaratensis*: Bhandari and Colin 1999: Khosla and Nagori, pp. 217, Pl. 2, Fig. 6.

2007b *Zonocypris gujaratensis*: Bhandari and Colin 1999: Khosla and Nagori, pp. 9, Pl. 2, Figs. 1–2.

2011b *Zonocypris gujaratensis*: Bhandari and Colin 1999: Khosla et al. pp. 237, Pl. 4, Figs. 1–2.

2017 *Zonocypris gujaratensis*: Bhandari and Colin 1999: Rathore et al. pp. 221, Figs. 3.7–3.8.

2019 *Zonocypris gujaratensis*: Bhandari and Colin 1999: Kapur et al. Figs. 3N–O.
2021 *Zonocypris gujaratensis*: Bhandari and Colin 1999: Kshetrimayum et al. pp. 12, 14, Figs. 5E–J.

Material A total of five moderately preserved carapaces.

Horizon, Age and Locality Upper Cretaceous (Maastrichtian) carbonaceous black shale unit of the intertrappean section exposed in the Shiraj (= Shriwas) well, Mohgaon-Kalan village, Chhindwara District, Madhya Pradesh, India.

Description Carapaces small, subovate to subtriangular in lateral view, rounded to globular in dorsal view. The greatest height at the mid position and the greatest length at the mid-height. Anterior margin is broader and more rounded than posterior margin, and posterior margin is less rounded and has a small arch. Dorsal margin is broad and arch shaped, strongly convex. Ventral margin is almost straight in lateral view. Hinge line is not clearly visible in dorsal view. Left valve is a little larger than right valve and overlaps around the entire periphery. The valve surface has strong spiral ribs extending over several whorls and, at intervals, dividing into secondary ribs (Khosla et al. 2011b). MPL/SK/SW/O/7355 measures length 0.480 mm, height 0.360 mm and width 0.330 mm.

Remarks In terms of overall shape, height, width and surface ornamentation, the present species differs from *Zonocypris spirula* (Whatley and Bajpai 2000a) and *Z. pseudospirula* (Whatley and Bajpai 2000b; Khosla et al. 2010). In dorsal view, they are fusiform and tumid, with different ornamentation than *Z. gujaratensis.* Furthermore, *Z. gujaratensis* differs from *Z. viriensis* (Khosla and Nagori 2005) in its' distinctive surface ornamentation. *Zonocypris viriensis* also has a distinctive ornamentation consisting of fine striations that are concentrically oriented in the periphery and irregularly present in the middle.

Distribution *Zonocypris gujaratensis* has been reported from the intertrappean beds of Anjar (Kachchh District), Gujarat (Bhandari and Colin 1999; Khosla and Nagori 2005), Mohgaon-Haveli (Chhindwara District) Manawar, Gujri (Dhar District), Madhya Pradesh (Khosla and Nagori 2007a; Rathore et al. 2017; Kshetrimayum et al. 2021), and Takli (Nagpur District), Maharashtra (Khosla and Nagori 2007b), and from the Lameta Formation of Bara Simla hill, Jabalpur (Khosla et al. 2011b).

4.3.21 Species *Zonocypris penchi* sp. nov. (Figs. 4.18L–O; Table 4.28)

Holotype MPL/SK/JML/O/6217, a left valve.

Material A total of 40 carapaces.

Horizon, Age and Locality Upper Cretaceous to Lower Palaeocene clayey limestone of unit 3 (JH 19 and 25) of the Jhilmili intertrappean beds, Chhindwara District, Madhya Pradesh, India.

Etymology Species named after Pench River, which flows through the study (Jhilmili) area.

Diagnosis A small carapace, subovate to subtriangular shaped in lateral view, with height at the central position, the greatest length behind the mid-height. Anterior margin is broad and concave, and posterior margin is short, less rounded and pointed. Carapace has reticulated ornamentation dominating the entire surface and concentric growth lines on the surface.

Description Carapaces are small, subovate to triangular shaped in lateral view; the maximum height at the middle, maximum length behind the mid-height. Anterior and posterior margins are equally rounded. Dorsal margin is fully arch shaped, and ventral margin is almost straight and gently arched in the middle portion. Right valve overlaps the left valve on the entire margin. Surface has strongly ornamented reticulate structure in the inflated part and slightly on the margins. Sexual dimorphic and internal structures are not visible.

Table 4.28 Dimensions of carapaces of *Zonocypris penchi* sp. nov.

	Dimensions (mm)		
Catalogue no.	Length	Width	Height
MPL/SK/JML/O/6217	0.265	0.223	0.259
MPL/SK/JML/O/6218	0.295	0.206	0.247
MPL/SK/JML/O/6219	0.311	0.213	0.246
MPL/SK/JML/O/6220	0.300	0.215	0.262
MPL/SK/JML/O/6221	0.271	0.226	0.261
MPL/SK/JML/O/6222	0.290	0.205	0.243
MPL/SK/JML/O/6223	0.305	0.221	0.243

Remarks Six species of *Zonocypris* were previously known from the Late Cretaceous of India. These species are *Z. spirula* (Whatley and Bajpai 2000a), *Z. viriensis* (Khosla and Nagori 2005), *Z. gujaratensis* (Bhandari and Colin 1999), *Z. pseudospirula* (Khosla et al. 2010), *Z. labyrinthicos* (Whatley et al. 2002b) and *Zonocypris* sp. *Zonocypris penchi* sp. nov. differs from these species by having a triangular shape and different surface ornamentation. The morphology and valve surface ornamentation of *Zonocypris penchi* sp. nov. is quite different from these species.

Zonocypris pseudospirula differs from *Z. penchi* sp. nov. by having a biconvex dorsal aspect, broad anterior margin, rounded posterior margin, subrectangular shape and coarse concentric ribs of ornamentation. *Zonocypris viriensis* differs from *Z. penchi* sp. nov. by having a markedly different ornamentation consisting of fine striations concentrically oriented in the peripheral region and irregularly present in the middle. *Zonocypris spirula* differs from *Z. penchi* sp. nov. by having a distinct valve surface ornamentation with a strongly tumid and fusiform dorsal valve. *Zonocypris gujaratensis* differs from *Z. penchi* sp. nov. by possessing a valve surface with strong spiral ribs extending over several whorls and, at intervals, dividing into secondary ribs.

4.3.22 Species Cypridopsis astralos (Whatley et al. 2002a) (Figs. 4.19A–D; Table 4.29)

Subfamily	Cypridopsinae (Kaufmann 1900)
Genus	*Cypridopsis* (Brady 1868)
Species	***Cypridopsis astralos*** (Whatley et al. 2002a)

2002a *Cypridopsis astralos*: Whatley et al. 2002a, pp. 174–176, Pl. 3, Figs. 14–19, Pl. 4, Figs. 1–7.

2011b *Cypridopsis astralos*: Whatley et al. 2002a: Khosla et al. pp. 240, Pl. 4, Figs. 11–13.

Material A total of five carapaces.

Horizon, Age and Locality Upper Cretaceous to Lower Palaeocene clayey limestone of unit 3 (JH 23–26) of the Jhilmili intertrappean beds, Chhindwara District, Madhya Pradesh, India.

Description Carapace medium-sized, elongate, subovate to subelliptical in lateral view, strongly inflated in dorsal view. The greatest height at or near the mid-length and the greatest length at the mid-height. Anterior margin is rounded, and posterior margin is subrounded to rounded. Dorsal margin is strongly convex, and ventral margin is slightly concave. Left valve is larger than right valve, overlapping the entire margins in dorsal and lateral views. Valve surface contains minute, weak and concentrically oriented reticulations and a small number of tubercles (Khosla et al. 2011b). Sexually dimorphic and internal structures are not distinguishable.

Table 4.29 Dimensions of carapaces of *Cypridopsis astralos* (Whatley et al. 2002a)

	Dimensions (mm)		
Catalogue no.	Length	Width	Height
MPL/SK/JML/O/6355	0.834	0.548	0.512
MPL/SK/JML/O/6356	0.830	0.546	0.511
MPL/SK/JML/O/6357	0.828	0.544	0.513
MPL/SK/JML/O/6358	0.829	0.495	0.510

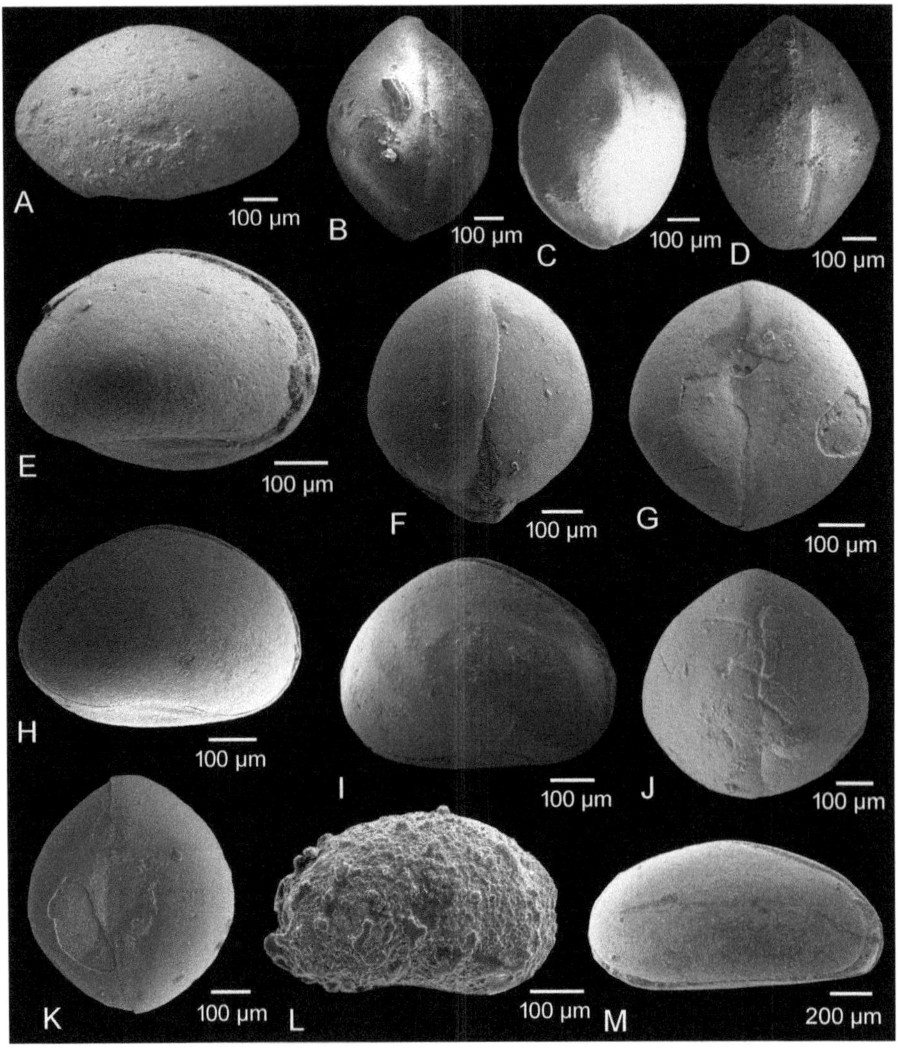

Fig. 4.19 (**A**–**D**) ***Cypridopsis astralos*** (Whatley et al. 2002a), Carapace, (**A**) Lateral view MPL/SK/JML/O/6355, left valve; (**B**) Dorsal view MPL/SK/JML/O/6356; (**C**) Dorsal view MPL/SK/JML/O/6357; (**D**) Ventral view MPL/SK/JML/O/6358. (**E**–**G**) ***Cypridopsis hyperectyphos*** (Whatley and Bajpai 2000a), Carapace (**E**) Lateral view MPL/SK/JML/O/5231, left valve; (**F**) Dorsal view MPL/SK/JML/O/5233; (**G**) Ventral view MPL/SK/JML/O/5232. (**H**–**K**) ***Cypridopsis elachistos*** (Whatley et al. 2002b), Carapace, (**H**) Lateral view MPL/SK/SW/O/5281, left valve; (**I**) Lateral view MPL/SK/GW/O/5287, right valve; (**J**) Dorsal view MPL/SK/SW/O/5282; (**K**) Ventral view MPL/SK/SW/O/5283. **L.** ***Candona* sp.** Carapace, (**L**) Lateral view MPL/SK/GP/O/5292, right valve; (**M**) ***Eucypris pelasgicos*** (Whatley and Bajpai 2000a), Carapace, lateral view MPL/SK/JML/O/5305, right valve

Remarks There are eight species of *Cypridopsis* – *C. elachistos* (Whatley et al. 2002b), *C. hyperectyphos* (Whatley and Bajpai 2000a), *C. huenei* (Khosla et al. 2011b), *C. ashui* (Khosla et al. 2011b), *C. dongargaonensis* (Khosla et al. 2005), *C. mohgaonensis* (Khosla and Nagori 2007a), *C. sahnii* (Khosla et al. 2005) and *C. wynnei* (Whatley and Bajpai 2000a). They are widely known from the several intertrappean localities of peninsular India. The dorsal outline of *C. astralos* is an important parameter that distinguishes this species from the other species mentioned above.

Distribution *Cypridopsis astralos* has been recorded from the intertrappean localities of Chandarki and Yanagundi (Gulbarga District), Karnataka (Whatley et al. 2002a), and the Bara Simla and Chotta Simla hills outcrops of the Lameta Formation (Jabalpur), Madhya Pradesh (Khosla et al. 2011b).

4.3.23 Species Cypridopsis hyperectyphos (Whatley and Bajpai 2000a) (Figs. 4.19E–G; Table 4.30)

2000a *Cypridopsis hyperectyphos*: Whatley and Bajpai, pp. 397–398, Pl. 4, Figs. 4–10.
2001 *Cypridopsis hyperectyphos*: Whatley and Bajpai 2000a: Bajpai and Whatley, pp. 96, Pl. 1, Figs. 6–8.
2002a *Cypridopsis hyperectyphos*: Whatley and Bajpai 2000a: Whatley et al. pp. 174, Pl. 3, Figs. 11–13.
2003a *Cypridopsis hyperectyphos*: Whatley and Bajpai 2000a: Whatley et al. pp. 80, Pl. 1, Figs. 12–13.
2005 *Cypridopsis hyperectyphos*: Whatley and Bajpai 2000a: Khosla et al. pp. 141, Pl. 2, Figs. 3–4.
2005 *Cypridopsis hyperectyphos*: Whatley and Bajpai 2000a: Khosla and Nagori, pp. 574, Pl. 1, Fig. 17.
2007b *Cypridopsis hyperectyphos*: Whatley and Bajpai 2000a: Khosla and Nagori, pp. 9, Pl. 2. Fig. 9.
2008 *Cypridopsis hyperectyphos*: Whatley and Bajpai 2000a: Sharma et al. pp. 178–180, Pl. 2, Figs. A–C.
2009a *Cypridopsis hyperectyphos*: Whatley and Bajpai 2000a: Khosla et al. pp.725, Pl. 2, Fig. 4.
2009 *Cypridopsis hyperectyphos*: Whatley and Bajpai 2000a: Sharma and Khosla, pp. 204, Pl. 3, Figs. A–B.
2010 *Cypridopsis hyperectyphos*: Whatley and Bajpai 2000a: Khosla et al. pp. 118, Figs. 3h–i.
2011a *Cypridopsis hyperectyphos*: Whatley and Bajpai 2000a: Khosla et al. pp. 236–237, Pl. 3, Figs. 8–11.

2011b *Cypridopsis hyperectyphos*: Whatley and Bajpai 2000a: Khosla et al. pp. 241–242, Pl. 5, Figs. 4–6.
2015 *Cypridopsis hyperectyphos*: Whatley and Bajpai 2000a: Khosla, pp. 348.
2017 *Cypridopsis hyperectyphos*: Whatley and Bajpai 2000a: Rathore et al. pp. 229–230, Figs. 3.9–3.11.
2018 *Cypridopsis hyperectyphos*: Whatley and Bajpai 2000a: Rathore, pp. 66.
2019 *Cypridopsis hyperectyphos*: Whatley and Bajpai 2000a: Kapur et al. pp. 1148–1149, Figs. 3D–E.

Material More than 265 carapaces and open valves.

Horizon, Age and Locality Upper Cretaceous to Lower Palaeocene clayey limestone of unit 3 (JH 16–21 and 23–26) of the Jhilmili intertrappean beds, and Upper Cretaceous to ?Lower Palaeocene hard clayey limestone unit of the Ghat Parasia intertrappean beds, Chhindwara District, Madhya Pradesh, India.

Description Carapaces medium-sized, subovate to triangular and of subglobular shape in lateral view and have an inflated to nearly circular outline in dorsal view. The maximum height is at or little posterior to mid-length, the maximum length is below the mid-height and the minimum width is at the posterior to mid-length. Posterior margin is less rounded and pointed than the anterior margin, which is more broadly rounded and concave; there is a definite decline from the maximum height to the postero-dorsal and antero-dorsal margins. The dorsal margin is more arched and strongly convex in lateral view, and the ventral margin is concave. A hinge line is clearly visible and slightly curved from the middle in dorsal view; a small depression is visible in ventral view but slightly concave in the middle. The left valve is larger than the right valve, overlapping along the periphery margins. The valve surface bears ornamentation consisting of concentrically arranged punctae. Sexually dimorphic features not observed, and other internal features are not visible.

Table 4.30 Dimensions of carapaces of *Cypridopsis hyperectyphos* (Whatley and Bajpai 2000a)

	Dimensions (mm)		
Catalogue no.	Length	Width	Height
MPL/SK/JML/O/5231	0.531	0.525	0.438
MPL/SK/JML/O/5232	0.520	0.500	0.409
MPL/SK/JML/O/5233	0.546	0.514	0.395
MPL/SK/JML/O/5234	0.517	0.492	0.344
MPL/SK/JML/O/5235	0.530	0.490	0.360
MPL/SK/JML/O/5236	0.460	0.486	0.345
MPL/SK/JML/O/5237	0.465	0.484	0.350
MPL/SK/JML/O/5238	0.463	0.491	0.351
MPL/SK/JML/O/5239	0.468	0.496	0.356
MPL/SK/JML/O/5240	0.529	0.510	0.359
MPL/SK/JML/O/5241	0.464	0.505	0.343

(continued)

Table 4.30 (continued)

Catalogue no.	Dimensions (mm)		
	Length	Width	Height
MPL/SK/JML/O/5242	0.550	0.470	0.358
MPL/SK/JML/O/5243	0.501	0.486	0.342
MPL/SK/JML/O/5244	0.460	0.494	0.353
MPL/SK/JML/O/5245	0.471	0.487	0.363
MPL/SK/JML/O/5246	0.463	0.481	0.341
MPL/SK/JML/O/5247	0.465	0.492	0.354
MPL/SK/JML/O/5248	0.473	0.429	0.347
MPL/SK/JML/O/5249	0.518	0.500	0.377
MPL/SK/JML/O/5250	0.482	0.472	0.342
MPL/SK/JML/O/5251	0.522	0.525	0.374
MPL/SK/JML/O/5252	0.465	0.437	0.363
MPL/SK/JML/O/5253	0.530	0.550	0.357
MPL/SK/JML/O/5254	0.528	0.545	0.355
MPL/SK/JML/O/5255	0.521	0.536	0.351
MPL/SK/JML/O/5256	0.485	0.484	0.347
MPL/SK/JML/O/5257	0.483	0.480	0.343
MPL/SK/JML/O/5258	0.468	0.506	0.348
MPL/SK/JML/O/5259	0.530	0.445	0.356
MPL/SK/JML/O/5260	0.478	0.480	0.349
MPL/SK/JML/O/5261	0.515	0.503	0.372
MPL/SK/JML/O/5262	0.519	0.532	0.342
MPL/SK/JML/O/5263	0.518	0.437	0.349
MPL/SK/JML/O/5264	0.506	0.491	0.366
MPL/SK/JML/O/5265	0.515	0.498	0.372
MPL/SK/JML/O/5266	0.470	0.427	0.345
MPL/SK/JML/O/5267	0.528	0.454	0.351
MPL/SK/JML/O/5268	0.493	0.474	0.351
MPL/SK/JML/O/5269	0.480	0.462	0.349
MPL/SK/JML/O/5270	0.556	0.459	0.342
MPL/SK/JML/O/5271	0.528	0.511	0.358
MPL/SK/JML/O/5272	0.552	0.472	0.369
MPL/SK/JML/O/5273	0.463	0.482	0.348
MPL/SK/JML/O/5274	0.468	0.457	0.346
MPL/SK/JML/O/5275	0.529	0.518	0.336
MPL/SK/GP/O/5276	0.496	0.486	0.343
MPL/SK/GP/O/5277	0.475	0.475	0.334
MPL/SK/GP/O/5278	0.489	0.457	0.344
MPL/SK/GP/O/5279	0.462	0.540	0.352
MPL/SK/GP/O/5280	0.559	0.572	0.376

Remarks Other than *C. hyperectyphos*, there are eight species of *Cypridopsis* – *C. elachistos* (Whatley et al. 2002b), *C. astralos* (Whatley et al. 2002a), *C. huenei* (Khosla et al. 2011b), *C. ashui* (Khosla et al. 2011b), *C. dongargaonensis* (Khosla et al. 2005), *C. mohgaonensis* (Khosla and Nagori 2007a), *C. sahnii* (Khosla et al. 2005) and *C. wynnei* (Whatley and Bajpai 2000a). They are known from several intertrappean localities of peninsular India. The typical morphology and ornamentation of *C. hyperectyphos* differ markedly from the other species of *Cypridopsis*.

Distribution Apart from Jhilmili, *Cypridopsis hyperectyphos* has been widely recorded from the intertrappean beds of Lakshmipur, Kora, and Anjar (Kachchh District), Gujarat, Yanagundi (Gulbarga District), Karnataka, Mamoni (Kota District), Rajasthan, Takli (Nagpur District) and Pinjaurni (Chandrapur District), Maharashtra (Whatley and Bajpai 2000a, 2002a, 2003a; Khosla and Nagori 2005; Khosla and Nagori 2007b; Sharma et al. 2008; Rathore 2018). This species has also been described from the Lameta Formation of Jabalpur, Madhya Pradesh (Khosla et al. 2011b).

4.3.24 Species Cypridopsis elachistos (Whatley et al. 2002b) (Figs. 4.19H–K; Table 4.31)

2002b *Cypridopsis elachistos*: Whately et al. pp. 109–110, Pl. 1, 2 Fig. 7–13, Fig. 1–2.
2007b *Cypridopsis elachistos*: Whatley et al. 2002b: Khosla and Nagori, pp. 9, Pl. II, Figs. 7–8.

Material More than 629 carapace and open valves.

Horizons, Age and Localities Upper Cretaceous (Maastrichtian) black carbonaceous shale unit of the intertrappean section exposed in the Shiraj (=Shriwas) well and Upper Cretaceous (Maastrichtian) black carbonaceous calcareous shale unit of the intertrappean section exposed in the Government well, Mohgaon-Kalan village, Chhindwara District, Madhya Pradesh, India.

Description Carapace small. In lateral view, it is subovate. In dorsal view, it is very inflated, with maximum width in the posterior third and a very truncated posterior. Carapace almost as long as it is wide and noticeably wider than it is tall. The anterior margin is narrow and asymmetrically rounded, with a strongly convex antero-ventral slope and a longer, almost straight antero-dorsal slope. The apex is below mid-height. At mid-height, the posterior margin is very bluntly pointed, with symmetrical postero-dorsal and postero-ventral slopes. Dorsal margin is convex, with the greatest height at mid-length. Valve tumidity obscures the ventral margin in lateral view; oral incurvature is shallow. Left valve is larger than the right valve and overlaps the entire margin. Surfaces can be smooth or have minute puncta. There was no sexual dimorphism observed. Internal features are not visible.

Table 4.31 Dimensions of carapaces of *Cypridopsis elachistos* (Whatley et al. 2002b)

Catalogue no.	Dimensions (mm)		
	Length	Width	Height
MPL/SK/SW/O/5281	0.390	0.363	0.263
MPL/SK/SW/O/5282	0.420	0.383	0.265
MPL/SK/SW/O/5283	0.373	0.372	0.278
MPL/SK/SW/O/5284	0.397	0.371	0.250
MPL/SK/SW/O/5285	0.381	0.369	0.248
MPL/SK/SW/O/5286	0.391	0.364	0.250
MPL/SK/GW/O/5287	0.388	0.370	0.276
MPL/SK/GW/O/5288	0.385	0.372	0.279
MPL/SK/GW/O/5289	0.379	0.376	0.266
MPL/SK/GW/O/5290	0.383	0.371	0.264

Remarks The very globular nature of carapaces and ornamentation consisting of fine and small punctae differentiate *Cypridopsis elachistos* from the other species of he genus *Cypridopsis,* namely *C. hyperectyphos* (Whatley and Bajpai 2000a), *C. huenei* (Khosla et al. 2011b), *C. ashui* (Khosla et al. 2011b), *C. dongargaonensis* (Khosla et al. 2005), *C. astralos* (Whatley et al. 2002a), *C. mohgaonensis* (Khosla and Nagori 2007a), *C. sahnii* (Khosla et al. 2005) and *C. wynnei* (Whatley and Bajpai 2000a), which are widely reported from several intertrappean localities of India.

Distribution This species has been previously described from the Mohgaon-Kalan intertrappean beds, Chhindwara District, Madhya Pradesh (Whatley et al. 2002b; Khosla and Nagori 2007b).

4.3.25 Species Candona sp. (Fig. 4.19L)

Subfamily	Candoninae (Daday 1900)
Genus	*Candona* (Baird 1845)
Species	***Candona* sp.**

Material Two moderately preserved carapaces.

Horizon, Age and Locality Upper Cretaceous to ?Lower Palaeocene reddish chert unit of the Ghat Parasia intertrappean beds, Chhindwara District, Madhya Pradesh, India.

Description Carapace small, subrectangular in outline, subovate, subspherical, elliptical to oval in shape in dorsal and lateral views. The maximum height at the mid-centre and maximum length at the mid-height. Anterior margin is broader and more rounded, and posterior margin is short and flat. Dorsal margin is arched, and ventral margin is almost straight and arched in the middle. Left valve is larger than right valve. The surface of the carapace is smooth and devoid of ornamentation. Sexually dimorphic and internal structures are not observable. MPL/SK/GP/O/5292 measures length 0.514 mm, width 0.323 mm and height 0.333 mm.

Remarks The morphology of the two carapaces described above indicates that they are *Candona* sp. The scarcity of the material prevents them from being identified at the species level. In India, the genus *Candona* is represented by *C. amosi* (Whatley et al. 2002a), *C. chuiensis* (Khosla et al., 2011b) and *C. phaseolus* (Kshetrimayum et al. (2021), which have been previously reported from the infra- and intertrappean beds of peninsular India.

4.3.26 Species Eucypris pelasgicos (Whatley and Bajpai 2000a) (Figs. 4.19M and 4.20A–N; Table 4.32)

Subfamily	Eucypridinae (Bronshein 1947)
Genus	*Eucypris* (Vávra 1891)
Species	***Eucypris pelasgicos*** (Whatley and Bajpai 2000a)

?1965 *Eucypris* sp.: Khanna and Mohan, Fig. 2, 1.

1990a *Candona altanulaensis*: Szczechura and Blaszyk 1970: Bhatia et al. Pl. 3, Fig. 3.

1996 *Candona altanulaensis*: Szczechura and Blaszyk 1970: Bhatia et al. pp. 297–302, Pl. 3, Fig. 3.

2000a *Eucypris pelasgicos*: Whatley and Bajpai, pp. 400–402, Pl. 5, Figs. 7–15.

2001 *Eucypris pelasgicos*: Whatley and Bajpai 2000a: Bajpai and Whatley, pp. 102–103, Pl. 2, Fig. 15; Pl. 3, Figs. 1–3.

2002b *Eucypris pelasgicos*: Whatley and Bajpai 2000a: Whatley et al. pp. 111–112, Pl. 2, Fig. 11.

2004 *Eucypris pelasgicos*: Whatley and Bajpai 2000a: Bajpai et al. pp. 152–154, Pl. 2, Figs. k-l.

2005 *Eucypris pelasgicos*: Whatley and Bajpai 2000a: Khosla et al. pp. 143, Pl. 3, Figs. 2–3.

2005 *Eucypris pelasgicos*: Whatley and Bajpai 2000a: Khosla and Nagori, pp. 574, Pl. 1, Fig. 12.

2007b *Eucypris pelasgicos*: Whatley and Bajpai 2000a: Khosla and Nagori, pp. 12, Pl. 2, Figs. 18–19.

2009a *Eucypris pelasgicos*: Whatley and Bajpai 2000a: Khosla et al. pp. 725, Pl. 2, Fig. 6.

2010 *Eucypris pelasgicos*: Whatley and Bajpai 2000a: Khosla et al. pp. 118, Figs. 4 a–b.
2011a *Eucypris pelasgicos* Whatley and Bajpai 2000a: Khosla et al. pp. 238, Pl. 3, Fig. 12.
2011b *Eucypris pelasgicos*: Whatley and Bajpai 2000a: Khosla et al. pp. 245, 246, Pl. 6, Figs. 7–8.
2017 *Eucypris pelasgicos*: Whatley and Bajpai 2000a: Rathore et al. pp. 223, Figs. 3.15–3.17.

Material More than 166 carapaces and open valves.

Horizon, Age and Locality Upper Cretaceous to Lower Palaeocene clayey limestone of unit 3 (JH 26) of the Jhilmili intertrappean beds, and Upper Cretaceous (Maastrichtian) black carbonaceous shale units of intertrappean sections exposed in the Shriwas (=Shiraj) and Government wells, Mohgaon-Kalan village, Chhindwara District, Madhya Pradesh, India.

Description Carapaces medium to large, subovate to elongate in lateral view, fusiform in dorsal view with a compressed anterior portion. The greatest height and length are at the mid-length. Anterior margin is more rounded, and the posterior margin is short and narrowly rounded. Dorsal margin is umbonate, and the ventral margin is nearly straight and curved postero-ventrally and antero-ventrally. Left valve overlaps the right valve on the entire periphery. The surface of the carapace is smooth. Sexually dimorphic and internal structures are not observable.

Table 4.32 Dimensions of carapace of *Eucypris pelasgicos* (Whatley and Bajpai 2000a)

	Dimensions (mm)		
Catalogue no.	Length	Width	Height
MPL/SK/JML/O/5305	0.810	0.410	0.463
MPL/SK/JML/O/5306	0.790	0.380	0.458
MPL/SK/JML/O/5307	0.801	0.390	0.461
MPL/SK/JML/O/5308	0.780	0.370	0.451
MPL/SK/SW/O/5309	0.720	0.360	0.430
MPL/SK/SW/O/5310	0.770	0.370	0.453
MPL/SK/SW/O/5311	0.800	0.390	0.440
MPL/SK/SW/O/5312	0.820	0.420	0.462
MPL/SK/GW/O/5313	0.750	0.320	0.420
MPL/SK/GW/O/5314	0.810	0.400	0.440
MPL/SK/GW/O/5315	0.790	0.360	0.360
MPL/SK/GW/O/5356	0.750	0.340	0.490
MPL/SK/GW/O/5357	0.730	0.330	0.480
MPL/SK/GW/O/5358	0.760	0.380	0.420
MPL/SK/GW/O/5359	0.810	0.430	0.480

Remarks Initially, this species was identified as *Candona altanulaensis* (Bhatia et al. 1990a) and later re-named as *Eucypris pelasgicos* by Whatley and Bajpai (2000a). It has a close resemblance to *E. intervolcanus*, but differs from it by its larger size and angular antero-ventral margin.

Distribution *Eucypris pelasgicos* has been widely reported from several intertrappean beds such as at Lakshmipur, Kora and Anjar (Kachchh District), Gujarat (Bajpai and Whatley 2001; Whatley and Bajpai 2000a; Khosla and Nagori 2005), Jhilmili (Chhindwara District), Khar (Khargaon District), Madhya Pradesh (Khosla et al. 2009a, 2011a; Rathore et al. 2017) and Takli (Nagpur District), Maharashtra. It has also been recorded from the Lameta Formation of the Nand-Dongargaon-Pisdura Basin (Maharashtra) and Jabalpur, Madhya Pradesh (Khosla and Nagori 2007b; Khosla et al. 2010, 2011b).

4.3.27 Species Eucypris sp. 1 (Figs. 4.20O–Q; Table 4.33)

Material Four moderately preserved carapaces.

Horizon, Age and Locality Upper Cretaceous to ?Lower Palaeocene black chert and hard clayey limestone units of the Ghat Parasia intertrappean beds, Chhindwara District, Madhya Pradesh, India.

Description Carapaces medium to large, subovate in lateral view, elliptical to oval in dorsal view. The greatest height and length at the mid-centre. Anterior and posterior margins are almost equally broad and rounded. Dorsal margin is arched and slightly concave, whereas ventral margin is concave. Left valve larger than right valve and distinctly overlaps the entire periphery in lateral view. The surface is smooth without ornamentation. Sexually dimorphic and internal features are not observable.

Table 4.33 Dimensions of carapace specimens of *Eucypris* sp. 1

	Dimensions (mm)		
Catalogue no.	Length	Width	Height
MPL/SK/GP/O/7338	0.940	0.565	0.624
MPL/SK/GP/O/7339	0.960	0.568	0.629
MPL/SK/GP/O/7340	0.935	0.579	0.618

Remarks On the basis of overall morphology and size, this species, *Eucypris* sp. 1, is tentatively identified as belonging to the genus *Eucypris*. The genus *Eucypris* is represented by five species – *E. pelasgicos* (Whatley and Bajpai 2000a), ?*E. verruculosa* (Whatley et al. 2002a), ?*E. intervolcanus* (Whatley and Bajpai 2000a), *E. catantion* (Whatley et al. 2003a) and *E. phulsagarensis* (Bajpai et al. 2004).

Fig. 4.20 (**A**–**N**) ***Eucypris pelasgicos*** (Whatley and Bajpai 2000a), Carapace, (**A**) Lateral view MPL/SK/SW/O/5309; right valve; (**B**) Lateral view MPL/SK/GW/O/5310, right valve; (**C**) Lateral view MPL/SK/SW/O/5311, right valve; (**D**) Lateral view MPL/SK/SW/O/5312, left valve; (**E**) Lateral view MPL/SK/GW/O/5313, left valve; (**F**) Lateral view MPL/SK/GW/O/5314, right valve; (**G**) Lateral view MPL/SK/GW/O/5315, left valve; (**H**) Lateral view MPL/SK/GW/O/5356, right valve; (**I**) Dorsal view MPL/SK/SW/0/5357; (**J**) Ventral view MPL/SK/GW/O/5358; (**K**) Dorsal view MPL/SK/JML/O/5306; (**L**) Ventral view MPL/SK/JML/O/5307; (**M**) Dorsal view MPL/SK/JML/O/5308; (**N**) Dorsal view MPL/SK/GW/O/5359. (**O**–**Q**) ***Eucypris* sp. 1.** Carapace, (**O**) Lateral view MPL/SK/GP/O/7338, left valve; (**P**) Lateral view MPL/SK/GP/O/7339, right valve; (**Q**) Dorsal view MPL/SK/GP/O/7340; (**R**) ?***Eucypris verruculosa*** (Whatley et al. 2002a), Carapace; (**R**) Lateral view MPL/SK/GW/O/7352, right valve

4.3.28 Species ?Eucypris verruculosa (Whatley et al. 2002a) (Fig. 4.20R)

2002a *?Eucypris verruculosa*: Whatley et al. 2002a, pp.177, Pl. 4, Figs. 8–9, 12–19.

Material Five moderately preserved carapaces.

Horizon, Age and Locality Upper Cretaceous (Maastrichtian) black calcareous carbonaceous shale unit of the intertrappean section exposed in the Government well, Mohgaon-Kalan village, Chhindwara District, Madhya Pradesh, India.

Description Carapace medium, elongate, subovate to subrectangular in lateral view, oval in dorsal view. The maximum height close to anterior margin and maximum width at the mid-length. Anterior margin is very well rounded and flat. Posterior margin is narrowly rounded and acuminate. Dorsal margin is arched, and inclined to the posterior, and ventral margin is rounded and arched in the centre. Anterior margin is very well rounded and flat. Posterior margin is narrowly rounded and acuminate. Left valve is slightly larger than right valve, overlapping the entire periphery. The surface is smooth and devoid of any ornamentation. Sexually dimorphic and internal features are not observable. MPL/SK/GW/O/7352 measures length 0.830 mm, width 0.319 mm and height 0.428 mm.

Remarks Certain morphological characteristics, such as a very well-rounded anterior margin and a posterior margin that is narrowly rounded and acuminate, led to the identification of carapaces recovered from the intertrappean beds found in the Government well as *?Eucypris verruculosa* (Whatley et al. 2002a). Based on these morphological characteristics, it differs from *E. pelasgicos* (Whatley and Bajpai 2000a), *E. intervolcanus* (Whatley and Bajpai 2000a) and *E. catantion* (Whatley et al. 2003a). Other species such as *E. phulsagarensis* (Bajpai et al. 2004), *E.* sp. 1 and *E.* sp. 2 have also been recovered from the Upper Cretaceous intertrappean beds in peninsular India.

Distribution Apart from Mohgaon-Kalan (Chhindwara District), Madhya Pradesh, this species has so far been reported from the intertrappean beds of Chandarki (Karnataka), Sichel hills (Telangana) and Takli (Nagpur District), Maharashtra (Whatley et al. 2002a, 2003a; Khosla and Nagori 2007a, b).

4.3.29 Species Cyclocypris amphibolos (Whatley et al. 2002a) (Figs. 4.21A–D; Table 4.34)

Family	Cyclocyprididae (Kaufmann 1900)
Genus	*Cyclocypris* (Brady and Norman 1889)
Type species	*Cypris globosa* (Sars 1863)
Species	***Cyclocypris amphibolos*** (Whatley et al. 2002a)

2002a *Cyclocypris amphibolos*: Whatley et al. 2002a, pp. 182–184, Pl. 6, Figs. 6–18.
2002c *Cyclocypris amphibolos*: Whatley et al. 2002a: Whatley et al. pp. 170–172, Pl. 2, Figs. 12–15.
2003a *Cyclocypris amphibolos*: Whatley et al. 2002a: Whatley et al. pp. 84, Pl. 1, Figs. 8–9.
2005 *Cyclocypris amphibolos*: Whatley et al. 2002a: Khosla et al. pp. 143–144, Pl. 3, Figs. 7–8.
2007a *Cyclocypris amphibolos*: Whatley et al. 2002a: Khosla and Nagori, pp. 219, Pl. 3, Figs. 8–11.
2007b *Cyclocypris amphibolos*: Whatley et al. 2002a: Khosla and Nagori, pp. 12–14, Pl. 3, Figs. 8–10.
2009a *Cyclocypris amphibolos*: Whatley et al. 2002a: Khosla et al. pp. 725. Pl. 2, Fig. 2.
2010 *Cyclocypris amphibolos*: Whatley et al. 2002a: Khosla et al. pp. 118, Figs. 4c–e.
2011a *Cyclocypris amphibolos*: Whatley et al. 2002a: Khosla et al. pp. 238–240, Pl. 3, Figs. 13–14.
2011b *Cyclocypris amphibolos*: Whatley et al. 2002a: Khosla et al. pp. 248–249, Pl. 6, Figs. 15–16.
2017 *Cyclocypris amphibolos*: Whatley et al. 2002a: Rathore et al. pp. 224–225, Figs. 4.1–4.3.

Material More than 50 well-preserved carapaces and valves.

Horizons, Age and Localities Upper Cretaceous to Lower Palaeocene clayey limestone of unit 3 (JH 25–26) of the Jhilmili intertrappean beds and Upper Cretaceous to ?Lower Palaeocene greenish chertified clay unit of the Ghat Parasia intertrappean beds, Chhindwara District, Madhya Pradesh, India.

Description Carapaces small to medium, somewhat irregularly subovate to subspherical in lateral view and elliptical to oval, and frequently fusiform, in dorsal view. The maximum height is at the mid-centre and maximum length at the mid-length. Anterior margin is broader and more rounded, and posterior margin is smaller or more narrowly rounded. Dorsal margin is anterodorsally convex, broad and sloping posteriorly, and the ventral margin is projected by valve tumidity and is less concave. Left valve is larger than right valve, overlapping the entire margins. The valve surface bears several fine papillae. Sexually dimorphic and internal features are not observable.

Table 4.34 Dimensions of carapaces of *Cyclocypris amphibolos* (Whatley et al. 2002a)

	Dimensions (mm)		
Catalogue no.	Length	Width	Height
MPL/SK/JML/O/5348	0.948	0.480	0.523
MPL/SK/JML/O/5349	0.718	0.457	0.501
MPL/SK/JML/O/5350	0.715	0.454	0.496
MPL/SK/GP/O/5351	0.693	0.438	0.501

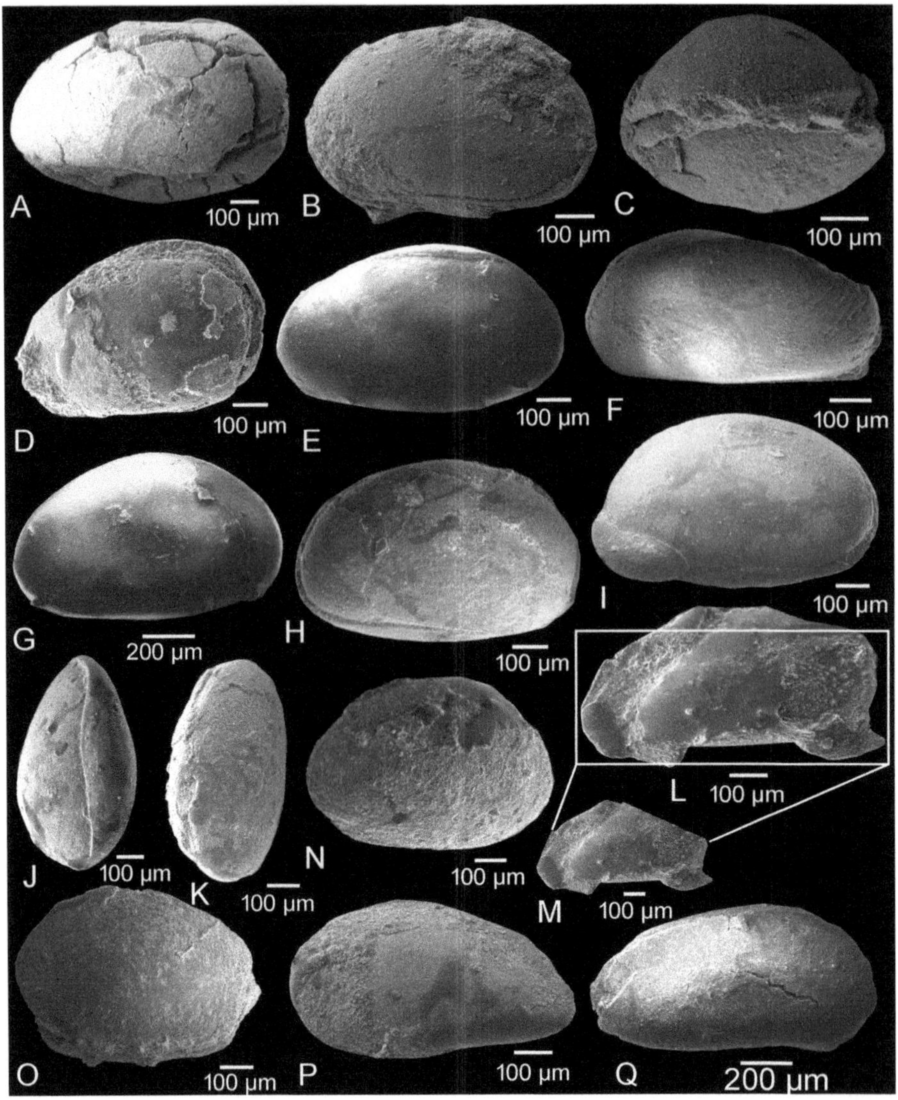

Fig. 4.21 (**A–D**) ***Cyclocypris amphibolos*** (Whatley et al. 2002a), Carapace; (**A**) Lateral view MPL/SK/JML/O/5348, right valve; (**B**) Lateral view MPL/SK/JML/O/5349, right valve; (**C**) Dorsal view MPL/SK/JML/O/5350; (**D**) Lateral view MPL/SK/GP/O/5351, right valve; (**E–K**) ***Cypria cyrtonidion*** (Whatley and Bajpai 2000a), Carapace; (**E**) Lateral view MPL/SK/JML/O/5291, left valve; (**F**) Lateral view MPL/SK/SW/O/5297, right valve; (**G**) Lateral view MPL/SK/GW/O/5299; right valve: (**H**) Lateral view MPL/SK/JML/O/5291, right valve; (**I**) Lateral view MPL/SK/JML/O/5293, right valve; (**J**) Ventral view MPL/SK/JML/O/5294; (**K**) Dorsal view MPL/SK/JML/O/5295; (**L–M**) ***Talicypridea pavnaensis*** (Khosla et al. 2005), (**L**) Enlarged view of MPL/SK/JML/O/8342; (**M**) Carapace, lateral view (MPL/SK/JML/O/8342), left valve showing the incurved ventral margin with horn-like structure which pointed towards right direction; (**N**) ***Cyprois rostellum*** (Whatley and Bajpai 2000a), female Carapace; Lateral view MPL/SK/SW/O/8320; right valve; (**O**) ***Cyprois* sp.** Carapace; Lateral view MPL/SK/GP/O/8344, left valve; (**P–Q**) ***Darwinula* sp.** Carapace; (**P**) Lateral view MPL/SK/GP/O/8346, right valve; (**Q**) Lateral view MPL/SK/GP/O/8347, left valve. (Figure **A**, Reproduced from Khosla et al. 2022 with permission from Wiley)

Remarks The present material of *Cyclocypris amphibolos* differs from other species of the genus *Eucypris* such as ?*E. verruculosa* (Whatley et al. 2002a), *E. pelasgicos* (Whatley and Bajpai 2000a), *E. intervolcanus* (Whatley and Bajpai 2000a), *E. catantion* (Whatley et al. 2003a), ?*E. phulsagarensis* (Bajpai et al. 2004), *Eucypris* sp. 1 and *Eucypris* sp. 2, as well as *Cyclocypris sahnii* (Bajpai and Whatley 2001), recovered from Upper Cretaceous intertrappean beds of peninsular India by characters such as irregularly subovate lateral outline, a dorsally regular fusiform outline, a symmetrically rounded anterior margin, a small, rounded posterior margin, a concave ventral margin and ornamentation consisting of fine papillae.

Distribution *Cyclocypris amphibolos* (Whatley et al. 2002a) has been described from the intertrappean beds of Yanagundi (Gulbarga District), Karnataka (Whatley et al. 2002a), Kora (Kachchh District), Gujarat (Whatley et al. 2002a), Mohgaon-Haveli and Jhilmili (Chhindwara District), Khar (Khargaon District), Madhya Pradesh (Khosla and Nagori 2007a; Khosla et al. 2009a, 2011a; Rathore et al. 2017) and Takli (Nagpur District), Maharashtra (Khosla and Nagori 2007a, b). It has also been recorded from the Dongargaon and Pisdura outcrops of the Lameta Formation (Chandrapur District), Maharashtra and Jabalpur, Madhya Pradesh (Khosla et al. 2005, 2010, 2011b).

4.3.30 Species Cypria cyrtonidion (Whatley and Bajpai 2000a) (Figs. 4.21E–K; Table 4.35)

Genus	*Cypria* (Zenker 1854)
Type species	*Cypris exculpta* (Fischer 1855)
Species	***Cypria cyrtonidion*** (Whatley and Bajpai 2000a)

1984 *Cyprois* sp.: Bhatia and Rana, pp. 33, Pl. 2, Fig. 12.
1988 *Cyprois* sp.: Mathur and Verma, pp. 173, Pl. 1, Figs. 1–2.
2000a *Cypria cyrtonidion*: Whatley and Bajpai, pp. 404, Pl. 6, Figs. 9–14.
2001 *Cypria cyrtonidion*: Whatley and Bajpai 2000a: Bajpai and Whatley, pp. 101–102, Pl. 2, Figs. 7–9,
2002a *Cypria cyrtonidion*: Whatley and Bajpai 2000a: Whatley et al. pp. 184, Pl. 6, Fig. 19.
2002b *Cypria cyrtonidion*: Whatley and Bajpai 2000a: Whatley et al. pp. 112–113, Pl. 2, Fig. 13.
2005 *Cypria cyrtonidion*: Whatley and Bajpai 2000a: Khosla et al. pp. 144, Pl. 3, Figs. 9–10.

2005 *Cypria cyrtonidion*: Whatley and Bajpai 2000a: Khosla and Nagori, pp. 574, Pl. 1, Fig. 10.
2007a *Cypria cyrtonidion*: Whatley and Bajpai 2000a: Khosla and Nagori, pp. 219–220, Pl. 3, Figs. 12–13.
2007b *Cypria cyrtonidion*: Whatley and Bajpai 2000a: Khosla and Nagori, pp. 14, Pl. 3, Figs. 11–12.
2008 *Cypria cyrtonidion*: Whatley and Bajpai 2000a: Sharma et al. pp. 182, Pl. 1, Fig. O.
2009a *Cypria cyrtonidion*: Whatley and Bajpai 2000a: Khosla et al. pp. 725, Pl. 2, Fig. 3.
2009 *Cypria cyrtonidion*: Whatley and Bajpai 2000a: Sharma and Khosla, pp. 202–204, Pl. 3, Figs. L–M.
2010 *Cypria cyrtonidion*: Whatley and Bajpai 2000a: Khosla et al. pp. 118, Figs. 4f–g.
2011a *Cypria cyrtonidion*: Whatley and Bajpai 2000a: Khosla et al. pp. 240–242. Pl. 3, Figs. 15–16.
2011b *Cypria cyrtonidion*: Whatley and Bajpai 2000a: Khosla et al. pp. 249–250, Pl. 7, Figs. 3–4.
2015 *Cypria cyrtonidion*: Whatley and Bajpai 2000a: Khosla, pp. 352, Fig. 5f.
2017 *Cypria cyrtonidion*: Whatley and Bajpai 2000a: Rathore et al. pp. 225, Figs. 4.4–4.6.
2018 *Cypria cyrtonidion*: Whatley and Bajpai 2000a: Rathore, pp. 6.
2021 *Cypria cyrtonidion*: Whatley and Bajpai 2000a: Kshetrimayum et al. pp. 8, 9, 11, Figs. 4J–M.

Material A total of 48 carapaces and open valves.

Horizons, Age and Localities Upper Cretaceous to Lower Palaeocene clayey limestone of unit 3 (JH 17 and 19) of the Jhilmili intertrappean beds and Upper Cretaceous (Maastrichtian) black carbonaceous shale units of intertrappean sections exposed in the Shiraj (=Shriwas) and Government wells, Mohgaon-Kalan village, Chhindwara District, Madhya Pradesh, India.

Description Carapace of medium size, subcircular to subquadrate in lateral view and semi-circular to subrectangular in dorsal view. The greatest height at the middle, the greatest length close to the ventral margin. Anterior margin is well rounded and concave, posterior margin identical to anterior margin but short. Dorsal margin is arched and well rounded, and ventral margin is almost straight from anterior to posterior. Posterior margin is concave, and anterior margin is concave. Left valve larger than right valve, overlapping the entire periphery expect dorsally where right valve overlaps the left valve. The surface is smooth without ornamentation. Sexually dimorphic and internal features are not observable.

Table 4.35 Dimensions of carapaces of *Cypria cyrtonidion* (Whatley and Bajpai 2000a)

	Dimensions (mm)		
Catalogue no.	Length	Width	Height
MPL/SK/JML/O/5291	0.510	0.258	0.510
MPL/SK/JML/O/5291.1	0.512	0.255	0.512
MPL/SK/JML/O/5293	0.513	0.257	0.513
MPL/SK/JML/O/5294	0.511	0.259	0.509
MPL/SK/JML/O/5295	0.515	0.262	0.510
MPL/SK/JML/O/5296	0.509	0.253	0.506
MPL/SK/SW/O/5297	0.514	0.254	0.516
MPL/SK/SW/O/5298	0.510	0.263	0.511
MPL/SK/GW/O/5299	0.498	0.246	0.505
MPL/SK/GW/O/5300	0.513	0.257	0.513

Remarks The recovered carapaces of *Cypria cyrtonidion* are quite different from *C. intertrappeana* (Bajpai and Whatley 2001) known from the Upper Cretaceous intertrappean beds of Kora (Kachchh District), Gujarat, in having smaller size together with a lack of marked postero-ventral overlap of the left valve, a postero-ventrally developed keel-like process, a broad, rounded anterior margin and a pointed posterior margin.

Distribution The species *Cypria cyrtonidion* has been reported from several inter-trappean localities – Lakshmipur, Kora and Anjar (Kachchh District), Gujarat (Whatley and Bajpai 2000a; Bajpai and Whatley 2001; Khosla and Nagori 2005), Chandarki and Yanagundi (Gulbarga District), Karnataka (Whatley et al. 2002a), Mohgaon-Kalan, Mohgaon-Haveli and Jhilmili (Chhindwara District), Gujri (Dhar District) and Khar (Khargone District), Madhya Pradesh (Whatley et al. 2002b; Khosla and Nagori 2007a; Khosla et al. 2009a; 2011a, b; Sharma and Khosla 2009; Khosla 2015; Rathore et al. 2017; Kshetrimayum et al. 2021), and Takli (Nagpur District) and Pinjaurni (Chandrapur District), Maharashtra (Rathore 2018). It has also been described from the infratrappean beds of Papro (Lalitpur District), Uttar Pradesh and the Lameta Formation of Dongargaon-Pisdura (Chandrapur District), Maharashtra (Sharma et al. 2008; Khosla et al. 2010, 2011b).

4.3.31 Species Talicypridea pavnaensis (Khosla et al. 2005) (Figs. 4.21L–M)

Subfamily	Talicyprideinae (Hou 1982)
Genus	*Talicypridea* (Khand 1977)
Species	***Talicypridea pavnaensis*** (Khosla et al. 2005)

2005 *Cypridea pavnaensis*: Khosla et al. 2005, pp. 144, Pl. 3, Figs. 11–14.
2011b *Talicypridea pavnaensis*: Khosla et al. 2005: Khosla et al. pp. 254–255, Pl. 8, Figs. 1–3.

Material One well-preserved carapace.

Horizon, Age and Locality Upper Cretaceous to Lower Palaeocene clayey limestone of unit 3 (JH 25) of the Jhilmili intertrappean beds, Chhindwara District, Madhya Pradesh, India.

Description Carapace medium to large, subovate to elongate in lateral view and lensoid in dorsal view. The maximum height at the mid-centre and the maximum length close to dorso-posterior margin. Anterior margin broad and less rounded. Dorsal margin is arched, anterodorsal margin is inclined, and postero-ventrally it is arched and slightly flat. Ventral margin is distinct, incurved with horn-like structure pointed towards the right, and the central part is almost straight. Anteroventral and postero-ventral angulations bulging downward beyond ventral margin, the former rounded. The left valve larger than the right valve, overlapping the posterior and dorsal margins. The carapace is smooth and lacks any ornamentation. Sexually dimorphic and internal features are not observable. One carapace, MPL/SK/JML/O/8342, measures length 0.772 mm, width 0.388 and height 0.443 mm.

Remarks Khosla et al. (2005) originally described this species from the Lameta Formation of the Nand-Dongargaon Basin, Chandrapur District, Maharashtra. Because the morphological details of the present carapace is very similar to those described from the Nand-Dongargaon Basin, it is identified as *Talicypridea pavnaensis*. This species resembles *T. jihgshanesis* (Ye in Hou et al. 1978) and *T. irregularis* (Hou in Hou et al. 1978) from the Upper Cretaceous of China (Hou et al. 1978). Furthermore, *T. pavnaensis* differs from *T. neustruevae* (Khand et al. 2007) from the Late Cretaceous of Mongolia in that it lacks indistinct wavy ornamentation on the valve surface.

Distribution Apart from Jhilmili, this species, *Talicypridea pavnaensis* (Khosla et al. 2005), is exclusively known from the Lameta Formation of the Nand-Dongargaon Basin, Chandrapur District, Maharashtra (Khosla et al. 2005, 2011b).

4.3.32 Species Cyprois rostellum (Whatley and Bajpai 2000a) (Fig. 4.21N; Table 4.36)

Family	Notodromadidae (Kaufmann 1900)
Genus	*Cyprois* (Zenker 1854)
Species	***Cyprois rostellum*** (Whatley and Bajpai 2000a)

2000a *Cyprois rostellum*: Whatley and Bajpai, pp. 406, Pl. 6, Figs. 15–19.
2001 *Cyprois rostellum*: Whatley and Bajpai 2000a: Bajpai and Whatley, pp. 108–109, Pl. 4, Figs. 4–10.
2007a *Cyprois rostellum*: Whatley and Bajpai 2000a: Khosla and Nagori, pp. 220, Pl. 3, Fig. 14–15.
2007b *Cyprois rostellum*: Whatley and Bajpai 2000a: Khosla and Nagori, pp. 14, Pl. 3, Figs. 15–16.
2008 *Cyprois rostellum*: Whatley and Bajpai 2000a: Sharma et al. pp. 182, Pl. 2, Figs. Q–R.
2011b *Cyprois rostellum*: Whatley and Bajpai 2000a: Khosla et al. pp. 256, Pl. 8, Figs. 10–11.

Material A total of five carapaces.

Horizons, Age and Localities Upper Cretaceous (Maastrichtian) black carbonaceous and calcareous shale units of intertrappean sections exposed in the Shiraj (=Shriwas) and Government wells, Mohgaon-Kalan village, Chhindwara District, Madhya Pradesh, India.

Description Carapaces medium to small, subcircular in lateral view, inflated in dorsal view. The maximum height at the mid-height, maximum length close to mid-height. Anterior margin is pointed and less rounded, and posterior margin is well rounded and broad. Dorsal margin is straight and arched, ventral margin is curved and short. A rostrum-like process projects from the cardinal angle. Left valve is slightly larger than right valve and overlaps anteroventrally. Valve is devoid of ornamentation. Sexually dimorphic and internal features are not observable.

Table 4.36 Dimensions of carapaces of *Cyprois rostellum* (Whatley and Bajpai 2000a)

	Dimensions (mm)		
Catalogue no.	Length	Width	Height
MPL/SK/SW/O/8320	0.269	0.165	0.257
MPL/SK/SW/O/8321	0.252	0.183	0.259
MPL/SK/GW/O/8322	0.325	0.179	0.346

Remarks A rostrum-like process projecting from the cardinal angle and other morphological features support identification of the present material as *Cyprois rostellum* (Whatley and Bajpai 2000a). These specimens are much smaller than those described from the Lameta Formation of the Jabalpur area by Khosla et al. (2011b).

Distribution This species has been reported from the intertrappean beds of Lakshmipur and Kora (Kachchh District), Gujarat, and Mohgaon-Kalan (Chhindwara District), Madhya Pradesh (Whatley and Bajpai 2000a; Bajpai and Whatley 2001; Khosla and Nagori 2007a, b). It is also known from the infratrappean beds of Papro (Lalitpur District), Uttar Pradesh (Sharma et al. 2008), and the Lameta Formation of Jabalpur, Madhya Pradesh (Khosla et al. 2011b).

4.3.33 Species *Cyprois* sp. (Fig. 4.21O; Table 4.37)

Material Two moderately preserved carapaces.

Horizon, Age and Locality Upper Cretaceous (Maastrichtian) to ?Lower Palaeocene reddish chert unit of the Ghat Parasia intertrappean beds, Chhindwara District, Madhya Pradesh, India.

Description Carapace small-medium, subspherical to subrectangular in lateral view and elliptical to subelliptical or oval in dorsal view. The maximum height at the mid-length and maximum length at the mid-height. Anterior and posterior margins are broad and equally arched. Dorsal margin is more rounded and arched. Ventral margin is almost straight and has a small arch in the centre. Left valve is larger than right valve and overlaps the entire periphery in lateral view. Valve surface is smooth and devoid of any ornamentation. Sexually dimorphic and internal features are not observable.

Table 4.37 Dimensions of carapaces of *Cyprois* sp.

	Dimensions (mm)		
Catalogue no.	Length	Width	Height
MPL/SK/GP/O/8344	0.796	0.405	0.558
MPL/SK/GP/O/8345	0.734	0.415	0.518

Remarks The carapaces of *Cyprois* sp. are considerably larger than *C. rostellum* Whatley et al. (2000a) recovered from the intertrappean beds of Shriwas (=Shiraj) and Government wells. Further, our material differs from *C. rostellum* by lacking a rostrum-like process and having long dorsal and ventral margins.

Distribution This species is only known from the Ghat Parasia intertrappean beds, Chhindwara District, Madhya Pradesh.

4.3.34 Species *Darwinula* sp. (Figs. 4.21P–Q; Table 4.38)

Family	Darwinulinidae (Brady and Norman 1889)
Genus	*Darwinula* (Brady and Robertson 1885)
Species	***Darwinula* sp.**

Material Five moderately preserved carapaces.

Horizon, Age and Locality Upper Cretaceous (Maastrichtian) to ?Lower Palaeocene reddish chert and fossiliferous clayey limestone units of the Ghat Parasia intertrappean beds, Chhindwara District, Madhya Pradesh, India.

Description Carapaces of medium size, subcylindrical, subovate to elongate in lateral view and subtriangular, elliptical in dorsal view. The greatest height close to anterior margin, the greatest length at the mid-height. Anterior margin is broader and more convex, and posterior margin is pointed. Dorsal margin is arched, inclined from the anterior and posterior margins, and ventral margin is curvy from the centre, and decreases in thickness towards the posterior margin. Left valve is slightly larger than right valve and overlaps the entire margin. Valve surface is smooth. Sexually dimorphic and internal features are not known.

Table 4.38 Dimensions of carapaces of *Darwinula* sp.

	Dimensions (mm)		
Catalogue no.	Length	Width	Height
MPL/SK/GP/O/8346	0.722	0.209	0.421
MPL/SK/GP/O/8347	1.122	0.429	0.589

Remarks The Ghat Parasia carapaces of *Darwinula* sp. are smaller than *D. torpedo* (Whatley et al. 2002a) described from the Yanagundi intertrappean beds, Gulbarga District, Karnataka.

Distribution *Darwinula* sp. is from the Ghat Parasia intertrappean beds, Chhindwara District, Madhya Pradesh.

4.4 Foraminiferans

4.4.1 Species Subbotina triloculinoides (Plummer 1926) (Figs. 4.22A–G; Table 4.39)

Kingdom	Protista (*Whittaker* 1969)
Phylum	Sarcodina (Schmarda 1871)
Class	Rhizopoda (von Siebold 1845)
Order	Foraminiferida (d'Orbigny 1826)
Superfamily	Globigerinoidea (Carpenter et al. 1862)
Family	Globigerinidae (Carpenter et al. 1862)
Genus	*Subbotina* (Brotzen and Pozaryska 1961)
Species	***Subbotina triloculinoides*** (Plummer 1926)

1926 *Globigerina triloculinoides*: Plummer, pp. 134, Pl. 8, Figs. l0a–b.
1928 *Globigerina pseudotriloba*: White, Pl. 27, pp. 194, Figs. 17a–b.
1952 *Globigerina stainforthi*: Bronnimann, pp. 23, Pl. 3, Figs. 10–12.
1957 *Globorotalia pusilla*: Bolli, pp.70, Pl. 15, Figs. 18–20.
1957a *Tylocidaris oedumi*: Loeblich and Tappan, pp.183, Pl. 40, Figs. 4a–c, Pl. 41, Figs. 2a–c; Pl. 42, Figs. 2a–c; Pl. 43, Figs. 5a–c; Pl. 45, Figs. 3a–c; Pl. 47, Figs. 2a–c; Pl. 52, Figs. 3–7; Pl. 62, Figs. 3–4.
1960 *Globigerina triloculinoides*: Bolli and Cita, pp. 13, Pl. 31, Figs. la–c.
1961 *Globigerina* (*Globigerina*) *microcellulosa*: Morozova, Pl. 1, Fig. 1411.
1961 *Subbotina triloculinoides*: Plummer 1926: Brotzen and Pozaryska, pp. 4160, Pl. 4, Fig. 2.
1962 *Globigerina triloculinoides*: Berggren, pp. 86, Pl. 14, Figs. la–2b.
1962 *Globigerina triloculinoides*: Hillebrandt, pp.119, Pl. 11, Figs. la–c.
1967 *Subbotina triloculinoides*: Plummer 1926: Belford, Pl. 1, Figs. 1–57.
1970 *Globigerina triloculinoides* Shutskaya, Pl. 18, Figs. la–c; Pl. 19, Figs. 3a–c; pp. 118, Pl. 21, Figs. 5a–c; Pl. 23, Figs. 12a–c.
1990 *Subbotina triloculinoides*: Plummer 1926: Stott and Kennett, pp. 559, Pl. 2, Fig. 12.
1979 *Subbotina triloculinoides*: Plummer 1926: Blow, Pl. 74, Fig. 6; Pl. 80, Fig. 1, Pl. 98, Fig. 7; Pl. 238, Fig. 5; Pl. 248, Fig. 1; Pl. 255, Fig. 9; pp. 128, Pl. 257, Fig. 9.
2009a *Subbotina triloculinoides*: Plummer 1926: Keller et al. pp. 46–47, Pl.1, Figs.1–7, Pl.2, Figs. 9–11.
2009b *Subbotina triloculinoides*: Plummer 1926: Keller et al. pp. 19, Pl.1, Fig. 8a–c, 25a–c.
2015 *Subbotina triloculinoides*: Plummer 1926: Khosla, pp. 349, 352, Fig. 5m.

Material More than 50 well-preserved tests.

Horizon, Age and Locality Upper Cretaceous to Lower Palaeocene clayey limestone of unit 3 (JH 19 and 25) of the Jhilmili intertrappean beds, Chhindwara District, Madhya Pradesh, India.

Description The tests are small to medium in size, with a low trochospiral pattern, tripartite, planispiral, bilaterally symmetrical and have four to five globular chambers (Keller et al. 2009a). The tests are tightly coiled. The last whorl is large and takes up to half of the test. The texture of the wall is cancellate and spinose. The spines appear at cancellate ridges. The test's periphery is very broadly round and distinctly lobate. A well-developed broad flap or lip covers a small umbilical depression and aperture. The aperture is slightly extra-umbilical and occasionally umbilical. The umbilicus is deep and narrow, and it is surrounded by an apertural lip. The wall texture of the test exhibits a weak to strong, moderate to coarse and honeycombed cancellate texture (Olsson et al. 2006).

Table 4.39 Dimensions of tests of *Subbotina triloculinoides* (Plummer 1926)

Catalogue no.	Dimensions (µm)		
	Length	Width	Thickness
MPL/SK/JML/F/2014	635	514	305
MPL/SK/JML/F/2002	423	330	297
MPL/SK/JML/F/2003	508	410	275
MPL/SK/JML/F/2004	524	298	284
MPL/SK/JML/F/2005	406	570	286
MPL/SK/JML/F/2006	516	302	319

Remarks *Subbotina triloculinoides* (Plummer 1926) recorded from the Jhilmili intertrappean beds, has globular chambers, a low trochospiral, tripartite test with cancellate and spinose wall texture and other characteristics. This species is restricted to the P1b–P4 zones and is commonly found in the Palaeocene (Olsson et al. 2006).

Distribution This species is globally found in low to high latitude areas (Olsson et al. 2006). In India, it is known from the intertrappean beds of the Government and Sunnamrayalu quarries of Rajahmundry, Andhra Pradesh (Malarkodi et al. 2010), subsurface sediments of Cretaceous-Palaeocene age from the Cauvery basin, Tamil Nadu (Nagendra et al. 2014), and the intertrappean beds of Jhilmili, Madhya Pradesh (Keller et al. 2009a, b).

4.4.2 Species *Globanomalina compressa* (Plummer 1926) (Fig. 4.22H; Table 4.40)

Genus	*Globanomalina* (Haque 1956)
Species	***Globanomalina compressa*** (Plummer 1926)

1926 *Globigerina compressa*: Plummer, pp.135, Pl. 8, Figs. 1a–c.
1953 *Globigerina compressa*: var. *compressa* Plummer 1926: Subbotina, pp. 63, Pl. 2, Figs. 2a, 6c.
1957 *Globorotalia compressa*: Plummer 1926: Bolli, pp. 77, Pl. 20, Figs. 21–23.
1960 *Globorotalia compressa*: Plummer 1926: Bolli and Cita, pp. 20, Pl. 32, Figs. 3a–c.
1962 *Globorotalia* (*Globorotalia*) *compressa*: Plummer 1926: Hillebrandt, pp. 125, Pl. 12, Figs. la–c.

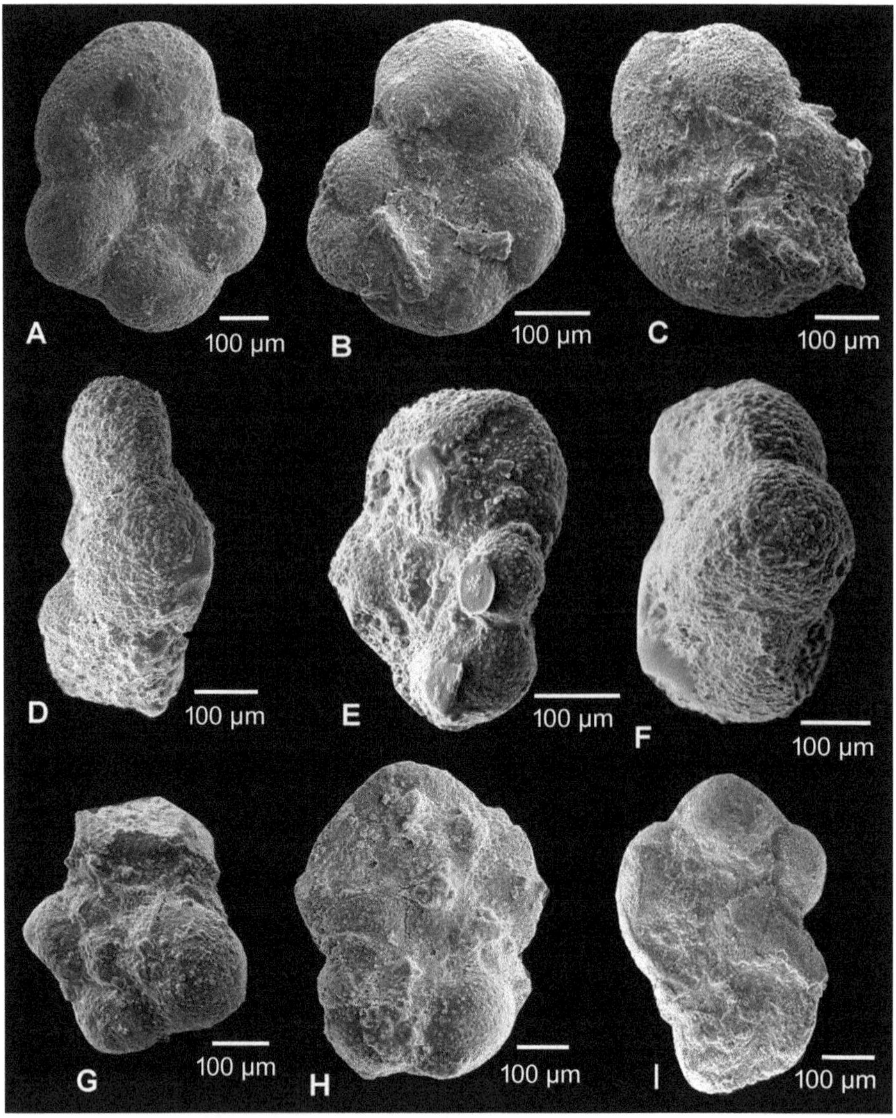

Fig. 4.22 (**A**–**G**) ***Subbotina triloculinoides*** (Plummer 1926), (**A**) Umbilical view MPL/SK/JML/F/2014; (**B**) Spiral view MPL/SK/JML/F/2002, cancellate spinose wall texture with calcite overgrowth of chambers in fig. **B**; (**C**) Umbilical view MPL/SK/JML/F/2003; (**D**) Side view MPL/SK/JML/F/2004, (**E**) Side view MPL/SK/JML/F/2005, deformed foraminifera; (**F**) Side view MPL/SK/JML/F/2006, (**G**) Umbilical view MPL/SK/JML/F/2006.i; **H** ***Globanomalina compressa*** (Plummer 1926), (**H**) Deformed umbilical view MPL/SK/JML/F/2040; (**I**)***Woodringina hornerstownensis*** (Olsson 1960); (**I**) Umbilical view MPL/SK/JML/F/2017. (Figure **A**, Reproduced from Khosla et al. 2022 with permission from Wiley)

1979 *Globorotalia (Turborotalia) compressa*: Plummer 1926: Blow, pp. 1062. Pl. 75, Figs. 10–11.
1983 *Globorotalia compressa*: Plummer 1926: Pujol, pp. 656, Pl. 2, Figs. 3–4.
1991 *Planorotalites compressus*: Plummer 1926: Huber, pp. 461, Pl. 3, Figs. 1–2.
1992 *Globanomalina compressa*: Plummer 1926: Berggren, pp. 563, Pl. 1, Figs. 14–16.
2009a *Globanomalina compressa*: Plummer 1926: Keller et al., Pl. 3. Fig. 12.
2015 *Globanomalina compressa*: Plummer 1926: Khosla, pp. 349, 352, Fig. 5q.

Material One moderately preserved test.

Horizon, Age and Locality Upper Cretaceous to Lower Palaeocene clayey limestone of unit 3 (JH 19) of the Jhilmili intertrappean beds, Chhindwara District, Madhya Pradesh, India.

Description Test (MPL/SK/JML/F/2040) is small, and it is trochospirally low, slightly angular, tightly coiled, rotaliform and compressed. It has five globular chambers and a moderately to severely imperforate peripheral margin. The chambers vary in size, increasing gradually towards the last (5th) chamber. The margin of the periphery is bluntly angular, lobate and moderately inflated. The umbilicus is extra-umbilical, the aperture is low and interio-marginal, and the lip surface is narrow, well-defined and arched. There is an umbilical depression. The wall has a typical perforate smooth structure.

Table 4.40 Dimensions of tests of *Globanomalina compressa* (Plummer 1926)

	Dimensions (μm)		
Catalogue no.	Length	Width	Thickness
MPL/SK/JML/F/2040	670	540	305

Remarks The test (MPL/SK/JML/F/2040) displays typical morphological features such as low trochospiral, compressed test, five chambers, smooth wall, imperforate peripheral margin and low aperture, which led its identification as *Globanomalina compressa* (Olsson et al. 1999).

Distribution Stratigraphically, this species ranges from P1c to P3 zones of late early Palaeocene to early late Palaeocene and is known from low to high latitudes worldwide (Olsson et al. 1999). In India, this species is known from Upper Cretaceous-Palaeocene intertrappean beds and Palaeocene sediments of the Pondicherry area (Cauvery basin), southern India (Keller et al. 2009a, b; Sharma and Malarkodi 2018).

4.4.3 *Species Woodringina hornerstownensis (Olsson 1960) (Figs. 4.22I, 4.23A; Table 4.41)*

Genus	*Woodringina* (Loeblich and Tappan 1957a, b)
Species	***Woodringina hornerstownensis*** (Olsson 1960)

1960 *Woodringina hornerstownensis*: Olsson, pp. 29, Pl. 4, Figs. 18–19.
1991 *Woodringina hornerstownensis*: D'Hondt, pp. 172, Pl. 2, Figs. 5–8.
1992 *Woodringina hornerstownensis*: Liu and Olsson, pp. 341, Pl. 1, Fig. 8.
1993 *Woodringina hornerstownensis*: MacLeod, pp. 63, Pl. 4, Figs. 6, 7, 11, 13.
1988 *Chiloguembelitria taurica*: Morozova: Keller, pp. 257, Pl. 3, Fig. 3.
2009a *Woodringina hornerstownensis*: Olsson: Keller et al. pp. 19, Fig. 8.
2015 *Woodringina hornerstownensis*: Olsson: Khosla, pp. 349.

Material Two well-preserved tests.

Horizon, Age and Locality Upper Cretaceous to Lower Palaeocene clayey limestone of unit 3 (JH–19 and 25) of the Jhilmili intertrappean beds, Chhindwara District, Madhya Pradesh, India.

Description The tests are small, elongated, biserial, tapering and twisted. They are nearly twice as long as broad and have three chambers in the first whorl and two chambers in the second whorl. Their wall surface is microperforated and has a guembelitriid texture. The chambers are more or less inflated, broad and wider than tall. They have a suture that is depressed. The aperture is small and asymmetrical. The anterior lip is thin and infolded.

Table 4.41 Dimensions of tests of ***Woodringina honerstownensis*** (Olsson 1960)

	Dimensions (μm)		
Catalogue no.	Length	Width	Thickness
MPL/SK/JML/F/2017	600	450	483
MPL/SK/JML/F/4008	400	300	306

Remarks The tests (MPL/SK/JML/F/2017 and MPL/SK/JML/F/4008) exhibit typical morphological characteristics such as biserial test, tapering, markedly twisted, wall microperforate and six chambers, identifying them as *Woodringina hornerstownensis*.

Distribution This species has a global distribution and ranges from the Pa to P3b Zones of the early Palaeocene to early late Palaeocene. It is known from high and low latitudes regions (Olsson et al. 1999). It is well known from the Deccan intertrappean beds of India (Keller et al. 2009a, b).

4.4.4 Species *Woodringina claytonensis* (Loeblich and Tappan 1957b) (Figs. 4.23B–E)

Genus	*Woodringina* (Loeblich and Tappan 1957b)
Species	***Woodringina claytonensis*** (Loeblich and Tappan 1957b)

1957b *Woodringina claytonensis*: Loeblich and Tappan, pp. 39, Figs. la–d.
1957a *Woodringina claytonensis*: Loeblich and Tappan, pp. 178. Pl. 40, fig. 6.
1991 *Woodringina claytonensis*: D'Hondt, pl. 1, Figs. 8, 11, 12; pp. 172. Pl. 2, Figs. 4, 12.
1992 *Woodringina claytonensis*: Liu and Olsson, pp. 341, Pl. 1, Figs. 4–6.
1993 *Woodringina claytonensis*: MacLeod, pp. 61. Pl. 3, Figs. 8–14. 2009a *Woodringina claytonensis*: Loeblich and Tappan: Keller et al. pp. 19, Fig. 8.
2009a *Woodringina claytonensis*: Loeblich and Tappan: Keller et al. Fig. 8, p.19.

Material Four well-preserved tests.

Horizon, Age and Locality Upper Cretaceous to Lower Palaeocene clayey limestone of unit 3 (JH 16, 17, 19, 25) of the Jhilmili intertrappean beds, Chhindwara District, Madhya Pradesh, India.

Description Tiny, small, triserial to biserial, and flaring tests. The first whorls have three chambers (triserial) and later two chambers (biserial). They have a twisted plane of biseriality and a few subglobular chambers that grow in size. The suture is constricted and divided into distinct chambers. The test wall is calcareous and perforated minutely. The aperture is low and asymmetrical, with a variable height ranging from low to high.

Dimensions: MPL/SK/JML/F (4010, 4011 and 4012), Maximum length 500 μm and width 425 μm.

Remarks The twisted plane of biseriality, asymmetrical aperture, initial triseriality followed by biseriality, and very minutely hispid wall are characters of these tests that identify them as *Woodringina claytonensis* (Olsson et al. 1999).

Distribution Stratigraphically, this species is known from the basal Pa to Plb zones of the Palaeocene worldwide. It is found commonly in sediments of low latitude open ocean environments and rarely in high latitude open marine environments (Olsson ct al. 1999). Based on stable isotope studies, it is inferred that this species preferred to live in warm surface water masses of oceans.

Fig. 4.23 **A.** ***Woodringina hornerstownensis*** (Olsson 1960), (**A**) Deformed spiral view MPL/SK/JML/F/4008. (**B–E**) ***Woodringina claytonensis*** (Loeblich and Tappen 1992), (**B**) Spiral view MPL/SK/JML/F/4010, (**C**) Umbilical view MPL/SK/JML/F/4011, (**D**) Umbilical view MPL/SK/JML/F/4012; (**E**) Enlarge view of **F**. (**F–H**) ***Hedbergella holmdelensis*** (Olsson 1964), (**F**) Umbilical view MPL/SK/JML/F/2045, (**G**) Umbilical view MPL/SK/JML/F/2046, (**H**) Edge view MPL/SK/JML/F/2047

4.4.5 *Species Hedbergella holmdelensis (Olsson 1964) (Figs. 4.23F–H; Table 4.42)*

Genus	*Hedbergella* (Bronnimann and Brown 1958)
Species	***Hedbergella holmdelensis*** (Olsson 1964)

1964 *Hedbergella holmdelensis*: Olsson, Pl. 1, pp. 160, Figs. la–c, 2a–c.
1994 *Hedbergella holmdelensis*: Olsson 1964: Huber, pp. 18, Pl. 2, Figs. 1–8.
2009a *Hedbergella holmdelensis*: Olsson 1964: Keller et al. pp. 47, 49, Pl. 2, Figs. 1–8.
2009b *Hedbergella holmdelensis*: Olsson 1964: Keller et al. pp. 19, Fig. 8.25d.
2015 *Hedbergella holmdelensis*: Olsson 1964: Khosla, pp. 349, 352, Fig. 20.

Material 20 well-preserved tests.

Horizon, Age and Locality Upper Cretaceous to Lower Palaeocene clayey limestone of unit 3 (JH 19) of the Jhilmili intertrappean beds, Chhindwara District, Madhya Pradesh, India.

Description Trochospirally, planispirally compressed and elongated tests are low to very low. They have five chambers, the last of which is ovate in shape, with a lobate central periphery and a rounded axial periphery. The umbilicus is small and deep. Chambers are globular in shape and gradually increasing in size. The spire appears to be somewhat raised. The wall is minutely perforated and hispid. The arch of the aperture is low to elevated, with a narrow lip. Sutures are depressed, curved and radial. The aperture is unclear. The texture of the wall is smooth and non-spinose.

Table 4.42 Dimensions of tests of *Hedbergella holmdelensis* (Olsson 1964)

	Dimensions (μm)		
Catalogue no.	Length	Width	Thickness
MPL/SK/JML/F/2045	262	244	249
MPL/SK/JML/F/2046	336	295	430
MPL/SK/JML/F/2047	418	291	251
MPL/SK/JML/F/2048	385	269	244
MPL/SK/JML/F/2049	414	286	241

Remarks *Hedbergella holmdelensis* differs from other species of foraminiferans by having cancellate spinose texture on the entire surface, a planispiral test and globular chambers (Olsson et al. 1999).

Distribution Stratigraphically, this species is known from the Lower Maastrichtian (Upper Cretaceous) to lower Zone P0 (Lower Palaeocene). It has a global distribution and occurs in marine sediments of high to low latitude environments (Olsson et al. 1999).

4.4.6 Species Guembelitria cretacea (Cushman 1933) (Fig. 4.24A; Table 4.43)

Family	Guembelitriidae (Montanaro-Gallitelli 1957)
Genus	*Guembelitria* (Cushman (1933)
Species	***Guembelitria cretacea*** (Cushman 1933)

1933 *Guembelitria cretacea*: Cushman, pp. 37, Pl. 4, Figs. 12a–b.
1961 *Guembelitria irregularis*: Morozova, pp. 11, Pl. 1, Fig. 9.
1970 *Guembelitria cretacea*: Cushman 1933: Olsson, pp. 601, Pl. 91, Figs. 4–5.
1973 *Guembelitria cretacea*: Cushman 1933: Smith and Pessagno, pp. 15, Pl. 1, Figs. 1–8.
1978 *Chiloguembelitria danica*: Hofker, Pl. 4, pp. 60, Fig. 14.
1988 *Chiloguembelitria danica*: Hofker 1978: Loeblich and Tappan, pp. 452. Pl. 4, Figs. 3–8.
1979 *Guembelitria* (?) *trifolia*: Morozova 1961: Blow, pp. 1384, Pl. 61, Fig. 9.
1989 *Guembelitria cretacea*: Cushman 1933: Keller, pp. 319, Figs. 1.1, 1.2.
1991 *Guembelitria cretacea*: Cushman 1933: D'Hondt, pp. 172, Pl. 1, Figs. 1, 3, 5, 6; Figs. 2, 4; Pl. 2, figs. 2–3.
1991 *Guembelitria cretacea*: Cushman 1933: D'Hondt and Keller, pp. 93, Pl. 3, Fig. 1.
1991 *Guembelitria irregularis*: Morozova 1961: D'Hondt, pp. 172, Pl. 1, Fig. 7.
1992 *Guembelitria cretacea*: Cushman 1933: Liu and Olsson, pp. 341, Pl. 1, Figs. 1–2.
2009b *Guembelitria cretacea*: Cushman 1933: Keller et al. pp. 46, Pl. 1, Fig. 9.
2015 *Guembelitria cretacea*: Cushman 1933: Khosla, pp. 350, Fig. 5n.

Material Five moderately preserved tests.

Horizon, Age and Locality Upper Cretaceous to Lower Palaeocene clayey limestone of unit 3 (JH 17, 18 and 20) of the Jhilmili intertrappean beds, Chhindwara District, Madhya Pradesh, India.

Description The tests are small, triserial, with 4–5 globular chambers and clearly depressed sutures. The aperture is large and asymmetrical, with a lip surrounding it. The aperture in the last whorl is semi-elliptical or semicircular towards the inner margin. They have a microperforated wall structure, and the surface of the tests reveals the presence of pore-mounds.

Table 4.43 Dimensions of a test of *Guembelitria cretacea* (Cushman 1933)

Catalogue no.	Dimensions (μm)		
	Length	Width	Thickness
MPL/SK/JML/F/2018	300	237	202

Remarks *Guembelitria cretacea* differs from other species of foraminiferans by having small and triserial test, aperture with a distinct lip, microperforate wall structure and globular chambers with depressed sutures (Olsson et al. 1999).

Distribution Stratigraphically, *Guembelitria cretacea* is known from the Maastrichtian (Upper Cretaceous) to Plb zone of Palaeocene. During the Maastrichtian, *Guembelitria cretacea* was generally restricted to near-shore marine environments and only rarely was found in the open oceans (Smith and Pessagno 1973; Olsson et al. 1999). It rapidly radiated in open oceans after the Cretaceous-Palaeogene boundary, and, during this interval, it was abundant at low and middle latitudes (Olsson et al. 1999, 2006). It attained a global distribution during the P0 and early Pa zone. In India, it is known from the intertrappean beds of peninsular India (Keller et al. 2009a, b).

4.4.7 Species Parasubbotina pseudobulloides (Plummer 1926) (Figs. 4.24B–F, and 4.26A–C; Table 4.44)

Genus	*Parasubbotina* (Olsson et al. 1992)
Species	***Parasubbotina pseudobulloides*** (Plummer 1926)

1926 *Globigerina pseudobulloides*: Plummer, pp. 133, Pl. 8, Figs. 9a–c.

1950 *Globigerina pseudobulloides*: Plummer 1926: Subbotina, pp. 106, Pl. 4, Figs. 8–10.

1957 *Globigerina pseudobulloides*: Plummer 1926: Troelsen, pp. 128, Pl. 30, Figs. 6a–c, 7a–c, 8a–c.

1957 *Globorotalia pseudobulloides*: Plummer 1926: Bolli, pp. 73, Pl. 17, Figs. 19–21.

1957a *Globorotalia pseudobulloides*: Plummer 1926: Loeblich and Tappan, Pl. 40, Figs. 3a–c, Pl. 42, Figs. 3a–c, Pl. 43, Figs. 3a–c, Pl. 44, Figs. 6a–c, pp. 192; Pl. 45, Figs. 1a–2c, Pl. 46, Figs. 6a–c.

1960 *Globigerina pseudobulloides*: Plummer 1926: Bolli and Cita, pp. 385. Pl. 33, Figs. 4a–c.

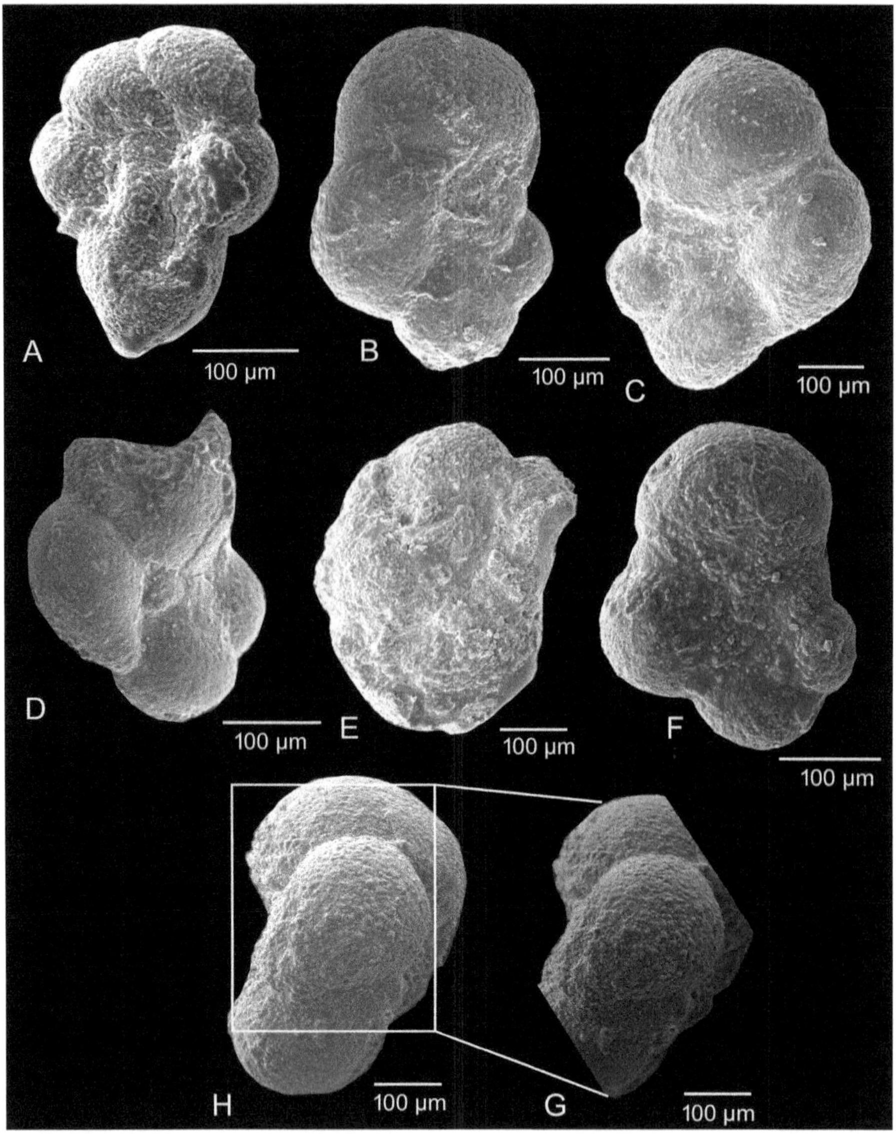

Fig. 4.24 A. ***Guembelitria cretacea*** (Cushman 1933), (**A**) Umbilical view MPL/SK/JML/F/2018. (**B–F**) Morphotypic variations of ***Parasubbotina pseudobulloides*** (Plummer 1926), (**B**) Umbilical view MPL/SK/JML/F/2019, (**C**) Spiral side view MPL/SK/JML/F/2020, (**D**) Umbilical view MPL/SK/JML/F/2021, (**E**) Deformed and mechanically compressed umbilical view MPL/SK/JML/F/2022, (**F**) Umbilical view MPL/SK/JML/F/2023. (**G–H**) ***Globigerinelloides aspera*** (Ehrenberg 1854), (**G**) Side view MPL/SK/JML/F/2024, (**H**) Enlarged view of G showing slightly perforate wall texture with calcite overgrowth

1963 *Globigerina pseudobulloides*: Plummer 1926: Gohrbandt, pp. 44, Pl. 1, Figs. 7–9.
1960 *Globorotalia pseudobulloides*: Plummer 1926: Olsson, pp. 46, Pl. 9, Figs. 19–21.
1962 *Globorotalia (Globorotalia) pseudobulloides*: Plummer 1926: Hillebrandt, pp. 124. Pl. 12, Figs. 2a–c.
1979 *Globorotalia (Turborotalia) pseudobulloides*: Plummer 1926: Blow, pp. 1096, Pl. 69, Figs. 2–3, Pl. 71, Figs. 4–5, Pl. 75, Figs. 2–3, Pl. 248, Figs. 6–8, Pl. 255, Figs. 1–6.
1992 *Globorotalia (Turborotalia) pseudobulloides*: Plummer 1926: Berggren, pp. 563, Pl. 1, Figs. 7–8.
1992 *Parasubbotina pseudobulloides*: Plummer 1926: Olsson et al. pp. 197, Pl. 3, Figs. 1–7.
2009a *Parasubbotina pseudobulloides*: Plummer 1926: Keller et al. pp. 47–49, Pl. 2, Fig. 12, Pl. 3, Figs. 1–11, 13–16.
2009b *Parasubbotina pseudobulloides*: Plummer 1926: Keller et al. Figs. 8.19d, 8.25a–d, p. 19.
2015 *Parasubbotina pseudobulloides*: Plummer 1926: Khosla, pp. 352, Fig. 5p.

Material 25 well-preserved tests.

Horizon, Age and Locality Upper Cretaceous to Lower Palaeocene clayey limestone of unit 3 (JH 19, 20, 24 and 25) of the Jhilmili intertrappean beds, Chhindwara District, Madhya Pradesh, India.

Description Rotaliform, very low trochospiral, medium-sized tests with five ventricose chambers. Chambers are globular, inflated and ovoid in shape, growing in size. The tests' periphery is lobate and rounded. Tests have a single, moderately sized, interio-marginal aperture. The aperture is curved and extends from the umbilicus to the margin. The umbilicus is narrow, deep and extra-umbilical. The test wall is thin and punctate, revealing a cancellate spinose structure with several spine holes (Olsson et al. 1999, 2006).

Table 4.44 Dimensions of tests of *Parasubbotina pseudobulloides* (Plummer 1926)

	Dimensions (μm)		
Catalogue no.	Length	Width	Thickness
MPL/SK/JML/F/2019	365	260	206
MPL/SK/JML/F/2020	557	430	215
MPL/SK/JML/F/2021	303	209	198
MPL/SK/JML/F/2022	484	353	210
MPL/SK/JML/F/2023	337	242	195
MPL/SK/JML/F/2035	505	471	244
MPL/SK/JML/F/2036	365	321	252
MPL/SK/JML/F/2037	300	257	243

Remarks *Parasubbotina pseudobulloides* can be easily differentiated from other species of planktic foraminiferans by possessing medium-sized tests, a last whorl consisting of five chambers, inflated and globular chambers and cancellate spinose wall structure (Plummer 1926; Olsson et al. 1999, 2006).

Distribution Stratigraphically, the occurrences of *Parasubbotina pseudobulloides* are restricted to Pa to P3a zones of the Palaeocene Epoch. It is known from sediments of low to high latitude environments worldwide (Olsson et al. 1999, 2006). In India, it is known from the Deccan intertrappean sediments of peninsular India (Keller et al. 2009a; Khosla 2015).

4.4.8 Species Globigerinelloides aspera (Ehrenberg 1854) (Figs. 4.24G–H; Table 4.45)

Superfamily	Globigerinacea (Carpenter et al. 1862)
Family	Planomalinidae (Bolli et al. 1957)
Genus	*Globigerinelloides* (Cushman and Ten Dam 1948)
Species	***Globigerinelloides aspera*** (Ehrenberg 1854)

1854 *Rotalia aspera*: Ehrenberg, Pl. 27, Figs. 57–58, pl. 28, fig. 42, pp. 24, pl. 31, fig. 44.
1854 *Phanerostomum asperum*: Ehrenberg, pp. 23, Pl. 30, Figs. 26a–b.
1879 *Globigerina aequilateralis*: Brady, pp. 517, Pl. 15, Figs. 14.
1910 *Globigerina aequilateralis*: Brady 1879: Heron-Allen and Earland, pp. 424, Pl. 8, Figs. 11–12.
1936 *Phanerostomum asperum*: Ehrenberg 1854: Brotzen, pp. 170, Pl. 13, Fig. 2.
1946 *Phanerostomum asperum*: Ehrenberg 1854: Schijfsma, pp. 94–96, Pl. 6, Fig. 8.
1951 *Phanerostomum asperum*: Ehrenberg 1854: Bandy, pp. 508, Pl. 75, Fig. 3.
1960 *Phanerostomum asperum*: Ehrenberg 1854: Belford, pp. 91, Pl. 25, Figs. 4–6.
1962 *Panomalina aspera*: Ehrenberg 1854: Barr, pp. 561–563, Pl. 69, Figs. 4a–b.
1963 *Phanerostomum asperum*: Ehrenberg 1854: Graham and Church, pp. 64–65, Pl. 7, Figs. 17a–c.
1963 *Planomalina (Globigerinelloides) aspera*: Ehrenberg 1854: Van Hinte, pp. 97, Pl. 12, Figs. 2a–3.
1964 *Planomalina alvarezi*: Olvera: Martin, pp. 84, Pl. 10, Figs. 8–9.
1965b *Planomalina (Globigerinelloides) aspera*: Ehrenberg 1854: Van Hinte, pp. 85, Pl.1, Figs. 2a–b.
1965 *Globigerina aspera*: Ehrenberg 1854: Takayanagi, pp. 201–202, Pl. 2, Figs. 9a–c.

1966 *Globigerinelloides aspera*: Ehrenberg 1854: Barr, pp. 50–504, Pl. 78, Figs. 4a, b.
2009a *Globigerinelloides aspera*: Ehrenberg 1854: Keller et al. pp. 19, Fig. 8.25f.
2015 *Globigerinelloides aspera*: Ehrenberg 1854: Khosla, pp. 349.

Material One well-preserved test.

Horizon, Age and Locality Upper Cretaceous to Lower Palaeocene clayey limestone of unit 3 (JH 19) of the Jhilmili intertrappean beds, Chhindwara District, Madhya Pradesh, India.

Description The test is small, and it has a planispiral shape with an equatorial aperture. It is bilaterally symmetrical and has five to six globular chambers. The rate of chamber size increase is slow. The test periphery is lobate and rounded, and the wall structure is spinose with circular pores. It has sutures that are depressed, radial and straight. Chambers have dense, dome-like pustules that have developed evenly (Georgescu 2012).

Table 4.45 Dimensions of a test of *Globigerinelloides aspera* (Ehrenberg 1854)

	Dimensions (μm)		
Catalogue no.	Length	Width	Thickness
MPL/SK/JML/F/2024	450	290	329

Remarks The morphological characters of the studied test resemble the specimens of this species described by Ehrenberg (1854) and Barr (1966) in test shape and form, nature of chambers and wall structure.

Distribution This species has a global distribution and is known from the Upper Cretaceous marine sediments of Poland, Europe, America, Asia, Australia and India (Gawor-Biedowa 1992; Keller et al. 2009a).

4.4.9 Species Globigerina (Eoglobigerina) pentagona (Morozova 1961) (Figs. 4.25A–C; Table 4.46)

Genus	*Eoglobigerina* (Morozova 1959)
Species	***Globigerina* (*Eoglobigerina*) *pentagona*** (Morozova 1961)

2009a *Globigerina* (*Eoglobigerina*) *pentagona*: Morozova 1961: Keller et al. Fig. 4.18a, Pl. 1, Figs. 10–12.
2015 *Globigerina* (*Eoglobigerina*) *pentagona*: Morozova 1961: Khosla, pp. 349. Fig. 5r.

Material 29 well-preserved tests.

Horizon, Age and Locality Upper Cretaceous to Lower Palaeocene clayey limestone of unit 3 of the Jhilmili (JH 16, 17, 19, 20, 24 and 25) intertrappean beds, Chhindwara District, Madhya Pradesh, India.

Description The tests are small, rounded and trochospiral in shape. They have two to two and a half whorls, with five equal-size globular chambers in each whorl. Chambers are inflated and growing in size. The periphery of the tests is rounded and scalloped, with a conical spiral side. The wall structure is smooth and perforated minutely. The test surface is gleaming. The dorsal surface is turreted, and the ventral margin is convex. The umbilicus is shallow, distinct and small. Sutures are deep and curved. The chambers are small and densely packed.

Table 4.46 Dimensions of tests of *Globigerina* (*Eoglobigerina*) *pentagona* (Morozova 1961)

	Dimensions (μm)		
Catalogue no.	Length	Width	Thickness
MPL/SK/JML/F/2015	528	406	187
MPL/SK/JML/F/2026	530	425	235
MPL/SK/JML/F/2027	518	398	188
MPL/SK/JML/F/2028	522	412	216
MPL/SK/JML/F/2029	498	396	218

Remarks The two to two-and-a-half whorls, inflated chambers, smooth and minutely perforated wall, turret-like dorsal surface are characters that place the studied tests in the species *Globigerina* (*Eoglobigerina*) *pentagona*. It differs from *G.* (*E.*) *tetragona* by possessing more chambers in the last whorl and a low spire. The conical form and high height of tests of *G.* (*E.*) *pentagona* differentiate it from *G.* (*E.*) *theodosica* and *G.* (*E.*) *edita* (Olsson et al. 1999, 2006).

Distribution Stratigraphically, this species is restricted to the Pa to P2 zones of the Palaeocene, and its fossils are found worldwide in marine sediments deposited in low and high latitude environments (Olsson et al. 1999, 2006). In India, it is known from the Deccan intertrappean sediments of peninsular India (Keller et al. 2009a; Khosla 2015).

4.4.10 *Foraminiferida Genus et Species indeterminate (Figs. 4.25D–G and 4.26D, E; Table 4.47)*

Order	Foraminiferida d'Orbigny (1826)
	Genus et Species indeterminate

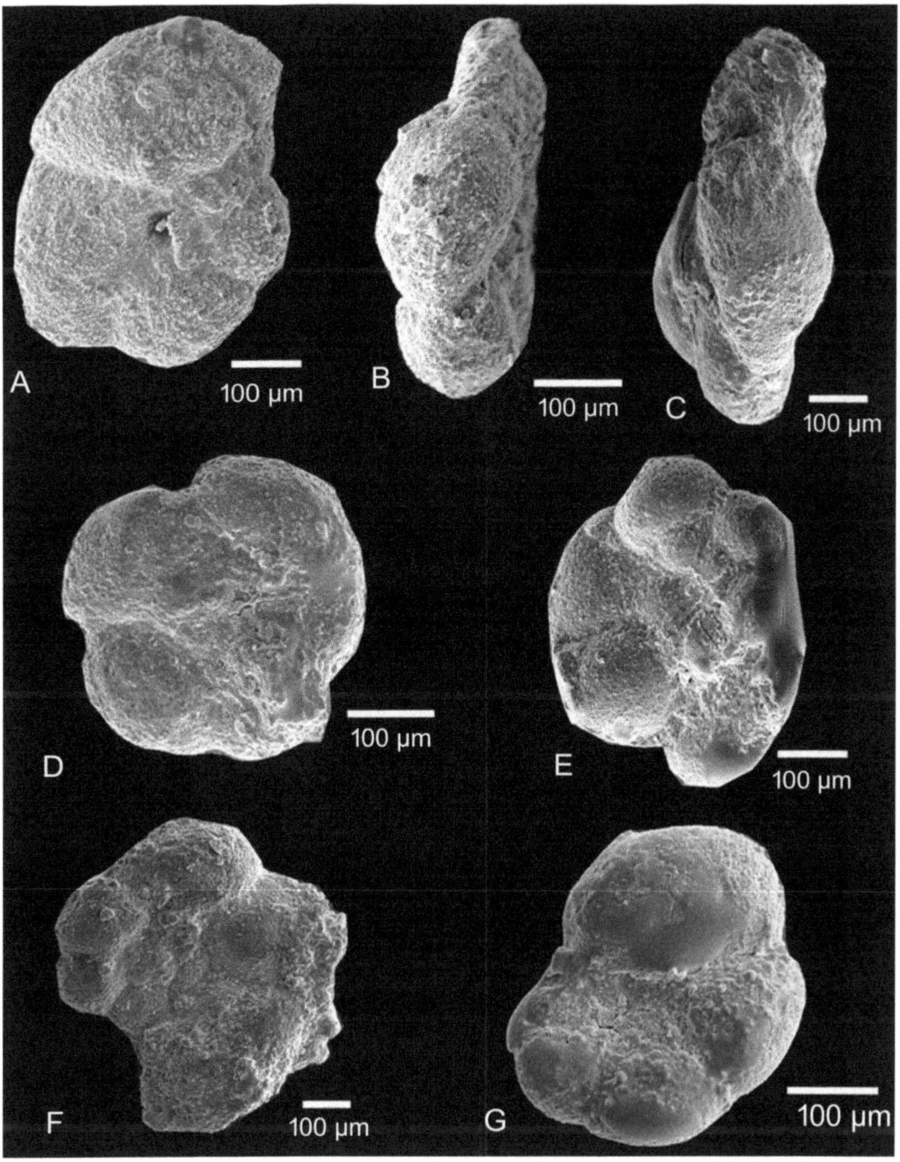

Fig. 4.25 (**A–C**) ***Globigerina*** (***Eoglogerina***) ***pentagona*** (Morozova 1961), (**A**) Umbilical view MPL/SK/JML/F/2015, (**B**) Side view MPL/SK/JML/F/2026, (**C**) Side view MPL/SK/JML/F/2027. (**D–G**) **Foraminiferida Genus et Species indeterminate**, (**D**) Umbilical view MPL/SK/JML/F/02030, (**E**) Umbilical view MPL/SK/JML/F/2031, (**F**) Umbilical view MPL/SK/JML/F/2032, (**G**) Umbilical view MPL/SK/JML/F/2033. (Figure **A**, Reproduced from Khosla et al. 2022 with permission from Wiley)

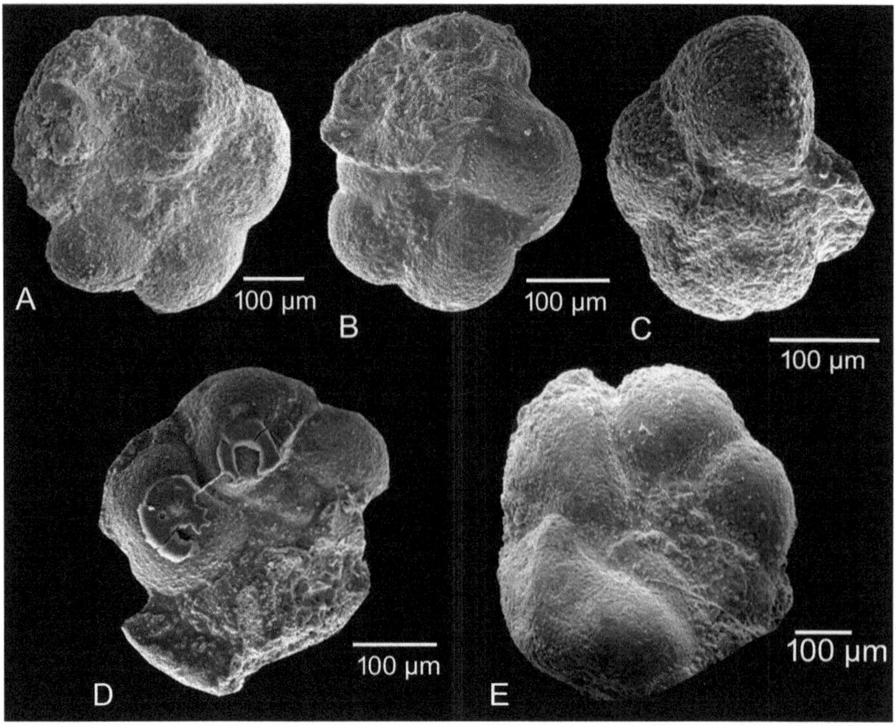

Fig. 4.26 (**A–C**) Morphotypic variations of deformed *Parasubbotina pseudobulloides* (Plummer 1926), (**A**) Umbilical view MPL/SK/JML/F/2035, (**B**) Umbilical view MPL/SK/JML/F/2036, (**C**) Umbilical view MPL/SK/JML/F/2037. (**D–E**) **Foraminiferida Genus et Species indeterminate**, (**d**) Umbilical view MPL/SK/JML/F/2038, (**E**) Umbilical view MPL/SK/JML/F/2039

Material Seven moderately preserved tests.

Horizon, Age and Locality Upper Cretaceous to Lower Palaeocene clayey limestone of unit 3 (JH 16, 17, 19, 20 and 25) of the Jhilmili intertrappean beds, Chhindwara District, Madhya Pradesh, India.

Description Small, circular, flat, trochospiral tests with variable (four to six) globular chambers. Their edges are broadly rounded and lobate. They have a spinose wall structure that is smooth and cancellate. The final test chamber is small. In some tests, there is a depression in the centre.

Table 4.47 Dimensions of indeterminate tests of foraminiferans

	Dimensions (μm)		
Catalogue no.	Length	Width	Thickness
MPL/SK/JML/F/2030	362	346	253
MPL/SK/JML/F/2031	469	357	261
MPL/SK/JML/F/2032	683	604	306
MPL/SK/JML/F/2033	358	280	263
MPL/SK/JML/F/2034	370	406	271
MPL/SK/JML/F/2038	352	360	252
MPL/SK/JML/F/2039	613	558	267

Remarks These tests bear characteristic features of foraminiferans and also appear quite different from *Subbotina triloculinoides* (Plummer 1926), *Globanomalina compressa* (Plummer 1926), *Hedbergella holmdelensis* (Olsson 1964), *Guembelitria cretacea* (Cushman 1933), *Parasubbotina pseudobulloides* (Plummer (1926) and *Globigerina* (*Eoglobigerina*) *pentagona* (Morozova 1961) in finer morphological details, such as shape and size of tests and arrangement of chambers.

4.5 Fishes

4.5.1 Species Igdabatis indicus (Prasad and Cappetta 1993) (Figs. 4.27A, B)

Class	Chondrichthyes (Huxley 1880)
Order	Myliobatiformes (Compagno 1973)
Superfamily	Myliobatoidea (Compagno 1973)
Family	Myliobatidae (Bonaparte 1838)
Genus	*Igdabatis* (Cappetta 1972)
Species	***Igdabatis indicus*** (Prasad and Cappetta 1993)

Material Isolated lateral teeth.

Horizons, Age and Localities Upper Cretaceous to ?Lower Palaeocene greenish limestone unit of the Ghat Parasia intertrappean beds and Upper Cretaceous (Maastrichtian) black carbonaceous shale unit of the intertrappean section exposed in the Shiraj (=Shriwas) well, Chhindwara District, Madhya Pradesh, India.

Description The occlusal surface of the teeth is convex and contains well-preserved small polygonal pits. Teeth have two lobes, which are separated by a groove in basal view. Lingual face of the teeth is convex and has a corrugated surface. The lobes are triangular in shape.

Dimensions MPL/SK/GP/CV/3001 and MPL/SK/GP/CV/3002 measure 1.4 and 1.6 in length and 0.80 and 0.85 in width, respectively

Remarks The morphological characters of the studied teeth resemble characters of lateral teeth of *Igdabatis indicus* (Prasad and Cappeta 1993) described by Prasad and Cappeta (1993) and Verma et al. (2017). Therefore, on this basis, these teeth are identified as *Igdabatis indicus*.

Distribution This species is widely known from the Upper Cretaceous Deccan infra- and intertrappean sediments of peninsular India and Palaeocene Fatehgarh Formation of Rajasthan (Prasad and Cappeta 1993; Khosla et al. 2004, 2009; Verma et al. 2016, 2017; Verma and Khosla 2018, 2019).

4.5.2 Species *Lepisosteus indicus* (Woodward 1908) (Figs. 4.27C–G)

Class	Actinopterygii (Von Klein 1885)
Subclass	Neopterygii (Regan 1923)
Order	Lepisosteiformes (Hay 1929)
Family	Lepisosteidae (Cuvier 1825)
Genus	*Lepisosteus* (Lacepède 1803)
Species	***Lepisosteus indicus*** (Woodward 1908)

Material Isolated teeth and scales.

Horizons, Age and Localities Upper Cretaceous to Lower Palaeocene clayey limestone of unit 3 of the Jhilmili intertrappean beds, Upper Cretaceous to ?Lower Palaeocene greenish limestone unit of the Ghat Parasia intertrappean beds and Upper Cretaceous (Maastrichtian) black carbonaceous calcareous shale unit of the intertrappean sections exposed in the Government and Shiraj (=Shriwas) wells, Chhindwara District, Madhya Pradesh, India.

Description The teeth are well-preserved and conical in shape. They have well-developed fine vertical ridges and grooves in the lower part. They have a smooth surface, translucent apical top and dark coloured basal part. The tooth edges are sharp and blade-like, descending from either side of the base to the apex of the basal part. The upper part is longer than the lower part. There are distinct straight lines from the translucent edge to the basal part.

The scales are well-preserved and vary in shape and size. They have a distinct peg- and socket-like articulation and show oval, sub-rectangular and rhombic shapes, depending on their position in the body. The caudal scales have a sub-rectangular outline with parallel ganoine ridges.

Dimension MPL/SK/GP/CV/3004 and MPL/SK/JML/CV/3005 measure 1.1 mm and 1.0 mm in height, respectively.

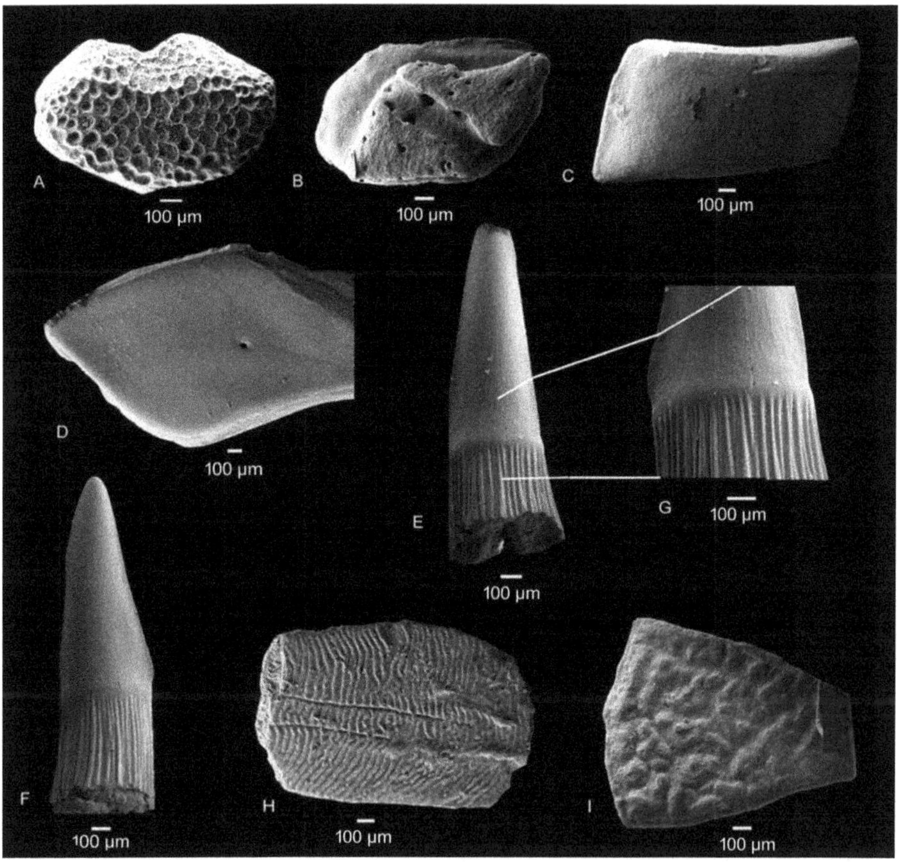

Fig. 4.27 (**A–I**) Fish remains from intertrappean beds of Chhindwara District, Madhya Pradesh. (**A**, **B**) ***Igdabatis indicus*** (Prasad and Cappeta 1993) from Ghat Parasia intertrappean beds, (**A**) MPL/SK/GP/CV/3001, Isolated lateral tooth in occusal view, (**B**) MPL/SK/GP/CV/3002, Isolated lateral tooth in basal view. (**C–G**) ***Lepisosteus indicus*** (Woodward 1908) from Jhilmili and Ghat Parasia intertrappean beds and intertrappean beds exposed within Government and Shriwas wells, Chhindwara District, Madhya Pradesh, (**C**) MPL/SK/GP/CV/3003, Isolated scale in basal view, (**D**) MPL/SK/GP/CV/3003, Isolated scale in external view, (**E**) MPL/SK/GP/CV/3004, Isolated tooth in lateral view, (**F**) MPL/SK/JML/CV/3005, Isolated tooth in lateral view and (**G**) Enlarged view of E. (**H–I**) **Osteoglossidae Genus et Species indeterminate from** Government well and Ghat Parasia intertrappean beds, (**H**) MPL/SK/GW/CV/3006, Isolated scale in external view and (**I**) MPL/SK/GP/CV/3007, Isolated scale in external view. Scale bar 100 μm. (Figure **E**, Reproduced from Kania et al. 2022 with permission from the Editor of the Himalayan Geology)

Remarks Woodward (1908) described *Lepisosteus indicus* (Woodward 1908) from the Upper Cretaceous Lameta Formation of Dongargaon, Maharashtra. The material investigated here shows close resemblance to *Lepisosteus indicus* (Woodward 1908).

Distribution *Lepisosteus indicus* is known from numerous infra- and intertrappean beds of peninsular India and the Palaeocene Fatehgarh Formation of Rajasthan (Prasad and Khajuria 1990; Kumar et al. 2005; Prasad et al. 2013).

4.5.3 Osteoglossidae Genus et Species indeterminate (Figs. 4.27H–I)

Division	Teleostei Müller (1845)
Subdivision	Osteoglossomorpha (Greenwood et al. 1966)
Order	Osteoglossiformes (Berg 1940)
Family	Osteoglossidae (Bonaparte 1832)
	Genus et Species indeterminate

Material Isolated scales.

Horizons, Age and Localities Upper Cretaceous to ?Lower Palaeocene greenish limestone and hard clayey limestone units of the Ghat Parasia intertrappean beds and Upper Cretaceous (Maastrichtian) black calcareous carbonaceous shale unit of the intertrappean section exposed in the Government well, Chhindwara District, Madhya Pradesh, India.

Description The present collection includes only scales, which are rhombic, quadrangular and polygonal in shape. The inner surface of the scales shows a concave depression with a raised rim bearing growth lines. The external surface shows diverse ornamentation from randomly distributed fine or coarse tubercles to linearly arranged tubercles. The inner surface is finely ornamented and forms a smooth internal surface. The external surface bears a coarse ornamentation.

Dimensions MPL/SK/GW/CV/3006 and MPL/SK/GW/CV/3007 measure 1.7 mm and 1.6 mm in width and 1.2 and 1.3 (maximum) in length, respectively.

Remarks The similar scales of Osteoglossiformes assigned to the genus *Phareodus* have been documented from the Upper Cretaceous infra- and intertrappean beds of peninsular India (Prasad and Sahni 1987; Khajuria and Prasad 1998).

Distribution The indeterminate scale remains of Osteoglossiformes are reported from several Deccan infra- and intertrappean sites, peninsular India (Prasad and Sahni 1987; Khajuria and Prasad 1998; Khosla et al. 2004; Verma et al., 2012; Khosla and Verma 2015).

4.6 Conclusions

1. A rich assemblage of charophytes consisting of *Platychara perlata* (Peck and Reker 1947), *Platychara raoi* (Bhatia and Mannikeri 1976), *Platychara sahnii* (Bhatia and Mannikeri 1976), *Platychara compressa* (Peck and Reker 1948), *Platychara* sp., *Platychara closasi* sp. nov., *Peckichara* cf. *varians* (Grambast 1957) *Nemegtichara* cf. *grambasti* (Bhatia et al. 1990b), ?*Grambastichara* sp., *Microchara shivarudrappai* sp. nov. and *Chara chhindwaraensis* sp. nov., including three new species, has been recovered for the first time from the Jhilmili section (District Chhindwara), Madhya Pradesh.
2. A total of 23 ostracod taxa (including three new species) represented by *Buntonia whittakerensis* sp. nov., *Neocyprideis raoi* (Jain 1978), *Limnocythere deccanensis* (Khosla et al. 2005), *Limnocythere martensi* sp. nov., *Frambocythere tumiensis anjarensis* (Bhandari and Colin 1999), *Gomphocythere strangulata* (Jones 1860), *Gomphocythere paucisulcatus* (Whatley et al. 2002b), *Gomphocythere dasyderma* (Whatley et al. 2002a), *Paracypretta subglobosa* (Sowerby 1840), *Paracypretta jonesi* (Bhatia and Rana 1984), *Paracypretta verruculosa* (Whatley et al. 2002a), *Strandesia jhilmiliensis* (Khosla et al., 2011a), *Stenocypris cylindrica* (Sowerby in Malcolmson 1840), *Periosocypris megistus* (Whatley et al. 2012), *Zonocypris spirula* (Whatley and Bajpai 2000a), *Zonocypris viriensis* (Khosla and Nagori 2005), *Zonocypris penchi* sp. nov., *Cypridopsis astralos* (Whatley et al. 2002a), *Cypridopsis hyperectyphos* (Whatley and Bajpai 2000a), *Eucypris pelasgicos* (Whatley and Bajpai 2000a), *Cyclocypris amphibolos* (Whatley et al. 2002a), *Cypria cyrtonidion* (Whatley and Bajpai 2000a) and *Talicypridea pavnaensis* (Khosla et al. 2005) were discovered from the Jhilmili intertrappean beds.
3. Nine taxa of foraminiferans represented by *Subbotina triloculinoides* (Plummer 1926), *Globanomalina compressa* (Plummer 1926), *Woodringina hornerstownensis* (Olsson 1960), *Woodringina claytonensis* (Loeblich and Tappan 1957a, b), *Hedbergella holmdelensis* (Olsson 1964), *Guembelitria cretacea* (Cushman 1933), *Parasubbotina pseudobulloides* (Plummer 1926), *Globigerinelloides aspera* (Ehrenberg 1854), *Globigerina* (*Eoglobigerina*) *pentagona* (Morozova 1961) and some unidentified forms were recovered from the Jhilmili intertrappean beds.
4. A new species of charophyte, *Platychara closasi* sp. nov., 10 taxa of ostracods consisting of *Limnocythere deccanensis* (Khosla et al. 2005), *Frambocythere tumiensis anjarensis* (Bhandari and Colin 1999), *Gomphocythere paucisulcatus* (Whatley et al. 2002b), *Periosocypris megistus* (Whatley et al. 2012), *Candona* sp., *Eucypris* sp. 1, *Cyclocypris amphibolos* (Whatley et al. 2002a), *Cyprois rostellum* (Whatley and Bajpai 2000a), *Cyprois* sp. and *Darwinula* sp. and three taxa of fishes comprising *Igdabatis indicus* (Prasad and Cappetta 1993), *Lepisosteus indicus* (Woodward 1908) and Osteoglossidae Genus et Species indeterminate (fishes) were recovered for the first time from the Ghat Parasia intertrappean beds (District Chhindwara, Madhya Pradesh).

5. Eight taxa of ostracod consisting of *Frambocythere tumiensis lakshmiae* (Whatley and Bajpai 2000a), *Gomphocythere strangulata* (Jones 1860), *Gomphocythere paucisulcatus* (Whatley et al. 2002b), *Gomphocythere* sp. 1, *Cypridopsis elachistos* (Whatley et al. 2002b) ?*Eucypris verruculosa* (Whatley et al. 2002a), *Cypria cyrtonidion* (Whatley and Bajpai 2000a), and *Cyprois rostellum* (Whatley and Bajpai 2000a) and two taxa of fishes, *Lepisosteus indicus* (Woodward 1908) and Osteoglossidae Genus et Species indeterminate, were recovered from the Government well (District Chhindwara, Madhya Pradesh).
6. Eight taxa of ostracods, *Frambocythere tumiensis lakshmiae* (Whatley and Bajpai 2000a), *Gomphocythere strangulata* (Jones 1860), *Gomphocythere paucisulcatus* (Whatley et al. 2002b), *Zonocypris labyrinthicos* (Whatley et al. 2002b), *Zonocypris gujaratensis* (Bhandari and Colin 1999), *Cypridopsis elachistos* (Whatley et al. 2002b), *Cypria cyrtonidion* (Whatley and Bajpai 2000a), and *Cyprois rostellum* (Whatley and Bajpai 2000a), and two fish taxa consisting of *Igdabatis indicus* (Prasad and Cappetta 1993) and *Lepisosteus indicus* (Woodward 1908), were recovered from the Shriwas well (District Chhindwara, Madhya Pradesh).

References

Agardh CA (1824) Systema Algarum. Literis Berlingianis, Lundae, pp 1–312

Apostolescu V (1956) Contribution a l'etude des ostracodes de l'Eocene inferieur (s.1) du bassin de Paris. Rev Inst Fr Pét 11:1327–1352

Apostolescu V (1961) Contribution B l'itude paliontologique (Ostracodes) et stratigraphique des basins Cretacts et Tertiaires de I' Afrique Occidentdle. Rev Inst Fr Pét 16:779–867

Baird W (1843) Notes on British Entomostraca. The Zoologist (Newman) 1:193–197

Baird W (1845) Arrangement of the British Entomostraca, with a list of species, particularly noticing those which have as yet been discovered within the bounds of the Club. Trans Berwickshire Naturalists' Club 2(139):145–158

Baird W (1850) The natural history of the British Entomostraca, vol 2. The Ray Society, London, pp 1–364

Baird W (1859) Description of some new recent Entomostraca from Nagpur, collected by the Rev. S. Hislop. Proc Zool Soc London 27:231–234

Bajpai S, Whatley RC (2001) Late Cretaceous non-marine ostracods from the Deccan intertrappean Beds, Kora: Western Kachchh, India. Rev Esp de Micropaleontol 33(1):91–111

Bajpai S, Sahni A, Jolly A, Srinivasan S (1990) Kachchh intertrappean biotas: affinities and correlation. In: Sahni A, Jolly A (eds) Cretaceous event stratigraphy and the correlation of the Indian nonmarine strata. A Seminar cum Workshop IGCP 216 and 245, Chandigarh, pp 101–105

Bajpai S, Mohabey DM, Kapur VV, Sharma R (2004) Late Cretaceous (Maastrichtian) freshwater ostracod fauna from Deccan intertrap sediments from Phulsagar, Mandla District, Madhya Pradesh. Gond Geol Magaz 19:147–157

Bandy OL (1951) Upper Cretaceous foraminifera from the Carlsbad area, San Diego County, California. J Paleontol 25:488–513

Barr FT (1962) Upper Cretaceous planktonic foraminifera from the Isle of Wight, England. Palaeontology 4(4):552–580

Barr FT (1966) The foraminiferal genus *Bolivinoides* from the Upper Cretaceous of the British Isles. Palaeontology 9:220–243

Barsotti G (1963) Paleocene ostracods of Libya (Sirte Basin) and their wide distribution. Rev Inst Fr Pét 18:1520–1528

Bassiouni MAA, Luger P (1990) Maastrichtian to Early Eocene Ostracoda from southern Egypt. Palaeontology, palaeoecology, palaeobiogeography and biostratigraphy. Ber Geowiss Abh Reihe A Geol Palaeontol 120(2):755–928

Bassiouni MAA, Morsi AM (2000) Paleocene-lower Eocene ostracodes from El Quss Abu said plateau (Farafra Oasis), Western Desert, Egypt. Palaeontogr Abt A 257:27–84

Belford DJ (1960) Upper Cretaceous foraminifera from the Toolonga calcilutite and Gingin chalk, Common wealth of Australia. Bull Bur Mineral Resour Geol Geophys 57:1–198

Belford DJ (1967) Paleocene planktonic foraminifera from Papua and New Guinea. Bull Bur Mineral Resour Aust 92:1–33

Benmansour S, Andreu B, Yahiaoui A (2016) The Campanian–Maastrichtian of the Aures Basin, Algeria: Paleobiogeographical distribution of ostracods. Cretac Res 58:86–107

Berg LS (1940) Classification of fishes, both recent and fossil. Travaux de l'Institut zoologique de l'Académie des Sciences de l'URSS 5(2):1–517

Berggren WA (1962) Some planktonic foraminifera from the Maestrichtian and type Danian stages of Southern Scandinavia. Stockholm Contrib Geol 9(1):1–106

Berggren WA (1992) Paleogene planktonic foraminifera magneto biostratigraphy of the southern Kerguelen Plateau (Sites 747–749). In: Wise SW, Schlich R et al (eds) Proceedings of the ocean drilling program, scientific results, vol 120. Ocean Drilling Program, College Station, pp 551–569

Bhandari A (1995) Ostracodes from the inter-trappean beds near Duddukuru, Andhra Pradesh and a note on their age and paleoecological significance. Indian J Pet Geol 4(1):89–107

Bhandari A (1998) Non-marine Ostracoda from the subsurface of the Cambay shale (Eocene) of Gujarat, India and their paleoecological significance. Rev Esp de Micropaleontol 30:1–10

Bhandari A, Colin JP (1999) Limnic ostracods of the intertrappean sediments (terminal Maastrichtian-basal Paleocene) of the Anjar region (Kachchh, Gujarat state), India: systematics, paleoecology, and paleobiogeographic affinities. J Micropaleontol 42(1):3–20

Bhatia SB, Mannikeri MS (1976) Some Charophyta from the Deccan intertrappean beds near Nagpur, Central India. Geophytology 6(1):75–81

Bhatia SB, Rana RS (1984) Palaeogeographic implications of the Charophyta and Ostracoda of the intertrappean beds of peninsular India. Mém Soc Geol Fr 147:29–35

Bhatia SB, Prasad GVR, Rana RS (1990a) Deccan volcanism, a Late Cretaceous event: conclusive evidence of ostracodes. In: Sahni A, Jolly A (eds) Cretaceous event stratigraphy and the correlation of the Indian nonmarine strata. A Seminar cum Workshop IGCP 216 and 245, Chandigarh, pp 47–49

Bhatia SB, Riveline J, Rana RS (1990b) Charophyta from the Deccan intertrappean beds near Rangapur, Andhra Pradesh, India. The Palaeobotanist 37:316–323

Bhatia SB, Prasad GVR, Rana RS (1996) Maastrichtian non-marine ostracodes from peninsular India: Palaeobiogeographic and age implications. Mem Geol Soc Ind 37:297–311

Blow WH (1979) The Cainozoic Globigerinida, vol 3. EJ Brill, Leiden, p 1452

Bolli HM (1957) The genera *Globigerina* and *Globorotalia* in the Paleocene-Lower Eocene Lizard Springs Formation of Trinidad, B.W.I. In: Loeblich AR et al (eds) Studies in foraminifera, vol 215. Bulletin – United States National Museum, pp 61–82

Bolli HM, Cita MB (1960) Globigerinee, Globorotaliedel Paleocene di Padernod' Adda (Italia). RivItal di Paleontol et Stratigr 66:361–408

Bolli HM, Loeblich AR, Tappan H (1957) Planktonic foraminiferal families Hantkeninidae, Orbulinidae, Globorotaliidae and Globotruncanidae. In: Loeblich AR, Tappan H, Beckmann JP, Bolli HM, Montanaro GE, Troelsen JC (eds) Studies in foraminifera, vol 215. Bulletin – United States National Museum, pp 3–50

Bonaparte CL (1832) Iconografiadelle fauna italica per le quattro classi degli animali vertebrate, Tomo III, Pesci. Roma, pp 1–556

Bonaparte CL (1838) Selachorum tabula analytica. Nuovi Ann Sci Nat Bologna 1(2):195–214

Brady GS (1868) Contributions to the study of the Entomostraca I. Ostracoda from the Arctic and Scandinavian Seas. Ann Mag Nat Hist 4(2):30–35

Brady GS (1879) A monograph of the Ostracoda of the Antwerp Crag. Trans Zool Soc London 10:379–409

Brady GS (1886) Notes on *Entomostraca* collected by Mr. A. Haly in Ceylon. J Linn Soc Lond Zool 19:293–317

Brady GS, Norman AM (1889) A monograph of the marine and freshwater Ostracoda of the North Atlantic and of North-Western Europe: section 1: Podocopa. Sci Trans R Dublin Soc Ser 2(4):63–270

Brady GS, Robertson D (1885) Genus *Darwinula*. In: Jones TR (ed) On the Ostracoda of the Purbeck formation; with notes on the Wealden species. Geological Society, London, The Q J Geol Soc London, p 346

Bronhstein ZS (1947) Fresh-water Ostracoda. Amerind Publishing Co, New Delhi. (Translation of Ostracoda Presnykh Vod Akademiya Nauk SSR, Zoologicheskii Institut)

Brönnimann PT (1952) Globigerinidae from the Upper Cretaceous (Cenomanian-Maestrichtian) of Trinidad, B.W.I. Bull Am Paleontol 34:5–71

Bronnimann PT, Brown NK (1958) *Hedbergella,* a new name for a Cretaceous planktonic foraminiferal genus. J Wash Acad Sci 48:15–17

Brotzen F (1936) Foraminiferen aus dem schwedischen untersten Senon von Eriksdal in Schonen. Arsb Sveriges Geol Undersökn Ser C 30(3):1–206

Brotzen F, Pozaryska K (1961) Foraminiferes du Paleocene et de l'Eocene inferieur en Pologne septentrionale; remarques paleogeographiques. Rev Micropaleontol 14:155–166

Cappetta H (1972) Les poisons cretaces et tertiaires du bassin des Iullemmeden (Republique du Niger). Palaeovertebrata 5:179–251

Carbonell G, Alzouma K, Dikouma M (1990) Les ostracodes Paléocènes du Niger: Taxinomie—Un témoignage de l'existenceé ventuelle de la mertrans saharienne? Geobios 23(6):671–697

Carbonnel G, Monciardini C (1995) Ostracodes Paléogènes au débouché de la mertrans saharienne (Mali-Niger). Trav Doc Lab Géol Lyon 135(1):3–123

Carpenter WB, Parker WK, Jones TR (1862) Introduction to the study of the foraminifera. Ray Society Publications, London, pp 1–139

Chanda S, Deb U, Mukherjee TN (1989) Study of fossil Charophyta from the intertrappean beds at Kateru area, Krishna-Godavari Basin, Andhra Pradesh, South India. In: First international symposium on extant and fossil Charophytes, Montpellier, pp 1–11

Charrière A, Haddoumi H, Mojon PO, Ferrière J, Cuche D, Zili L (2009) Miseen évidence par charophytes et ostracodes de l'âge Paléocène des depots discordants sur les rides anticlinales de la régiond' Imilchil (Haut Atlas, Maroc): consequences paléogéographiques et structurales. C R Palevol 8(1):9–19

Colin JP (1980) Sur la morphologie, la systématique, la biogéographie et l'évolution des ostracodes limniques Timiriaseviinae (Limnocytheridae). Paléobiol Continent 11:1–52

Ciccioli PL, Ballent S, Tedesco A, Barreda V, Limarino CO (2005) Hallazgo de depositos cretacicosen La Precordillera de La Rioja (formacion cienaga del Rio Huaco). Rev Asoc Geol Argentina 60(1):122–131

Colin JP, Moghaddam HV, Safari A, Grai SS (2012) Presence of *Frambocythere* Colin, 1980, (limnic ostracode) in the Maastrichtian of the Zagros Mountains, Iran: a newly recognized link between southern Europe and the Far East. Carnets de Géol Mém Brest Lett 2012/02(CG2012-L02):173–181

Compagno LJV (1973) Interrelationships of living elasmobranchs. Zool J Linnean Soc 53:15–61

Cushman JA (1933) Some new foraminiferal genera. Contrib Cushman Lab Foram Res 8:4–7

Cushman JA, Ten Dam A (1948) *Globigerinelloides*, a new genus of Globigerinidae. Contrib Cushman Lab Foram Res 24:42–43

Cuvier G (1825) Recherches sur les Ossements fossiles, où l'on rérablit les charactères de plusieurs animaux dont les revolutions du globe ont dètruit les espèes, vol 5, no 2, Dufour and D'Ocagne, Paris, pp 1–547

d'Hondt S (1991) Phylogenetic and stratigraphic analysis of earliest Paleocene triserial and biserial planktonic foraminifera. J Foraminifer Res 21:168–181

d'Hondt SL, Keller G (1991) Some patterns of planktic foraminiferal assemblage turnover at the Cretaceous-Tertiary boundary. Mar Micropaleontol 17(1–2):77–11

d'Orbigny AD (1826) Tableau méthodique de la classe des Céphalopodes. Ann Sci Nat 7:96–169

Daday JJ (1900) Ostracoda Hungarie. A Magyarorszagi Kagylosrakok Maganrajza, pp 1–319

de Sowerby JC (1840) Explanations of the plates and wood-cuts. In: Malcolmson JC (ed) On the fossils of the eastern portion of the Great Basaltic District of India, vol 2, no 5. Trans Geol Soc London, pp 511–567

Difley R, Ekdale AA, Gillette DD (1999) Biostratigraphic aspects of the Cretaceous-Tertiary (KT) boundary interval at North Horn Mountain, Emery County, Utah. Vertebrate Paleontology in Utah. Utah Geological Survey Miscellaneous Publications, pp 389–398

Doweld AB (2015) *Chara nikolskae*, a new name for *Chara elliptica* Nikolskaja (1984) non Fritzsche (1924) and *Aclistochara elliptica* (Fritzsche) comb. nov. (*Characeae*). Phytotaxa 226(2):199–200

Ehrenberg CG (1854) Mikrogeologie. Das Erden und Felsen schaffende Wirken des unsichtbar kleinen selbstständigen Lebens auf der Erde. Leopold Voss, Leipzig, pp xxviii + 374

El Sogher A (1996) Late Cretaceous and Paleocene Ostracodes from the Waha limestone and Hagfa shale formations of the Sirt Basin, Libya. In: Salim MJ, Mouzughi AJ, Hammuda OS (eds) The geology of Sirt Basin, Libya, vol 1. Elsevier, Amsterdam, pp 287–382

Elewa AMT, Abdelhady AA (2020) Migration routes of the genus *Buntonia* Howe, 1935 (Ostracoda, Crustacea) during the Late Cretaceous–Paleogene in Africa. Arab J Geosci 13:1182. https://doi.org/10.1007/s12517-020-06203-y

Feist M, Colombo F (1983) La limite Cretace-Tertiaire dans le nord-est de l'Espagne, du 1016 point de vue des charophytes. Géol Méditerr 10:303–326

Fischer S (1855) Beitragzur Kenntniss der Ostracoden. Abh der Akad München Mathem Physikal CI 7:637–666

Foster CA, Swain FM, Petters SW (1983) Late Paleocene Ostracoda from Nigeria. Rev Esp de Micropaleontol 15(1):103–166

Fouch TD, Hanley JH, Forester RM, Keighin CW, Pitman JK, Nichols DJ (1987) Chart showing lithology, mineralogy, and paleontology of the nonmarine North Horn Formation and Flagstaff Member of the Green River Formation, Price Canyon, Central Utah, Miscellaneous investigations series. United States Geological Survey, pp 1–8

Fritzsche CH (1924) Neue Kreidefaunen aus Südamerika (Chile, Bolivia, Peru, Columbia). Neu Jb Miner Geol Paläontol 50:1–56

Galeeva LI (1955) Ostracods of the Cretaceous deposits of the Mongolian People's Republic. State Scientific and Technical Publishing, Moskva, pp 1–97

Gawor-Biedowa E (1992) Campanian and Maastrichtian foraminifera from the Lublin Upland, Eastern Poland. Palaeontol Pol 52:3–187

Georgescu MD (2012) Restudy of the type specimens of *Phanerostomum asperum* Ehrenberg 1854 (Late Cretaceous, Foraminiferida). Micropaleontology 58(3):291–303

Gohrbandt K (1963) Apercu sur la subdivision du Paleoecene et de l'Eocene le plus inferieur dans I' Halvetikum au nord de Salzbourg (Auteiche) base sur des foraminiferes planctoniques. Colloque sur l Paleogene. Mem Bur Rech Geol Min 28:599–605

Graham JJ, Church CC (1963) Campanian foraminifera from the Stanford University campus (California School of Earth Sciences), vol 8, no 1, Stanford University, pp 1–107

Grambast L (1956) La plaque basale de la gyrogonite. C R des Séances de L'Acad des Sci Paris 242:2585–2587

Grambast L (1957) Ornamentation de la gyrogonite et systematique chez les charophytes fossiles. Rev Gén Bot 64:339–362

Grambast L (1959) Tendances évolutives dans le phylum des Charophytes. C R des Séances de L'Acad des Sci Paris 249:557–559

Grambast L (1962) Classification de l'embranchement des Charophytes. Naturalia Monspeliensia, Montpellier, Série Botanique 14:63–86
Grambast L (1971) Remarques phylogenetiques et biocronologiques sur les Septorella du Cretace terminal de Provence et les charophytes associes. Paléobiol Cont 2(2):1–38
Grambast L (1974) Phylogeny of the Charophyta. Taxon 23:463–481
Grambast L, Grambast N (1954) Sur la position systématique de quelques Charophytes tertiaires. Rev Gén Bot 61:665–671
Grambast L, Martinez M, Mattauer M, Thaler L (1967) *Perutherium altiplanense*, nov. gen., nov. sp., premier mammifere Mesozoique d'Amerique du Sud. C R Acad des Sci Ser D 264:707–710
Greenwood PH, Rosen DE, Weitzman SH, Myers GS (1966) Phyletic studies of teleostean fishes, with a provisional classification of living forms. Bull Am Mus Nat Hist 131(4):339–456
Grekoff N (1969) On the stratigraphic value and paleogeographic relationships of some Cretaceous, Paleocene and Lower Eocene ostracods from eastern Algeria. In: Proceedings of the third African micropaleontology colloquium, Cairo, pp 227–248
Groves J (1933) Charophyta. Fossilium Catalogus II, pars 19:1–74
Haque AFM (1956) The foraminifera of the Ranikot and the Laki of the Namal Gorge, Salt Range. Mem Geol Surv Pak, Paleontol Pakist 1:1–300
Hartmann G (1964) Asiatische Ostracoden. Systematiche und zoogeographische Untersuchungen. Internationale Revue der gesamten Hydrobiologie. Systematische Beihefte 3:1–155
Hay WRA (1929) Second bibliography and catalogue of the fossil Vertebrata of North America. Publi Carnegie Instit Wash 309:1–2003
Helmdach FF (1978) Nicht marine ostracodes aus der Spanischer Oberkreide. Berl Geowiss Abh 3:71–78
Heron-Allen E, Earland A (1910) On the recent and fossil foraminifera of the shore-sands of Selsey Bill, Sussex – 5. The Cretaceous Foraminifera. J R Microsc Soc London 1910:401–426
Hislop S (1860) On the Tertiary deposits, associated with trap-rock, in the East Indies– with descriptions of the fossil shells. Q J Geol Soc London 16:154–166
Hofker J (1978) Analysis of a large succession of samples through the Upper Maastrichtian and the Lower Tertiary of drill hole 47.2, Shatsky rise, Pacific, Deep Sea drilling project. J Foram Res 8:46–75
Holifield BR (1964) Lower Tertiary Charophyta of North America (MA thesis). University of Missouri, Columbia, p 130
Horn af Rantzien H (1951) On the fossil Charophyta of Latin America. Sven Bot Tidskr 45:658–677
Horn af Rantzien H (1959) Morphological types and organ-genera of Tertiary Charophyte fructifications. Stockh Contrib Geol 4:45–197
Hou YT (1982) On the taxonomy and origin of cristid-ostracods. Acta Palaeontol Sin 2(1):73–82
Hou Y, Junde H, Chunhui Y (1978) The Cretaceous-Tertiary ostracods from the marginal region of the Yangtze-Han River plain in Central Hubei. Mem Nanjing Instit Geol Paleontol 9:129–206
Howe HV (1962) Ostracod taxonomy. Louisiana State University Press, Baton Rouge
Howe HV, Chambers J (1935) Louisiana Jacksone Eocene Ostracoda. Louisiana Geol Surv Geol Bull 5:1–65
Howe HV (1935) Louisiana Jackson Eocene Ostracoda. Louis Geoll Sur Geol Bull 5:1–65.
Huang RJ (1988) Charophytes of Nanxiong Basin, Guangdong and its Cretaceous-Tertiary boundary. Acta Micropaleontol Sin 27(4):457–474
Huang R, Wang Z (1978) Triassic Charophytes of Shaanxi. Acta Palaontol Sin 17 (3):267–276
Huber BT (1991) Maestrichtian planktonic foraminifer biostratigraphy and the Cretaceous/ Tertiary boundary at hole 738C (Kerguelen Plateau, Southern Indian Ocean). In: Proceedings of the Ocean drilling program scientific results, vol 119. Ocean Drilling Program, College Station, pp 451–465
Huber BT (1994) Ontogenetic morphometrics of some Late Cretaceous trochospiral planktonic foraminifera from the Austral Realm. Smithson Contrib Paleobiol 77:1–85
Huxley TH (1880) On the applications of the laws of evolution to the arrangement of the Vertebrata and more particularly of the Mammalia. Proc Zool Soc London 43:649–662

Jaillard E, Cappetta H, Ellenberger P, Feist M, Grambast-Fessard N, Lefranc JP, Sige B (1993) Sedimentology, palaeontology, biostratigraphy and correlation of the Late Cretaceous Vilquechico Group of southern Peru. Cretac Res 14:623–661

Jain SP (1978) Ostracodes des "inter-trappean beds" (Eocene Infeerieur) de Kateru, Rajahmundry, Côte est de l' Inde. Rev de Microplaeontol 21(2):51–58

Jones T (1860) Notes on the fossil Cypride. Q J Geol Soc London 16:154–189

Kapur VV, Khosla A, Tiwari N (2019) Paleoenvironmental and paleobiogeographical implications of the microfossil assemblage from the Late Cretaceous intertrappean beds of the Manawar area, District Dhar, Madhya Pradesh, Central India. Hist Biol 31(9):1145–1160

Karczewska J, Ziembińska-Tworzydlo M (1969) Upper Cretaceous Charophyta from the Nemegt Basin, Gobi Desert. Paleontol Pol 21:121–144

Karczewska J, Ziembińska-Tworzydlo M (1972) Lower Tertiary Charophyta from the Nemegt Basin, Gobi Desert. Acta Paleontol Pol 27:51–81

Kaufmann A (1900) Neue Ostracoden aus der Schweiz. Zool Anz 23:131–133

Keller G (1989) Extended Cretaceous/Tertiary boundary extinctions and delayed population change in planktonic foraminifera from Brazos River, Texas. Paleoceanography 4:287–332

Keller G, Khosla SC, Sharma R, Khosla A, Bajpai S, Adatte T (2009a) Early Danian planktic foraminifera from Cretaceous-Tertiary intertrappean beds at Jhilmili, Chhindwara District, Madhya Pradesh, India. J Foraminifer Res 39(1):40–55

Keller G, Adatte T, Bajpai S, Mohabey DM, Widdowson M, Khosla A, Sharma R, Khosla SC, Gertsch B, Fleitmann D, Sahni A (2009b) K-T transition in Deccan traps of Central India marks major marine seaway across India. Earth Planet Sci Lett 282:10–23

Kesling RV (1951) The morphology of ostracod moult stages. University of Illinois Press, Urbana

Khajuria CK, Prasad GVR (1998) Taphonomy of a Late Cretaceous mammal-bearing microvertebrate assemblage from the Deccan intertrappean beds of Naskal, peninsular India. Palaeogeogr Palaeoclimatol Palaeoecol 137:153–172

Khand Y (1977) Novye vidy ostrakod is porgranchichykh sloev verkhnego mela I paleogena Zaaltayskoy Gobi, Mongolskoy Narodnoy Respubliki. In: *Mesozoic and Cenozoic faunas, floras and biostratigraphy of Mongolia*, vol 4. The Joint Soviet-Mongolian Paleontological Expedition. Transaction, pp 106–111

Khanna SN, Mohan M (1965) Intertrappean beds of Kutch, western India. Bull Geol Soc India 2:48–51

Khosla A (2014) Upper Cretaceous (Maastrichtian) charophyte gyrogonites from the Lameta Formation of Jabalpur, Central India: palaeobiogeographic and palaeoecological implications. Acta Geol Pol 64(3):311–323

Khosla A (2015) Palaeoenvironmental, palaeoecological and palaeobiogeographical implications of mixed fresh water and brackish marine assemblages from the Cretaceous–Palaeogene Deccan intertrappean beds at Jhilmili, Chhindwara District, Central India. Rev Mex de Cien Geol 32(2):344–357

Khosla A, Sahni A (2000) Late Cretaceous (Maastrichtian) ostracodes from the Lameta Formation, Jabalpur Cantonment area, Madhya Pradesh, India. J Palaeont Soc India 45:57–78

Khosla A, Verma O (2015) Paleobiota from the Deccan volcano-sedimentary sequences of India:paleoenvironments, age and paleobiogeographic implications. Hist Biol 27(7):898–914

Khosla SC, Nagori ML (2002) Ostracodes from the inter-trappean beds (Early Paleocene) of the east coast of India. Paleontol Res 6(2):191–210

Khosla SC, Nagori ML (2005) A restudy of ostracode fauna from the intertrappean beds of the Anjar, Kachchh District, Gujarat. J Geol Soc India 66(5):573–580

Khosla SC, Nagori ML (2007a) Ostracoda from the Intertrappean beds of Mohgaon-Haveli, Chhindwara District, Madhya Pradesh. J Geol Soc India 69:209–221

Khosla SC, Nagori ML (2007b) A revision of the Ostracoda from the Intertrappean beds of Takli, Nagpur District, Maharashtra. J Paleontol Soc India 52:1–15

Khosla A, Prasad GVR, Verma O, Jain AK, Sahni A (2004) Discovery of a micromammal-yielding Deccan intertrappean site near Kisalpuri, Dindori District, Madhya Pradesh. Curr Sci 87(3):380–383

Khosla A, Sertich JJW, Prasad GVR, Verma O (2009) Dyrosaurid remains from the intertrappean beds of India and the Late Cretaceous distribution of Dyrosauridae. J Vertebr Paleontol 29(4):1321–1326
Khosla SC, Nagori ML, Mohabey DM (2005) Effect of Deccan volcanism on non-marine Late Cretaceous Ostracode fauna: a case study from Lameta Formation of Dongargaon area (Nand-Dongargaon basin), Chandrapur District, Maharashtra. Gondwana Geol Mag 8:133–146
Khosla SC, Nagori ML, Jakhar SR, Rathore AS (2009a) Mixed marine, brackish water and non-marine microfaunal association in the inter-trappean beds (Early Palaeocene) of Jhilmili, Chhindwara District, Madhya Pradesh. J Geol Soc India 73:724–732
Khosla SC, Nagori ML, Jakhar SR, Rathore AS, Kumari M (2009b) A restudy of the Ostracoda *Cypris cylindrica* Sowerby (in Malcolmson, 1840) from the Deccan intertrappean beds (Late Cretaceous) of Lakshmipur, Kachchh, Gujarat. J Geol Soc India 74:579–584
Khosla A, Kania S, Lucas SG, Verma O (2022) Charophytes from the Cretaceous–Palaeogene transition in the Jhilmili intertrappean beds of Central India. Geol J 57(11):4412-4438 https://doi.org/10.1002/gj.4528
Khosla SC, Nagori ML, Jakhar SR, Rathore AS (2010) Stratigraphical and palaeoecological implications of the Late Cretaceous ostracods from the Lameta Formation of Pisdura, Chandrapur District, Maharashtra, India. Gond Geol Mag 25:115–124
Khosla SC, Nagori ML, Jakhar SR, Rathore AS (2011a) Early Danian lacustrine–brackish water Ostracoda from the Deccan intertrappean beds near Jhilmili, Chhindwara District, Madhya Pradesh, India. Micropaleontology 57(3):223–245
Khosla SC, Rathore AS, Nagori ML, Jakhar SR (2011b) Non-marine Ostracoda from the Lameta Formation (Maastrichtian) of Jabalpur, Madhya Pradesh and Nand-Dongargaon Basin, Maharashtra, India: their correlation, age and taxonomy. Rev Esp de Micropaleontol 43:209–260
Khosla A, Chin K, Alimohammadin H, Dutta D (2015) Ostracods, plant tissues, and other inclusions in coprolites from the Late Cretaceous Lameta Formation at Pisdura, India: Taphonomical and palaeoecological implications. Palaeogeogr Palaeoclimatol Palaeoecol 418:90–100
Khosla A, Chin K, Verma O, Alimohammadin H, Dutta D (2016) Paleobiogeographical and paleoenvironmental implications of the freshwater Late Cretaceous ostracods, charophytes and distinctive residues from coprolites of Lameta Formation at Pisdura, Chandrapur District (Maharashtra), Central India. In: Khosla A, Lucas SG (eds) Cretaceous period: biotic diversity and biogeography, vol 71. New Mexico Museum of Natural History & Science, Albuquerque, pp 173–184
Klie W (1938) Krebstiere oder Crustacea-III: Ostracoda. Gustav Fischer, Jena, Muschelkrebse
Knowlton FH (1888) Description of a new fossil species of the genus *Chara*. Bot Gaz 13:156–157
Krömmelbein K, Weber R (1971) Ostracoden des "Nordost-Brasilianischen Wealden". Beihefte zum Geologischen Jahrbuch, Beiheft 115:1–93
Kshetrimayum DS, Parmar V (2019) Late Cretaceous (Maastrichtian) faunal assemblage from a new intertrappean locality, Uthawali, District Dhar, Madhya Pradesh, Central India. Researcher 15(1):13–26
Kshetrimayum DS, Parmar V, Lourembam RS, Prasad GVR (2021) A diversified Ostracoda (Crustacea) assemblage from the Upper Cretaceous intertrappean beds of Gujri, Dhar District, Madhya Pradesh, India. Cretac Res 124:104784. https://doi.org/10.1016/j.cretres.2021.104784
Kumar K, Rana RS, Paliwal BS (2005) Osteoglossid and lepisosteid fish remains from the Paleocene Palana Formation, Rajasthan, India. Palaeontology 48(6):1187–1209
Lacépède BGE (1803) Histoire naturelle des poissons, vol 4. Plassan, Paris
Latreille PA (1806) Genera crustaceorum et insectorum secundumordinem naturalem in familia disposita, iconibus exemplis queplurimis explicata. Paris Argentorati 1:1–302
Li S, Gao QQ, Zhang YY, Qu HY, Wang MY, Wan XQ (2013) Late Cretaceous-early Paleocene charophytes from the Songliao Basin, North China, SK1(N) core. Acta Micropalaeontol Sin 30(1):1–16
Li S, Wang Q, Zhang H, Lu H, Martín-Closas C (2016) Charophytes from the Cretaceous–Paleogene transition in the Pingyi Basin (Eastern China) and their Eurasian correlation. Cretac Res 59:179–200

Li S, Wang QF, Zhang HC, Wan XQ, Martín-Closas C (2019) Charophytes from the Cretaceous–Paleocene boundary in the Songliao Basin (North-Eastern China): a Chinese biozonation and its calibration to the Geomagnetic Polarity Time Scale. Pap Palaeontol 5:47–81

Lindley J (1836) A natural system of botany. Longman, London

Linnaeus C (1758) Systema Naturae per regna tria naturae, secundum classes, ordines, genera, species, cum characteribus, differentiis, synonymis, locis. Editio decima, reformata 1:1–824

Liu C, Olsson RK (1992) Evolutionary adaptive radiation of microperforate planktonic foraminifera following the K/T mass extinction event. J Foram Res 22:328–346

Loeblich AR, Tappan H (1957a) Planktonic foraminifera of Paleocene and Early Eocene age from the Gulf and Atlantic coastal plains. Bull U S Natl Mus 215:173–198

Loeblich AR, Tappan H (1957b) *Woodringina,* a new foraminiferal genus (Heterohelicidae) from the Paleocene of Alabama. J Wash Acad Sci 47:39–40

Loeblich AR, Tappan H (1964) Treatise on invertebrate Paleontology, part C: Protista 2, Sarcodina, chiefly "The camoebians" and Foraminiferida. Geol Soc Am Univ Kansas Press, Lawrence

Loeblich AR, Tappan H (1988) Foraminiferal genera and their classification, vol 2. Van Nostrand and Reinhold Company, New York, pp 1–970

Loeblich AR, JR, Tappen H (1992) Present status of Foraminiferal Classification. In: Takayanagi, Y, Saito T (eds) Studies in benthic foraminifera, Tokyo: Tokai University Press, pp, 93–102

MacLeod N (1993) The Maastrichtian-Danian radiation of triserial and biserial planktic foraminifera: testing phylogenetic and adaptational hypotheses in the (micro) fossil record. Mar Micropaleontol 21(1–3):47–100

Madler K (1955) Zur taxionomie der tertiaren charophyten. Geol Jahrb 70:265–328

Malarkodi N, Keller G, Fayazudeen PJ, Mallikarjuna UB (2010) Foraminifera from the early Danian intertrappean beds in Rajahmundry quarries, Andhra Pradesh. J Geol Soc India 75:851–863

Malcolmson JG (1840) On the fossils of the Eastern portion of the Great Basaltic District of India. Trans Geol Soc London 52(5):537–575

Mandelstam MI (1960) Ostracods: Nadsemeystvo Cytheracea. In: Orlov YA (ed) Osnovy Paleontologii, 8. Gosudarstvennoe Nauchno-Technicheskoe Isdatelstvo Literaturiy Po Gelogii I Ochrane Nedra, Moskva, pp 365–411

Martin L (1964) Upper Cretaceous and lower Tertiary foraminifera from Fresno County, California. Jahrb Geol Bundesanst 9:1–128

Massieux M, Tambareau Y, Villatte J (1981a) Charophytes thanétiennes et ilerdiennes de la Couverture occidentale du Massif de Mouthoumet (Aude). Géol Méditerr 8(1):1–17

Massieux M, Tambareau Y, Villatte J (1981b) Characees Paleocenes et eocenes du versant nord des Pyrenees. Rev Micropaleontol 24(2):69–82

Mathur AK (1972) Some ostracodes from the Tatrot Formation near Pinjaur. Himal Geol 2:388–396

Mathur AK, Verma KK (1988) Freshwater ostracodes from the intertrappean beds of southeastern Rajasthan. Geol Surv India Spec Publ 11(2):169–174

Migula W (1897) The Characeae of Germany, Austria and Switzerland. In: Rabenhorst X (ed) Kryptogamic Flora vol 5. Kummer E, Leipzig, pp 1–765

Mohabey DM, Udhoji SG, Verma KK (1993) Palaeontological and sedimentological observations on non-marine Lameta Formation (Upper Cretaceous) of Maharashtra, India: their palaeontological and palaeoenvironmental significance. Palaeogeogr Palaeoclimatol Palaeoecol 105:83–94

Montanaro-Gallitelli E (1957) A revision of the foraminiferal family Heterohelicidae. Bull U S Natl Mus 215:133–154

Moore RC, Benson RH, Berdan JM, van den Bold WA, Hanai T (1961) Treatise on invertebrate Paleontology, Arthropoda. Geological Society of America and University Kansas Press, Kansas

Morozova VG (1959) Stratigraphy of the Danian-Montian deposits of the Crimea according to the foraminfera. Dokl Akad Nauk SSSR 124:1113–1116

Morozova VG (1961) Datsko-Montskie planktonnye foraminifery yuga SSSR (Danian-Montian Planktonic Foraminifera of the Southern USSR). Paleontol Zh 2:8–19

Morsi AMM, Speijer RP, Stassen P, Steurbaut E (2011) Shallow marine ostracode turnover in response to environmental change during the Paleocene-Eocene thermal maximum in Northwest Tunisia. J Afr Earth Sci 59:243–268
Müller J (1845) Über den Bau und die Grenzen der Ganoiden, und über das natürliche System der Fische. Arch Naturgeschichte 11(1):91–141
Müller GW (1894) Die Ostracoden des Golfes von Neapel und der angrenzenden Meeres-Abschnitte. Fauna und Flora des Golfes von Neapel. Friedlander and Sohn (Berlin), pp 1–404
Müller GW (1898) The ostracods. Scientific results of the trips to Madagascar and East Africa in the years 1889–95 by Dr A Voeltzkow Dep Senckenb. Nat Res Ges 21:255–296
Musacchio EA (1972) Carofitasd el cretacio inferiore n sedimentitas chubuenses al este de la Herreira, Chubut. Ameghiniana 9:354–356
Nagendra R, Sathiyamoorthya P, Reddy AN (2014) Foraminiferal biostratigraphy and paleobathymetric trends across the Cretaceous–Tertiary boundary in Neyveli area, Cauvery basin, India. Stratigr Geol Correl 22(7):708–716
Neale JW (1969) The taxonomy, morphology and ecology of recent Ostracoda. Oliver & Boyd, Edinburgh
Neale JH (1976) The genus *Finmarchinella* Swain 1963 (Crustacea: Ostracoda) and its species. Bull Br Mus (Nat Hist) Zool 27(2):83–93
Olsson RK (1960) Foraminifera of latest Cretaceous and earliest Tertiary age in the New Jersey coastal plain. J Paleontol 34:1–58
Olsson RK (1964) Late Cretaceous planktonic foraminifera from New Jersey and Delaware. Micropaleontology 10:157–188
Olsson RK (1970) Planktonic foraminifera from base of Tertiary, millers ferry, Alabama. J Paleontol 44:598–604
Olsson RK, Hemleben C, Berggren WA, Liu C (1992) Wall texture classification of planktonic foraminifera genera in the Lower Danian. J Foram Res 22:195–213
Olsson RK, Berggren WA, Hemleben C, Huber BT (1999) Atlas of Paleocene Planktonic Foraminifera. Smithson Contrib Paleobiol 85:1–252
Olsson RK, Hemleben C, Huber BT, Berggren WA (2006) Taxonomy, biostratigraphy, and phylogeny of Eocene *Globigerina*, *Globoturborotalita*, *Subbotina*, and *Turborotalita*. In: Pearson PN, Olsson RK, Huber BT, Hemleben C, Berggren WA (eds) Atlas of Eocene planktonic foraminifera, vol 41. Cushman Found Spec Publ, pp 111–168
Peck RE (1959) Stratigraphic distribution of charophyta and non-marine ostracodes. In: Itermountain association of petroleum Geologists guidebook, 10th annual field conference, pp 115–121
Peck RE, Forester RM (1979) The genus *Platychara* from the Western Hemisphere. Rev Palaeobot Palynol 28:223–236
Peck RE, Reker CC (1947) Cretaceous and lower Cenozoic charophyta from Peru. Novit Am Mus Nat Hist 1369:1–6
Peck RE, Reker CC (1948) Eocene Charophyta from North America. J Paleontol 22(1):85–90
Plummer HJ (1926) Foraminifera of the midway formation in Texas. Univ Tex Bull 2644:1–206
Prasad GVR, Sahni A (1987) Coastal-plain microvertebrate assemblage from the terminal Cretaceous of Asifabad, peninsular India. J Paleontol Soc Ind 32:5–19
Prasad GVR, Khajuria K (1990) A record of microvertebrate fauna from the inter-trappean beds of Naskal, Andhra Pradesh. J Palaeontol Soc India 35:151–161
Prasad GVR, Cappetta H (1993) Late Cretaceous selachians from India and the age of the Deccan traps. Palaeontology 36(1):231–248
Prasad GVR, Verma O, Flynn JJ, Goswami A (2013) A new Late Cretaceous vertebrate fauna from the Cauvery Basin, South India: implications for Gondwanan paleobiogeography. J Vertebr Paleontol 33(6):1260–268
Puckett TM, Colin JP, Mitchell S (2012) New species and genera of Ostracoda from the Maastrichtian (Late Cretaceous) of Jamaica. Micropaleontology 58(5):397–455

Pujol C (1983) Cenozoic planktonic foraminiferal biostratigraphy of the southwestern Atlantic (Rio-Grande Rise)- deep-sea drilling project leg-72. Initial Rep Deep Sea Drill Proj 72:623–673
Rao KS, Rao SRN (1939) Fossil charophyta of the Deccan intertrappeans near Rajahmundry, India. Mem Geol Surv India Palaeontol Indica 29:1–14
Rathore AS (2018) Late Cretaceous (Maastrichtian) non-marine Ostracod fauna from the intertrappean beds of Pinjaurni, Chandrapur district, Maharashtra. JK Knowledge Initiative 2(1):64–73
Rathore AS, Grover P, Verma V, Lourembam RS, Prasad GVR (2017) Late Cretaceous (Maastrichtian) non-marine ostracod fauna from Khar, a new intertrappean locality, Khargaon District, Madhya Pradesh, India. Paleontol Res 21(3):215–229
Regan CT (1923) The skeleton of *Lepisosteus*, with remarks on the origin and evolution of the lower neopterygian fishes. Proc Zool Soc London 93:445–461
Reyment RA (1963) Studies on Nigerian Upper Cretaceous and Lower Tertiary Ostracoda, part 2: Danian, Paleocene and Eocene Ostracoda. Stockhlom Contrib Geol 10:1–286
Reyment RA (1981) The Ostracoda of the Kalambaina Formation (Paleocene), northwestern Nigeria. Bull Geol Inst Univ Upsala NS 9:51–65
Reyment RA, Reyment ER (1980) The Paleocene trans-Saharan transgression and its ostracod fauna. In: Salem MJ, Busrewfl MT (eds) The geology of Libya. Academic, London, pp 245–254
Sahni A, Khosla A (1994) A Maastrichtian ostracode assemblage (Lameta Formation) from Jabalpur Cantonment, Madhya Pradesh, India. Curr Sci 67(6):456–460
Salahi D (1966) Ostracodes du Crétacé supérieur et du Tertiaire en provenance d'un sondage de la région de Zelten (Libye). Rev Inst Fr Pét 21(1):3–32
Sarr R (1995) Étude biostratigraphique et paléoenvironnementale des séries d'âge Crétacé terminal à Eocène moyen du Sénégal occidental. Systématique et migration des Ostracodes. Unpublished thèse de doctorat (inédite), Université Dakar, pp 1–335
Sarr R (2015) Paleobiogeographic implications of Paleocene ostracods from Senegal and related West and North Africa basins. Rev Micropaleontol 58(4):333–349
Sars GO (1863) Beretning om en i sommeren 1862 foretagen zoologisk reise I Christianias og Trondhjems stifter. Nyt Magazin for Naturvidenskaberne 12:193–252
Sars GO (1889) On some freshwater Ostracoda and Copepoda raised from dried Australian mud. Forhandlinger I Christiania. Videnskabs-Selskabet 8:2–79
Sars GO (1924) The freshwater Entomostraca of the Cape Province (Union of South Africa). Ostracoda. Ann S Afr Mus 20:105–193
Sars GO (1925) An account of the Crustacea of Norway with short descriptions and figures of all the species. Ostracoda 9(3–10):73–208
Schijfsma E (1946) The foraminfera from the Hervian (Campanian) of southern Limburg. Mededeel. Mededelingen van de Geologische Stichting 7:1–174
Schmarda LC (1871) Zoologie II, Ausgabe. Band I, Wilhelm Braumüller, Wien
Schudack ME (1993) Die Charophyten im Oberjura und Unterkreide Westeuropas. Mit einer phylogenetischen Analyse der Gesamtgruppe. Berl Geowiss Abh Reihe A 8:1–209
Shahin A (2000) Paleogene ostracodes of Gebel Withr, southwestern Sinai, Egypt: palaeontology, biostratigraphy and palaeobiogeography. J Afr Earth Sci 31(2):285–315
Shahin A (2005) Maastrichtian to Middle Eocene ostracodes from Sinai, Egypt: systematics, biostratigraphy and paleobiogeography. Rev de Paléobiol 24(2):749–779
Sharma R, Khosla A (2009) Early Palaeocene ostracoda from the Cretaceous-Tertiary (K-T) Deccan intertrappean sequence at Jhilmili, district Chhindwara, Central India. J Palaeontol Soc India 54(2):197–208
Sharma CSB, Malarkodi S (2018) *Globanomalina* from the Palaeocene sediment of Pondicherry area, Cauvery Basin, Southern India. Int J Geol Earth Sci 4(1):67–71
Sharma R, Bajpai S, Singh MP (2008) Freshwater ostracoda from the Paleocene-age Deccan intertrappean beds of Lalitpur (Uttar Pradesh), India. J Paleontol Soc India 53(2):81–87
Shivarudrappa TV (1989) Charophytes from the intertrappeans of Gurmatkal (Karnataka, India) and their stratigraphical significances. Proceedings of the first international symposium on extant and fossil charophytes, Montpellier, Abstract Book, p 54

Shutskaya EK (1970) Stratigrafiya, foraminifery i paleogeografiya nizhnego paleogena Kryma, predkavkaz'ya i zapadnoi chadsti srednei azii [Stratigraphy, foraminifera and paleogeography of the lower Paleogene in the Crimea, Precaucasus and the Western Part of Central Asia]. Trudv, Vsesoyuznyi Nauchno-lssledovadetel 'skii Geologorazvedochnyi Neftyanoi Institut (VNIGNI) 70:1–256

Smith GM (1938) Cryptogamic botany, vol II. Bryophytes and Pteridophytes. McGraw-Hill, New York

Smith CC, Pessagno EA (1973) Planktonic foraminifera and stratigraphy of the Corsicana Formation (Maestrichtian) North-central Texas. Spec Publ Cushman Found Foraminifer Res 12:1–68

Srinivasan S, Sahni A, Bajpai S (1992) Fossil Charophyta from the Deccan intertrappean beds of Gurmatkal, Gulbarga District, Karnataka. Curr Sci 63:396–398

Srinivasan S, Bajpai S, Sahni A (1994) Charophytes from Deccan intertrappean beds of peninsular India: implications for age and correlation of Deccan volcanic. Geobios 27:559–571

Stankevitch ES (1974) Ostracoda from Upper Cretaceous deposits ofcontinental water-basins in the northern Gobi (Mongolia), In: Problemy issledovaniya drevnkh ozer evrazii, Leningrad, p. 57–70

Stott LD, Kennett JP (1990) Antarctic Paleogene planktonic foraminifer biostratigraphy: ODP Leg 113, Sites 689 and 690. Proc Ocean Drill Program Sci Results 113:549–569

Stuhlmann F (1888) Preliminary report of a trip to East Africa with the support of the Royal Academy of Sciences to study freshwater fauna. Meet Rep R Prussian Acad Sci 49:1255–1269

Subbotina NN (1950) Mikrofauna i stratigrafiya Elburganskogo Gorozonta Goriathego Klyitcha. Trudy Vsesoyuznego Neftyanogo Nauchno-Issledovatel 'skogo Geologo-Razvedochnogo Instituta (VNIGRI), Mikrofauna SSSR, Sbornik 51:5–112

Subbotina NN (1953) Iskopaemye foraminifery SSSR (Globigerinidy, Khantkenininidy i Globorotaliidy). Trudy Vsesoyznogo Neftyanogo Nauchno-Issledovatel 'skogo Geologo-Razvedochnogo Instituta (VNIGRI) 76:1–296

Sylvester-Bradley PC (1948) The ostracode genus "Cythereis". J Paleontol 22:792–797

Szczechura J, Blaszyk J (1970) Fresh-water ostracodes from the Nemegt Formation (Upper Cretaceous) of Mongolia. Palaeontol Pol 38:65–21

Takayanagi Y (1965) Upper Cretaceous planktonic foraminifera from the Putah Creek subsurface section along the Yolo-Solano County line, California, Second series (Geology) 36. Tohoku University, pp 161–237

Tian JX, Chen J, Lu NH, Chen C, Wang H, Li T, Li S, Wang QF, Zhang HC, Wan XQ (2021) Discovery of charophyte flora across the Cretaceous-Paleocene transition in the Jiaolai basin. Palaeoworld 30(3):538–550

Troelsen JC (1957) Some planktonic foraminifera of the type Danian and their stratigraphic importance. Bull US Natl Mus 215:125–131

Vaillant S (1719) Caracteres de quatorze genres de plantes. Mem Acad R Sci Paris 1719:17–20

Van Hinte JV (1963) Zur Stratigraphie und Mikropaläontologie der Oberkreide und des Eozäns des Krappfeldes (Kärnten). Jahrb Geol Bundesanst Sonderband 8:1–147

Vávra W (1891) Kritisches Verzeichniss der Ostracoden Böhmens. Ceska Spolecnost Nauk Prague, pp 159–168

Vávra W (1895) Die von Dr F Stuhlmann gesammelten Süsswasser-Ostracoden Zanzibar's. Beiheft zum Jahrbuch der Hamburgischen Wissenschaftlichen Anstalten 12:1–23

Verma O, Khosla A (2018) Application of internet technology in assembling literature for palaeontological research. Iran J Sci Technol Trans A Sci 4:1715–1723

Verma O, Khosla A (2019) Developments in the stratigraphy of the Deccan Volcanic Province, peninsular India. Compt Rend Geosci 351:461–476

Verma O, Prasad GVR, Khosla A, Parmar V (2012) Late Cretaceous gondwanatherian mammals of India: Distribution, interrelationships and biogeographic implications. J Paleontol Soc India 57:95–104

Verma O, Khosla A, Goin FJ, Kaur J (2016) Historical biogeography of the Late Cretaceous vertebrates of India: Comparison of geophysical and paleontological data. In: Khosla A, Lucas SG (eds) Cretaceous period: Biotic diversity and biogeography, vol 71. New Mexico Museum of Natural History & Science, Albuquerque, pp 317–330

Verma O, Khosla A, Kaur J, Prasanth M (2017) Myliobatid and pycnodont fish from the Late Cretaceous of Central India and their paleobiogeographic implications. Hist Biol 29(2):253–265

Vicente A, Martín-Closas C, Arz JA, Oms O (2015) Maastrichtian basal Paleocene charophyte biozonation and its calibration to the Global Polarity Time Scale in the southern Pyrenees (Catalonia, Spain). Cret Res 52:268–285

von Hillebrandt A (1962) Das Paleozan und seine foraminiferen fauna im Becken von Reichenhall und Salzburg. Bayerische Akademie der Wissenschaften 108:1–181

von Klein A (1885) Beiträge zur Bildung des Schädels der Knochenfische. II. Jh Ver vaterländ Naturk Württemberg 41:107–261

Von Siebold CT (1845) Lehrbuch der Vergleichenden Anatomie der Wirbellosen Thiere [Heft 1]. Veit and Comp, Berlin, pp 1–208

Von Siebold CT (1848) Lehrbuch der vergleichenden Anatomie. Teil 1. Lehrbuch der vergleichenden Anatomie der wirbellosen Thiere. Veit, Berlin, p 679

Wang Z (1978) Paleogene charophytes from the Yangtze-Han River basin. Characeae. Mem Nanjing Inst Geol Paleontol Acad Sin 9:101–120

Wang QF (2004) Fossil Charophytes from late Cretaceous to Eocene of North Jiangsu Basin, China (PhD dissertation). University of Beijing, Beijing, China

Wang Z, Lu HN, Zhao CB (1985) In: Wang TQ (ed) Cretaceous charophytes from Songliao Basin and adjacent areas (pp. 1–94). Harbin, China: Heilongjiang Scientific and Technical Publishing House (in Chinese with English abstract).

Whatley RC, Bajpai S (2000a) A new fauna of Late Cretaceous non-marine Ostracoda from the Deccan intertrappean beds of Lakshmipur, Kachchh (Kutch District), Gujarat, western India. Rev Esp de Micropaleontol 32(3):385–409

Whatley RC, Bajpai S (2000b) Further nonmarine Ostracoda from the Late Cretaceous intertrappean deposits of the Anjar region, Kachchh, Gujarat, India. Rev Micropaleontol 43(1):173–178

Whatley RC, Bajpai S, Srinivasan S (2002a) Upper Cretaceous nonmarine Ostracoda from intertrappean horizons in Gulbarga district, Karnataka state, South India. Rev Esp de Micropaleontol 34(2):163–186

Whatley RC, Bajpai S, Srinivasan S (2002b) Upper Cretaceous intertrappean nonmarine Ostracoda from Mohgaonkala (Mohgaon-Kalan), Chhindwara District, Madhya Pradesh state, Central India. J Micropaleontol 21:105–114

Whatley RC, Bajpai S, Whittaker JE (2002c) New record and new species of Upper Cretaceous ostracoda for Indian intertrappean deposits. Boll Soc Paleontol Ital 41(2–3):163–173

Whatley RC, Bajpai S, Whittaker J (2003a) Freshwater ostracoda from the Upper Cretaceous intertrappean beds at Mamoni (Kota district), southeastern Rajasthan, India. Rev Esp de Micropaleontol 35(1):75–86

Whatley RC, Bajpai S, Whittaker JE (2003b) Indian intertrappean ostracoda in the collections of the Natural History Museum, London. Cretac Res 24(1):73–88

Whatley RC, Bajpai S, Whittaker JE (2003c) The identity of the non-marine ostracod *Cypris subglobosa* Sowerby from the intertrappean deposits of Peninsular India. Palaeontology 46(6):1281–1296

Whatley RC, Khosla SC, Rathore AS (2012) *Periosocypris megistus* n. gen. and n. sp.: a new gigantic non-marine cyprid ostracod from the Maastrichtian Lameta Formation of India. J Palaeontol Soc Ind 57(2):113–117

White MP (1928) Some index foraminifera of the Tampico embayment area of Mexico. J Paleontol 2(4):280–317

Whittaker RH (1969) New concepts of kingdoms of organisms. Science 163:150–159

Woodward AS (1908) On some fish remains from the Lameta Beds at Dongargaon, Central Province. Mem Geol Surv India, Palaeontol Indica NS 3:1–6

Yang JL, Wang QF, Lu HN (2005) Discovery of Cretaceous and Paleocene charophyte floras from well Hu-2 in the southern edge of Junggar Basin. Acta Micropaleontol Sin 22(3):251–268

Zenker W (1854) Monographie der Ostracoden. Arch Naturg 20(1): 1–87

Chapter 5
Palaeoecological, Palaeoenvironmental and Age Implications of the Cretaceous-Palaeogene Microbiota-Bearing Deccan Intertrappean beds of the Chhindwara District, Madhya Pradesh, India

5.1 Introduction

Fossils and lithologies (rock-types) serve as good indicators for making both palaeoecological and palaeoenvironmental reconstructions (Raup and Stanley 1985; Dodd and Stanton 1990; Bottjer 2016; Verma 2018). The physical and biological aspects of sedimentary rocks are very useful to record palaeoecological and palaeoenvironmental observations. For example, the presence of fossils and their associations in sedimentary rocks serve as tools to identify different types of palaeoenvironments such as non-marine, marine or terrestrial. The presence of brachiopods, corals, echinoderms, foraminiferans, trilobites and cephalopods in the sedimentary rocks serves as excellent indicators of marine palaeoenvironments. However, some forms of bivalves and gastropods are inhabitants of freshwater environments, so their presence in strata is only useful to infer palaeoenvironment, when their association with other organisms is to be established (Weller 1960; Jones 2011; Verma 2018). The fossils of plants, fungi, ostracods, insects, amphibians, reptiles, birds and mammals are good indicators of non-marine environments (Jones 2011). The nature of the shell chemistry of charophytes, algae and ostracods is used to infer ancient freshwater environments. The benthic foraminiferans are useful to know temperature, depth and nature of substratum. The planktic foraminiferans are useful to find marine incursions into the freshwater sedimentary successions. Charophytes are also useful in making palaeoecological and palaeoenvironmental observations. Their presence in sediments indicates freshwater and brackish water or supra-tidal environments (Soulié-Märsche 1994).

Cretaceous sedimentary rocks are widely distributed in India (Vaidyanadhan and Ramakrishnan 2010). Their geological and palaeontological research began in the early 1850s (Oldham 1856). Following that, rich fossil assemblages were reported from many Cretaceous basins, including the Cauvery, Gondwana, Narmada,

A. Khosla et al., *Microbiota from the Late Cretaceous-Early Palaeocene Boundary Transition in the Deccan Intertrappean Beds of Central India*, Topics in Geobiology 54, https://doi.org/10.1007/978-3-031-28855-5_5

Saurashtra, Kachhch, Barmer and Jaisalmer, as well as sedimentary sequences associated with the Deccan volcanics in peninsular India (e.g. Krishnan 1949; Wadia 1966; Stoliczka 1873; Kumar 1996; Naqvi 2005; Khosla and Verma 2015; Verma 2015; Khosla et al. 2016; Verma et al. 2016; Khosla and Lucas 2020). Fossils were also reported from the Cretaceous deposits of the Himalayan region (Krishnan 1949; Wadia 1966; Stoliczka 1873; Kumar 1996; Naqvi 2005). In the last few decades, the Cretaceous-Palaeogene boundary has been demarcated at few places in India such as: (i) the Um Sohryngkew section of Meghalaya, Northeast India and (ii) within the Deccan intertrappean beds, peninsular India (Bhandari et al. 1994; Keller et al. 2009a, b; Khosla 2015). In fact, palaeoecological and palaeoenvironmental investigations are important to document the effects of the mass extinctions that took place across various boundaries of the geological time scale, notably the Cretaceous-Palaeogene boundary mass extinction (Khosla and Lucas 2021).

It is noted that palaeoecological and palaeoenvironmental investigations based on microfossils on the Upper Cretaceous-Lower Palaeocene rocks in India have received attention in the last few decades (e.g. Khosla 2015 and references therein). The Deccan intertrappean beds are exposed in numerous places along the north-western, north-eastern, eastern, south-eastern and southern margins of the traps in peninsular India. In addition to the record of the Cretaceous-Palaeogene boundary, these trap sediments have yielded a rich biotic assemblage comprising invertebrates, vertebrates, plants and microfossils (Khosla and Sahni 2003; Verma et al. 2012; Khosla and Verma 2015; Verma and Khosla 2018, 2019). It is also noted that palaeoecologically and palaeoenvironmentally, only a few intertrappean sites have been studied so far (e.g. Khosla and Sahni 2003; Khosla 2015; Khosla and Verma 2015; Kania et al. 2022; Khosla et al. 2022). Based on microfossils such as charophytes, ostracods, foraminiferans and fishes, this study presents inferences regarding the palaeoecology and palaeoenvironment of the intertrappean beds exposed in the Chhindwara region of Madhya Pradesh, Central India.

5.2 Charophytes

Charophytes are fresh and brackish water green algae that have a fossil record dating back to the Silurian Period. Most charophytes calcified to some extent, and the group is best known for fossils of the calcified female fructification, known as the gyrogonite. Charophytes are a group of land plant ancestors that are restricted to water bodies such as lakes (temporary or permanent) and swamps. They grow submerged in shallow, quiet and smoothly flowing fresh and brackish water environments (Mccourt et al. 1996; Sanjuan and Alqudah 2018; Khosla et al. 2022). Their calcified fructifications (gyrogonites) typically fossilise and have been found in freshwater deposits ranging from Silurian to Recent (Feist-Castel 1975, 1977; Feist and Grambast-Fessard 1991; Conkin and Conkin 1992; Riveline et al. 1996; Soulié-Märsche 1999; Zhamangara and Lucas 1999; Feist et al. 2005a, b; Lucas 2018; Sanjuan and Alqudah 2018).

The living charophytes are characteristically found in freshwater or non-marine environments, with the exception of a few that can endure some salinity but are not suited to growing in an open marine environment (Villalba-Breva and Martín-Closas 2011, 2013; Schubert et al. 2015; Vicente et al. 2016a, b, 2019). They grow in almost all inland waters, specifically slow-flowing rivers and streams, lake and pond bottoms (both impermanent and permanent), and can survive in sand and silt substrates as well as marl-rich lacustrine environments (Barbosa et al. 2021). They proliferate as "charophyte meadows" in alkaline aquatic environment with clear, nutrient-rich water at depths ranging from 1 to 10 m (Garcia 1994; Khosla et al. 2022).

Over the last four decades, many researchers have considered the study of Late Cretaceous charophytes to be an important biostratigraphic, palaeoecologic and palaeoenvironmental tool for delineating the Cretaceous-Palaeogene boundary and assessing associated environmental impacts in North and South America (e.g. Peck and Forester 1979; Jaillard et al. 1993), Europe (Feist 1979; Masriera and Ullastre 1988; Galbrun et al. 1993; Feist et al. 2005a, b; Villalba-Breva and Martín-Closas 2013; Villalba-Breva et al. 2012; Vicente et al. 2016a, b) and Asia (Karczewska and Ziembińska-Tworzydlo 1983; Bhatia and Rana 1984; Wang et al. 1985; Van Itterbeeck et al. 2005; Khosla 2014; Li et al. 2019, 2020a, b; Tian et al. 2021; Kania et al. 2022; Khosla et al. 2022). In comparison to global records, charophytes from Upper Cretaceous deposits in India have been largely ignored, with only a few charophyte records documented from the infra- and intertrappean horizons of peninsular India to date (Bhatia and Mannikeri 1976; Bhatia and Rana 1984; Bhatia et al. 1990a, b; Srinivasan et al. 1992, 1994; Khosla 2014; Kapur et al. 2019). As a consequence, no attempt has been made to evaluate their significance for demarcating the Cretaceous-Palaeogene boundary and reconstructing the palaeoecology, palaeoenvironment and palaeobiogeography of the northward drifting Indian plate before to the India–Asian docking (Khosla et al. 2022). This study of Indian charophyte records will help us better understand the palaeoecology and palaeoenvironments of the Late Cretaceous-Early Palaeocene charophytic flora.

The Jhilmili and Ghat Parasia intertrappean beds of Chhindwara District of Madhya Pradesh contain prolific charophyte fossil assemblages, which are represented by *Platychara perlata* (Peck and Reker 1947), *Platychara raoi* (Bhatia and Mannikeri 1976), *Platychara sahnii* (Bhatia and Mannikeri 1976), *Platychara compressa* (Peck and Reker 1948), *Platychara* sp., *Platychara closasi* sp. nov., *Peckichara* cf. *varians* Grambast (1957), *Nemegtichara* cf. *grambasti* (Bhatia et al. 1990b),?*Grambastichara* sp., *Microchara shivarudrappai* sp. nov. and *Chara chhindwaraensis* sp. nov. (Table 5.1). These charophytes are found in a 60 cm thick clayey and nodular limestone horizon that is underlain by claystone and overlain by laminated claystone of the Unit 3 of the Jhilmili section and hard clayey limestone horizon of the Ghat Parasia intertrappeans (Table 5.1; Fig. 5.1). The Jhilmili and Ghat Parasia charophytes have not received any considerable attention from micropalaeontologists to date. Furthermore, these sections have been poorly studied with regard to the sedimentology, palaeoecology and taphonomy of the charophyte-bearing intertrappean beds. For the first time, the present study deals with the

Table 5.1 Palaeoecological and palaeoenvironmental implications of charophyte taxa

Division	Taxon	Section	Palaeoecology and palaeoenvironments	References
Charophyta	*Platychara perlata* (Peck and Reker 1947)	Jhilmili	Non-marine water, permanent lakes and temporary ponds near floodplain, tolerates some salinity and low energy environment	Khosla (2014) and Vicente et al. (2019)
	Platychara raoi (Bhatia and Mannikeri 1976)	Jhilmili	Non-marine water, permanent lakes and temporary ponds near floodplain, tolerates some salinity and low energy environment	Khosla (2014, 2015)
	Platychara sahnii (Bhatia and Mannikeri 1976)	Jhilmili	Non-marine water, permanent lakes and temporary ponds near floodplain, tolerates some salinity and low energy environment	Khosla (2014)
	Platychara compressa (Peck and Reker 1948)	Jhilmili	Non-marine water, permanent lakes and temporary ponds near floodplain, tolerates some salinity and low energy environment	Khosla (2014) and Vicente et al. (2019)
	Platychara sp.	Jhilmili	Non-marine water, permanent lakes and temporary ponds near floodplain, tolerates some salinity and low energy environment	Khosla (2014), Vicente et al. (2019) and this study
	Platychara closasi sp. nov.	Ghat Parasia	Non-marine water, permanent lakes and temporary ponds near floodplain, tolerates some salinity and low energy environment	Khosla (2014), Vicente et al. (2019) and this study
	Peckichara cf. *varians* (Grambast 1957)	Jhilmili	Non-marine water, shallow freshwater lakes, brackish water and low energy environment	Khosla (2014, 2015) and Li et al. (2016)
	Nemegtichara cf. *grambasti* (Bhatia et al. 1990b)	Jhilmili	Non-marine water and low energy environment	Khosla (2014, 2015)
	?*Grambastichara* sp.	Jhilmili	Non-marine water and low energy environment	Bhatia et al. (1990b) and this study
	Microchara shivarudrappai sp. nov.	Jhilmili	Non-marine water, temporary ponds in floodplain, brackish water and low energy environment	This study
	Chara chhindwaraensis sp. nov.	Jhilmili	Non-marine water, tolerates salinity and low energy environment	Khosla (2014, 2015) and this study

Modified after Khosla (2015), Kania et al. (2022) and Khosla et al. (2022)

Fig. 5.1 Field photographs showing panoramic views of (**A**) the Jhilmili intertrappean beds and (**B**) the Ghat Parasia intertrappean beds, Chhindwara District, Madhya Pradesh, Central India (after Kania et al. 2022). Dotted line with arrow shows position of Cretaceous-Palaeogene (K-Pg) boundary transition where charophytes were collected from the Jhilmili intertrappean beds. (Reproduced from Kania et al. 2022 with permission from the Editor of the Himalayan Geology)

charophyte flora from Upper Cretaceous-Lower Danian freshwater/lacustrine to brackish-marine deposits of the Jhilmili and Ghat Parasia areas of Central India and presents new insights on palaeoecology and palaeoenvironments.

Sedimentological, taphonomical and palaeoecological studies were carried out in the Jhilmili intertrappean beds in order to establish the palaeoenvironment of the charophyte assemblages and to investigate the degree to which they were affected by palaeoecological changes. Sedimentary studies carried out by Keller et al. (2009a, b), Sharma and Khosla (2009) and Khosla (2015) have provided a basis for the present study. Lithologically, the Jhilmili intertrappean beds are mainly composed of alternating purple siltstone, red clayey siltstone, laminated clayey limestone and packstones, representing fluvio-lacustrine and brackish water conditions. Six lithological units have been recorded and are described below, from base to top. The base of the section is marked by the 3.5 m thick Deccan traps. Above this, there is a 1.5 m thick purple siltstone (palaeosol). The lower half of the palaeosol in Unit 2 contains rotated quartz grains, slickenslide structures, manganese streaks and carbonate nodules in large numbers, which have been considered the major features of the palaeosol (Keller et al. 2009a, b; Khosla 2015; Kania et al. 2022; Khosla et al. 2022).

Keller et al. (2009a, b) and Khosla (2015) considered that the purple palaeosols (JH 14) might be the product of the weathering of basalts. The upper part of the palaeosol (ca. 1.5 m thick) is mainly composed of red clayey siltstone containing calcareous concretions, rounded quartz grains, carbonate clasts and nodules indicative of significant water level fluctuations and indicating somewhat humid conditions (Keller et al. 2009a, b). The reddish colour of the palaeosol in Unit 2 is connected to the presence of iron oxides and hydroxides (Keller et al. 2009a, b). Humid and dry conditions have been assigned to the topmost 30 cm of Unit 2 due to the presence of oxy-vertisols and slickenside structures (Keller et al. 2009a, b).

The uppermost parts of the palaeosol are intercalated with coarse-grained sandstones, which are devoid of any horizontal bedding or planar cross-bedding. The coarse-grained sediments have been identified as tempestites. The presence of the clay mineral smectite in this unit also points to semi-arid to sub-humid climatic conditions (Keller et al. 2009a, b). The third unit (JH16, JH19, JH22 and JH23) is 60 cm thick and is mainly formed of laminated clayey and nodular limestone containing fresh water charophytes (JH 17 and JH 19), ostracods and brackish marine planktic foraminiferans (JH16, JH18, JH20–22, JH24 and JH25). Small- to medium-sized gyrogonites of charophytes are very abundant in white, light yellow and pink-brownish coloured clayey limestone (JH17 and JH19), including *Platychara perlata*, *Platychara raoi*, *Platychara sahnii*, *Platychara compressa*, *Platychara* sp., *Peckichara* cf. *varians*, *Nemegtichara* cf. *grambasti*, ?*Grambastichara* sp., *Microchara shivarudrappai* sp. nov. and *Chara chhindwaraensis* sp. nov. Good preservation of these fossils is considered evidence of autochthony in shallow fluvio-lacustrine facies with large amounts of terrigenous input (Kania et al. 2022; Khosla et al. 2022).

Charophytes were recovered from the clayey limestone facies of Unit 3 in Jhilmili. These facies have produced a high concentration of gyrogonites, demonstrating that they were deposited in situ in a supratidal environment (Soulié-Märsche 1989). The occurrence of limestone bands clearly indicates a drier climate with the development of water bodies (ponds and lakes) with ostracods and algae (Khosla et al. 2022). The charophytes are found in association with ostracods and planktic foraminiferans, which were probably transported by a short-lived marine incursion. The association of charophytes with other biota further suggests an environment with fluctuating salinity, with both freshwater and brackish influences (Vicente et al. 2016a, b). A similar situation occurs in two charophyte-bearing localities in the Upper Cretaceous of Spain, the Lower Red Unit (Maastrichtian) of the Coll de

Nargó (Vallcebre Basin, Eastern Pyrenees) and the El Barrancdela Possa sections (Vicente et al. 2015, 2016a, b). Vicente et al. (2015, 2016a, b) further suggested that small- to medium-sized charophytes, especially *Microchara,* developed in shallow, ephemeral floodplain ponds/lakes impacted repeatedly by fine terrigenous inputs with variable water tables and subject to drying events. Indeed, similar extant charophytes suggest that species with gyrogonites of small size in floodplain ponds may be adapted to withstand very shallow and precarious conditions (Kania et al. 2022; Khosla et al. 2022).

Most charophytes prefer a non-marine environment, but a few can tolerate some salinity, such as *Platychara, Peckichara* and *Chara* (Villalba-Breva and Martín-Closas 2011, 2013; Schubert et al. 2015; Vicente et al. 2016a, b, 2019). The genera *Microchara* and *Platychara* preferred to live in permanent lakes on the floodplain of a fluvial system. The genus *Microchara* across the Cretaceous-Palaeocene transition in the Jiaolai Basin (eastern China) was inferred to have favoured lakes or temporary ponds, especially in floodplain deposits (Li et al. 2020a; Tian et al. 2021; Kania et al. 2022; Khosla et al. 2022).

Platychara is considered an indicator of local palustrine and lacustrine conditions in a floodplain environment (Khosla 2014). *Peckichara* has also been considered to have favoured shallow freshwater lakes (Li et al. 2016). Several workers (Feist-Castel 1975; Soulié-Märsche 1989; Vicente et al. 2019) also considered that *Microchara* and *Platychara* of Maastrichtian to Danian age in the south Pyrenean basin (Spain) preferred to dwell in temporary ponds, especially in floodplain deposits. Apart from a freshwater habitat, Vicente et al. (2016a, b, 2019) suggested that both *Microchara* and *Peckichara* could also be indicative of brackish water assemblages. As a whole, *Platychara*, *Peckichara*, *Microchara*, *Chara* and *Nemegtichara* are charophytes that inhabited non-marine (freshwater), low energy, permanent lakes/lacustrine to floodplain channels (Khosla 2014, 2015).

Peckichara and *Nemegtichara* have also been considered to have favoured shallow freshwater lakes (Bhatia et al. 1990b; Li et al. 2016). Much better documented is the presence of *Peckichara* in Late Cretaceous freshwater enduring lakes of the Pyrenees (Villalba-Breva et al. 2012). *Peckichara* typically favoured a freshwater environment (Villaba-Breva et al. 2012). *Chara* has varied ecological preferences and lived in an assortment of environments, for instance freshwater and brackish water bodies, coastal ponds, marshes and lakes (Li et al. 2020a, b; Saber et al. 2021; Tian et al. 2021).

Nemegtichara cf. *grambasti* and *Chara chhindwaraensis* sp. nov. gyrogonites indicate shallow fluvio-lacustrine facies with high terrigenous input. They are found in sediments dense with planktic foraminiferans and ostracods, which were most likely transported by a short-lived marine incursion. Vicente et al. (2015) reported a similar situation in the Late Cretaceous of the Pyreenes (Catalonia), which was interpreted as indicating postmortem transport of planktic foraminiferans by flotation during high tides in a mudflat by comparison to similar situations in extant environments. The association of charophytes with marine microfossils suggests an environment with fluctuating salinity, with influences from both fresh and brackish water (Vicente et al. 2016a, b; Khosla et al. 2022). A similar situation exists in two

Upper Cretaceous charophyte-bearing locations in Spain namely, the Grey Unit at El Barranc de la Posa (Vallcebre Basin, Eastern Pyrenees) and Lower Red Unit (Maastrichtian) at Coll de Nargó (Vallcebre Basin, Eastern Pyrenees) (Vicente et al. 2015; Vicente et al. 2016a, b). Certainly, experiments with extant charophytes indicate that species producing small-sized gyrogonites may be adapted to withstand extremely shallow and perilous conditions (Sanjuan et al. 2015, 2020; Khosla et al. 2022).

As a result, the discovery of non-marine aquatic charophytes in Unit 3 of the Jhilmili intertrappean section of clayey limestone is likely evidence that a freshwater or lake environment temporarily prevailed over a shallow marine environment at low tide intervals.

The hard clayey limestone of the Ghat Parasia intertrappean beds is where the newly discovered species of *Platychara, closasi, was discovered. This section also produced partial remnants of the gar Lepisosteus indicus (Woodward* 1908) and ostracods *Frambocythere tumiensis anjarensis (Bhandari and Colin* 1999) and *Periosocypris megistus. A non-marine aquatic charophyte is Platychara (Khosla* 2015). The gar, *Lepisosteus indicus, is also thought to be a species of freshwater environment ranging from fluvial to lacustrine. The ostracod taxa, Frambocythere tumiensis anjarensis and Periosocypris megistus, have been considered as inhabitants of freshwater lakes/ponds in a low energy environment. (Grande* 2010; Whatley et al. 2012). Therefore, the presence of the aforementioned biotic components suggests that the limestone was deposited in a fluvial to lacustrine environment. (Kania et al. 2022). *Igdabatis indicus, a Late Cretaceous myliobat that is thought to have lived in a brackish to shallow marine coastal habitat, has left behind dental remains in the underlying greenish limestone unit (Prasad and Cappetta* 1993; Verma et al. 2017). The Ghat Parasia section contains a marine biotic signature, hence it is suggested that future palaeontological research on this signature should be more concentrated concentrated.

5.3 Ostracods

Ostracods are a group of tiny invertebrates belonging to the phylum Arthropoda. They have a long geological history, ranging from the Ordovician to Recent. Their shells are made of two valves or carapaces, which are usually from 0.1 mm to 32 mm in size (Mahalakshmi and Hussain 2015). They live at various water depths and altitudes in aquatic environments and are found in a diverse range of aquatic habitats, such as oceans, seas, lagoons, lakes and other fresh water bodies (Jones 2011). Ostracods are ecologically very crucial in their ecosystems because they regulate many ecological factors such as nutrient cycles, water clarity, food sources and also serve an important bio-indicator of the ecosystem health. Thus, their fossil remains are important components in palaeoecological studies because of abundance of their calcareous valves of various species in the deposits of marine, brackish and freshwater environments, and, also, they act as an ecological indicator of the aquatic environment, tsunamis or severe storms. They also serve as an effective tool for the bio-zonation of marine and non-marine environments because of their diversity, small size, mode of preservation and widespread distribution (Dodd and Stanton 1990; Jones 2011; Whatley 2012; Bottjer 2016; Khosla et al. 2013; Khosla 2015; Khosla and Verma 2015).

The intertrappean beds exposed at Jhilmili and Ghat Parasia, and within the Shiraj and Government wells, Chhindwara District, Madhya Pradesh, have yielded a rich assemblage of ostracods (Table 5.2). This assemblage is represented by *Buntonia whittakerensis* sp. nov., *Neocyprideis raoi* (Jain 1978), *Limnocythere deccanensis* (Khosla et al. 2005), *Limnocythere martensi* sp. nov., *Frambocythere tumiensis anjarensis* (Bhandari and Colin 1999), *Frambocythere tumiensis lakshmiae* (Whatley and Bajpai 2000a), *Gomphocythere strangulata* (Jones 1860), *Gomphocythere paucisulcatus* (Whatley et al., 2002b), *Gomphocythere dasyderma* (Whatley et al. 2002a), *Gomphocythere* sp. 1, *Paracypretta subglobosa* (Sowerby 1840), *Paracypretta jonesi* (Bhatia and Rana 1984), *Paracypretta verruculosa* (Whatley et al. 2002a), *Strandesia jhilmiliensis* (Khosla et al. 2011a), *Stenocypris cylindrica* (Sowerby in Malcolmson 1840), *Periosocypris megistus* (Whatley et al. 2012), *Zonocypris spirula* (Whatley and Bajpai 2000a), *Zonocypris viriensis* (Khosla and Nagori 2005), *Zonocypris labyrinthicos* (Whatley et al. 2002b), *Zonocypris gujaratensis* (Bhandari and Colin 1999), *Zonocypris penchi* sp. nov., *Cypridopsis astralos* (Whatley et al. 2002a), *Cypridopsis hyperectyphos* (Whatley and Bajpai 2000a), *Cypridopsis elachistos* (Whatley et al. 2002b), *Candona* sp., *Eucypris pelasgicos* (Whatley and Bajpai 2000a), *Eucypris* sp. 1,? *Eucypris verruculosa* (Whatley et al. 2002a), *Cyclocypris amphibolos* (Whatley et al. 2002a), *Cypria cyrtonidion* (Whatley and Bajpai 2000a), *Talicypridea pavnaensis* (Khosla et al. 2005), *Cyprois rostellum* (Whatley and Bajpai 2000a), *Cyprois* sp. and *Darwinula* sp.

Buntonia whittakerensis sp. nov. is a marine/brackish water form, which reinforces the interpretation of the existence of a shallow marine environment at Jhilmili. The genus *Buntonia* is widely documented from the Cenozoic of Africa, and it has been inferred that it lived in deeper and warm offshore waters at various depths, ranging from 25 to 41 m, with different temperature ranges, from 16.0 to 30.6 °C and 6.3 to 11.0 °C during the summer and winter seasons, respectively (Keen 2004; Khosla et al. 2011a, b; Ozawa and Tanaka 2019).

Neocyprideis raoi (Jain 1978) is another brackish water ostracod taxon indicating shallow marine environmental conditions during the deposition of the clayey limestone of unit 3 at the Jhilmili section (Sharma and Khosla 2009; Khosla et al. 2011a; Khosla 2015). It was earlier discovered in Rajahmundry's intertrappean deposits near the southeast coast, Andhra Pradesh (Jain 1978). It is inferred that *Neocyprideis raoi* lived with other fresh water ostracods in the lake at times of low mesohaline salinities (Sharma and Khosla 2009; Khosla et al. 2011a). This genus also lived in near shore environments (Khosla 2015). Therefore, it has been inferred that only the topmost portion of unit 3 was deposited in marine-brackish settings, and the brackish ostracods and planktic foraminiferans in middle unit 3 were the

Table 5.2 Palaeoecological and palaeoenvironmental preferences of ostracod taxa

Class	Taxon	Section	Palaeoecology and palaeoenvironments	References
Ostracoda	*Buntonia whittakerensis* sp. nov.	Jhilmili	Brackish-marine water with shallow water depth	Khosla et al. (2011a), Khosla (2015) and this study
	Neocyprideis raoi (Jain 1978)	Jhilmili	Brackish-marine water, near shore water	Sharma and Khosla (2009), Khosla (2015) and this study
	Limnocythere deccanensis (Khosla et al. 2005)	Jhilmili and Ghat Parasia	Non-marine, low energy and temporary pools	Sharma and Khosla (2009), Khosla (2015) and this study
	Limnocythere martensi sp. nov.	Jhilmili	Non-marine, low energy and temporary pools	This study
	Frambocythere tumiensis anjarensis (Bhandari and Colin 1999)	Jhilmili and Ghat Parasia	Non-marine, low energy aquatic, poor swimmer, epibenthonic walker, crawler	Sharma and Khosla (2009), Khosla et al. (2011b), Khosla (2015) and this study
	Frambocythere tumiensis lakshmiae (Whatley and Bajpai 2000a)	Shiraj (=Shriwas) and Government wells	Non-marine, low energy aquatic, poor swimmer, epibenthonic walker, crawler	Sharma and Khosla (2009), Khosla et al. (2011a, b), Khosla (2015) and this study
	Gomphocythere strangulata Jones (1860)	Shiraj (=Shriwas) well, Government well and Jhilmili	Non-marine, low energy aquatic, poor swimmer, epibenthonic walker, crawler	Sharma and Khosla (2009), Khosla (2015) and this study
	Gomphocythere paucisulcatus (Whatley et al. 2002b)	Shiraj (=Shriwas) well, Government well, Jhilmili and Ghat Parasia	Non-marine, low energy aquatic, poor swimmer, epibenthonic walker, crawler	Khosla et al. (2011a, b), Khosla (2015) and this study
	Gomphocythere dasyderma (Whatley et al. 2002a)	Jhilmili	Non-marine, low energy aquatic, poor swimmer, epibenthonic walker, crawler	Khosla (2015) and this study
	Gomphocythere sp. 1	Government well	Non-marine, low energy aquatic, poor swimmer, epibenthonic walker, crawler	Khosla et al. (2011a), Khosla (2015) and this study
	Paracypretta subglobosa (Sowerby 1840)	Jhilmili	Non-marine, low energy aquatic, active swimmer	Sharma and Khosla (2009), Khosla (2015) and this study

	Paracypretta jonesi (Bhatia and Rana 1984)	Jhilmili	Non-marine, low energy aquatic, active swimmer	Khosla et al. (2011a), Khosla (2015) and this study
	Paracypretta verruculosa (Whatley et al. 2002a)	Jhilmili	Non-marine, low energy aquatic, active swimmer	Khosla et al. (2011a, b), Khosla (2015) and this study
	Strandesia jhilmiliensis (Khosla et al. 2011a)	Jhilmili	Non-marine, low energy aquatic, active swimmer, lived in shallow freshwater environment	Khosla et al. (2011a) and this study
	Stenocypris cylindrica (Sowerby in Malcolmson 1840)	Jhilmili	Non-marine, low energy aquatic, active swimmer, fresh water environment	Khosla et al. (2011a, b), Khosla (2015) and this study
	Periosocypris megistus Whatley et al. (2012)	Jhilmili and Ghat Parasia	Non-marine, low energy aquatic, active swimmer, lived in temporary and permanent water bodies	Khosla et al. (2011a, b), Khosla (2015) and this study
	Zonocypris spirula (Whatley and Bajpai 2000a)	Jhilmili	Non-marine, low energy aquatic, active swimmer, well ornamented and indicates sluggish flow conditions	Sharma and Khosla (2009), Khosla et al. (2011a), Khosla (2015) and this study
	Zonocypris viriensis (Khosla and Nagori 2005)	Jhilmili	Non-marine, low energy aquatic, active swimmer, well ornamented and indicates sluggish flow conditions	Khosla et al. (2011a), Khosla (2015) and this study
	Zonocypris labyrinthicos Whatley et al. (2002b)	Shiraj (=Shriwas) well	Non-marine, low energy aquatic, active swimmer, well ornamented and indicates sluggish flow conditions	Khosla (2015) and this study
	Zonocypris gujaratensis (Bhandari and Colin 1999)	Shiraj (=Shriwas) well	Non-marine, low energy aquatic, active swimmer, well ornamented and shows sluggish conditions	Khosla et al. (2011a b), Khosla (2015) and this study
	Zonocypris penchi sp. nov.	Jhilmili	Non-marine, low energy aquatic, active swimmer, well ornamented and indicates sluggish flow conditions	This study
	Cypridopsis astralos (Whatley et al. 2002a)	Jhilmili	Non-marine, low energy water, active swimmer, lived in permanent ponds and lakes	Sharma and Khosla (2009), Khosla et al. (2011a, b), Khosla (2015) and this study
	Cypridopsis hyperectyphos (Whatley and Bajpai 2000a)	Jhilmili	Non-marine, low energy water, active swimmer, lived in permanent ponds and lakes	Sharma and Khosla 2009, Khosla et al. (2011b), Khosla (2015) and this study

(continued)

Table 5.2 (continued)

Class	Taxon	Section	Palaeoecology and palaeoenvironments	References
	Cypridopsis elachistos (Whatley et al. 2002b)	Shiraj (=Shriwas) and Government wells	Non-marine, low energy water, active swimmer, lived in permanent ponds and lakes	Sharma and Khosla (2009), Khosla (2015) and this study
	Candona sp.	Ghat Parasia	Freshwater, temporary pools, waters with slightly increased salinity	Khosla (2015), Cohuo et al. (2017) and this study
	Eucypris pelasgicos (Whatley and Bajpai 2000a)	Jhilmili	Non-marine, low energy aquatic water, temporary pools, active swimmer	Khosla et al. (2011b), Khosla (2015) and this study
	Eucypris sp. 1	Ghat Parasia	Non-marine, low energy water, temporary pools, active swimmers	Khosla (2015) and this study
	?*Eucypris verruculosa* (Whatley et al. 2002a)	Government well	Non-marine, low energy water, temporary pools, active swimmers	Khosla (2015) and this study
	Cyclocypris amphibolos (Whatley et al. 2002a)	Jhilmili and Ghat Parasia	Non-marine, low energy water, cool water with high oxygen content	Khosla et al. (2011a, b), Khosla (2015) and this study
	Cypria cyrtonidion (Whatley and Bajpai 2000a)	Shiraj (=Shriwas) well, Government well and Jhilmili	Non-marine, temporary pools, moderate to active swimmer	Khosla et al. (2011a, b), Khosla (2015) and this study
	Talicypridea pavnaensis (Khosla et al. 2005)	Jhilmili	Non-marine, moderate to active swimmer, epifaunal, detritivore grazer	Khosla et al. (2011b), Khosla (2015) and this study
	Cyprois rostellum (Whatley and Bajpai 2000a)	Shiraj and Government wells	Non-marine, actively mobile, detritivore, grazer, nekton-benthonic, live at pool and lake margins	Van Itterbeeck et al. (2007) and El Hajj et al. (2021)
	Cyprois sp.	Ghat Parasia	Non-marine, actively mobile, detritivore, grazer, nekton-benthonic, lived at pool and lake margins	Van Itterbeeck et al. (2007), El Hajj et al. (2021)
	Darwinula sp.	Ghat Parasia	Non-marine, low energy water, active swimmer, lived in permanent water bodies	Sharma and Khosla (2009), Khosla et al. (2011a), Khosla (2015) and this study

Modified after Khosla (2015)

after effect of momentary irregular intrusions of marine waters such as storms, high tides and/or erosion of brackish marine sediments that were deposited earlier.

The two species *Limnocythere deccanensis* (Khosla et al. 2005) and *L. martensi* sp. nov. were found in the Jhilmili and Ghat Parasia intertrappean beds (Table 5.2). Their presence indicates a non-marine environment of low energy and temporary pools. The living representatives of *Limnocythere* are found in numerous water bodies such as rivers, streams, ponds, swamps and lakes (Wang et al. 2021). Palaeoecologically, the genus *Limnocythere* lived only in a non-marine lacustrine environment and was associated with non-swimmer cytheraceans. Mckenzie (1981) reported that some of the extant species of the genus are endobenthonic and spend a considerable part of the day within the sediments and some part of the day resting on it. A very few species of *Limnocythere* have been reported from temporary pools, but most of its representatives prefer to live in a permanent water body. *Limnocythere* has been abundantly reported from lagoonal and lacustrine environments of the Imichil region of Morocco (Charriere et al. 2009). In Australia, they live in saline and freshwater environments, where temperatures mostly lie near 15 °C, 20 °C and 25 °C (Yin et al. 1999). In Laguna Babicora, North America, they are found well adapted to alkaline conditions in a lake environment and live in oligo- to mesohaline conditions with temperature ranges from 5.6 °C to 21 °C (Feist et al. 2002). In Quaternary deposits of Patagonia and Argentina, this genus was found to live in a sodic water environment (rich in chlorine, sulphate and bicarbonate) with mesohaline conditions (Cusminsky et al. 2005; Ramon et al. 2012; Ramos et al. 2017). In Europe, it is found in large numbers and preferred mesohaline aquatic environments (Meisch 2000). It is inferred that *L. deccanensis* and *L. martensi* lived in shallow marine and brackish water environments with moderate temperature and low salinity. Their weak surface ornamentation indicates the existence of high-stress conditions, probably due to the variations in physico-chemical factors that would have been created by the volcanic flows of the Deccan Volcanic Province.

The two subspecies *Frambocythere tumiensis anjarensis* (Bhandari and Colin 1999) and *F. t. lakshmiae* (Whatley and Bajpai 2000a) have been recovered from the Jhilmili and Ghat Parasia sites and also from the Shiraj and Government wells (Table 5.2). Their heavy surface ornamentation favours that they were poor swimmers and epibenthonic walkers/crawlers (Khosla 2015). Thus, the recovered forms of *Frambocythere* are poor swimmers to non-swimmers and crawlers, which usually preferred non-marine and low energy aquatic water environments (Sharma and Khosla 2009; Khosla et al. 2011a; Khosla 2015). The presence of their heavily ornamented carapaces in the intertrappean beds of the Jhilmili and Ghat Parasia sites and the Shiraj (=Shriwas) and Government wells clearly indicates increasing alkalinitic conditions in the aquatic environment.

Four forms of the genus *Gomphocythere* – *G. strangulata* (Jones 1860), *G. paucisulcatus* (Whatley et al., 2002b), *G. dasyderma* (Whatley et al. 2002a) and *Gomphocythere* sp. 1 – were recovered from the intertrappean beds of the Shiraj (= Shriwas) and Government wells, and the Jhilmili and Ghat Parasia sections (Table 5.2). *Gomphocythere* commonly lives on sediments with a mixture of clay, sand and gravel in an aquatic environment having moderate oxygen content, slight alkalinity and salinity, and brackish to freshwater conditions (Sharma and Khosla 2009; Khosla et al. 2011a, b; Khosla, 2015; Külköylüoğlu et al. 2015). They are poor swimmers, epibenthonic walkers and crawlers and usually prefer to live in a low energy aquatic environment of permanent waters (Mckenzie 1971).

Three species of *Paracypretta* – *P. subglobosa* (Sowerby 1840), *P. jonesi* (Bhatia and Rana 1984) and *P. verruculosa* (Whatley et al. 2002a) – were found in the Jhilmili intertrappean beds (Table 5.2). These species are active swimmers and commonly prefer to live in non-marine and low energy aquatic water environments (Whatley and Bajpai, 2000b; Sharma and Khosla 2009; Khosla et al. 2011a, b; Khosla 2015). Sometimes, they tolerate slightly saline environmental conditions and are found in both temporary and semi-permanent water bodies (Martens et al. 1996; Khosla 2015). In Jhilmili, carapaces of *Paracypretta* are large in size, indicating that they had sufficient time and suitable environmental conditions to reach their mature stage.

A single species of *Strandesia*, *S. jhilmiliensis* (Khosla et al. 2011a), was recovered from Jhilmili. This species is an active swimmer and usually prefers to thrive in a non-marine, low energy aquatic environment of shallow permanent water bodies like ponds and lakes (Khosla et al. 2011a; Khosla 2015). Another single form, of the genus *Stenocypris*, *S. cylindrica* (Sowerby in Malcolmson 1840) was recorded from Jhilmili. This species is also an active swimmer and prefers to live in permanent non-marine water bodies like ponds and lakes. It is a common component of epigenal and low energy aquatic environments (Khosla et al. 2011a, b; Khosla 2015). The species *Periosocypris megistus* (Whatley et al. 2012), which has been recovered from the Jhilmili and Ghat Parasia intertrappean beds, is one of the megaspecies of ostracods. This species is an active swimmer and thrives in a non-marine, low energy aquatic environment, particularly in temporary and permanent water bodies like lakes or ponds (Khosla 2015).

Zonocypris is a diverse genus during the present study because five species of the genus – *Zonocypris spirula* (Whatley and Bajpai 2000a), *Zonocypris viriensis* (Khosla and Nagori 2005), *Zonocypris labyrinthicos* (Whatley et al. 2002b), *Zonocypris gujaratensis* (Bhandari and Colin 1999) and *Zonocypris penchi* sp. nov. – have been recovered from the Jhilmili and Shiraj (=Shriwas) well sections. The living forms of *Zonocypris* are detritivorous, commonly occur in shallow fluctuating lakes and are found to be adapted to slightly alkaline environmental conditions (Mazzini 2011). The above-mentioned species are active swimmers, and the presence of heavily ornamentation on carapaces indicates sluggish water conditions (Khosla 2015). These species normally live in non-marine, low energy aquatic environments (Sharma and Khosla 2009; Khosla et al. 2011a; Khosla 2015). Furthermore, the greater calcification of their heavily ornamented carapaces shows increasing alkalinity in the environment.

Three forms (*Cypridopsis astralos* Whatley et al. 2002a, *C. hyperectyphos* Whatley and Bajpai 2000a and *C. elachistos* Whatley et al. 2002b) of *Cypridopsis* were recovered from the Jhilmili, Shiraj well and Government well sections. These forms are active swimmers in a freshwater environment. It has been reported that *Cypridopsis* has the capacity to tolerate a wide range of salinity (Külköylüoğlu

2003) and prefers to thrive in non-marine and low energy environments of permanent water bodies. The species *Candona* sp. belongs to the family Candonidae. It may be noted that Candonidae is one of the diversified families of ostracods; its representatives show variable size, distinct shell morphology and lack swimming setae, indicating their benthic life (Cohuo et al. 2017). It has been inferred that *Candona* has wide ecological tolerances that permit it to colonise a variety of habitats across a wide spatial range (Cohuo et al. 2017). Thus, it is a non-marine organism that prefers to live in temporary pools having waters with slightly increased salinity.

Three species belonging to the genus *Eucypris* – *E. pelasgicos* (Whatley and Bajpai 2000a),?*E. verruculosa* (Whatley et al. 2002a) and *Eucypris* sp. 1 – were found in the Jhilmili, Ghat Parasia and Government well sections (Table 5.2). These are active swimmers and normally occur in a non-marine and low energy aquatic environment, preferably in temporary water bodies (Khosla 2015). Another species, *Cyclocypris amphibolos* (Whatley et al. 2002a), was recovered from both the Jhilmili and Ghat Parasia sections. The genus *Cyclocypris* is a non-marine active swimmer and prefers to live in low energy aquatic environments having cooler water with high oxygen content. It has the capability to tolerate a wide range of salinity. *Cypria cyrtonidion* (Whatley and Bajpai 2000a) is found widely distributed, as it was recovered from the Shiraj well, Government well and Jhilmili sections. It is a moderate to active swimmer and prefers to live in a non-marine low energy environment. *Talicypridea pavnaensis* Khosla et al. (2005) was only recovered from the Jhilmili section. It is a moderate to active swimmer and prefers to live in non-marine fluvio-lacustrine conditions.

Cyprois rostellum (Whatley and Bajpai 2000a) and *Cyprois* sp. were recovered from the Shiraj (=Shriwas), Government and Ghat Parasia sections (Table 5.2). *Cyprois* is an active and strong freshwater swimmer and non-marine ostracod that likes to live on the margins of pools and lakes with abundant aquatic vegetation and prefers to stay close to the substrate (El Hajj et al. 2021).

Only one form of the genus *Darwinula*, *Darwinula* sp., was recovered from the Ghat Parasia section. It is a non-marine to brackish water ostracod that prefers low-energy aquatic environments such as lakes, ponds and slow-moving streams. It lives in shallow water bodies up to a depth of 20 m, but it is more abundant up to a depth of 1 m. *Darwinula* species can be found in a wide range of habitats, including lakes, ponds, slow-moving streams/rivers, coastal lagoons, springs, wetlands, marshes, hot springs, rice fields and peat bogs (Khosla 2015; Yavuzatmaca and Külköylüolu 2019).

Overall, the Jhilmili ostracod assemblage is dominated by the presence of two species, *Zonocypris viriensis* and *Limnocythere deccanensis* (JH 17 and JH 18). The presence of ostracod genera in the Jhilmili assemblage: *Limnocythere, Zonocypris, Darwinula, Eucypris, Gomphocythere, Frambocythere, Paracandona, Paracypretta, Periosocypris, Strandesia, Stenocypris, Cypria, Centrocypris, Cyclocypris* and *Cypriodopsis*, indicates that lakes/rivers were present, and the biota thrived in freshwater, low-energy aquatic lacustrine environments, swamps and ponds (Khosla 2015; Khosla et al. 2022). Palaeoecologically, the ostracod assemblages show a mixture of active and poor swimmers. Active swimmers are represented by *Stenocypris, Zonocypris, Cypria, Cypridopsis, Paracypretta, Heterocypris, Cyclocypris, Strandesia* and *Eucypris*, whereas poor swimmers are represented by *Darwinula* and the three ostracod genera *Gomphocythere, Frambocythere* and *Limnocythere* belonging to the family Limnocytheridae (Sharma and Khosla 2009; Khosla et al. 2011a; Khosla 2015). Various workers (McKenzie 1971; Sharma and Khosla 2009; Khosla et al. 2011a, b; Khosla 2015; Kapur et al. 2019) concluded that two of the ostracod genera, *Gomphocythere* and *Frambocythere,* are clearly epibenthonic walkers. These authors also speculated that some *Limnocythere* species might be endobenthonic. Though a few types of *Limnocythere* have been found in temporary pools, the majority of Limnocytheridae require permanent water (Khosla 2015). The typical *Cypridopsis* rarely lived in streams and rivers, preferring to thrive in permanent waters such as freshwater swamps and ponds, whereas *Cyclocypris* also lived in small ponds and occasionally in large lakes (Khosla 2015).

The presence of genera such as *Limnocythere*, *Gomphocythere*, *Periosocypris*, *Candona*, *Eucypris*, *Cyclocypris*, *Cyprois* and *Darwinula* in the Ghat Parasia favours fluvio-lacustrine, low energy depositional environmental settings. The intertrappean sediments of the Shiraj and Government wells yielded a smaller number of ostracods compared to the Jhilmili and Ghat Parasia assemblages. The presence of the genera *Frambocythere*, *Gomphocythere*, *Zonocypris*, *Cypridopsis*, *Eucypris*, *Cypris* and *Cyprois* in the Shiraj (=Shriwas) and Government wells suggests fluvio-lacustrine low energy depositional environmental settings. It is to be noted that no brackish and shallow marine ostracod taxa were recovered from Ghat Parasia, the Shiraj (=Shriwas) well or the Government well during the present study.

5.4 Foraminiferans

Foraminifera are marine (planktic and benthic) microorganisms and are very sensitive to various environmental conditions such as light, salinity, temperature and depth, so their fossil remains serve as a very important tool to reconstruct palaeoecological and palaeoenvironmental conditions of the past (Olsson et al. 1999, 2006; Keller et al. 2009a, b; Sharma and Khosla 2009). The variation in the morphologies of foraminiferan tests and the evolution of their wall structure together indicate changes in various environmental conditions of the geological past in which they once had lived. In fact, different species of foraminiferans live in different realms of marine environments; therefore, they are used to find palaeoshorelines, infer depth of water columns, map palaeolocations of tropics, reconstruct global ocean temperature changes and determine global sea level fluctuations (Olsson et al. 1999, 2006).

Out of the investigated sections – Ghat Parasia, Jhilmili, Shiraj (=Shriwas) well and Government well – a rich assemblage of foraminiferans was recovered solely from the Jhilmili section (Table 5.3). This assemblage is represented by *Subbotina triloculinoides* (Plummer 1926), *Globanomalina compressa* (Plummer 1926),

Table 5.3 Palaeoecological and palaeoenvironmental inferences of foraminiferans

Order	Taxon	Section	Palaeoecology and palaeoenvironments	References
Foraminiferida	*Subbotina triloculinoides* (Plummer 1926)	Jhilmili	Brackish-marine waters of low to high latitudes across the globe	Olsson et al. (1999, 2006), Keller et al. (2009a, b), Sharma and Khosla (2009) and this study
	Globanomalina compressa (Plummer 1926)	Jhilmili	Brackish-marine waters of low to high latitudes across the globe	Olsson et al. (1999, 2006), Keller et al. (2009a, b), Khosla (2015) and this study
	Woodringina hornerstownensis (Olsson 1960)	Jhilmili	Brackish-marine waters of low to high latitudes across the globe, lived in intense and stressful conditions	Olsson et al. (1999, 2006), Keller et al. (2009a, b), Sharma and Khosla (2009), Khosla (2015) and this study
	Woodringina claytonensis (Loeblich and Tappan 1957b)	Jhilmili	Brackish-marine waters of low latitude open oceans and rarely found in high latitude open marine environments	Olsson et al. (1999, 2006), Keller et al. (2009a, b), Sharma and Khosla (2009), Khosla (2015) and this study
	Hedbergella holmdelensis (Olsson 1964)	Jhilmili	Brackish-marine waters of low to high latitudes across the globe	Olsson et al. (1999, 2006), Keller et al. (2009a, b), Khosla (2015) and this study
	Guembelitria cretacea (Cushman 1933)	Jhilmili	Brackish-marine water, disaster opportunist, persisted in intense and stressful conditions, mostly occurred in shallow continental shelf environment and rarely in the open oceans	Smith and Pessagno (1973), Olsson et al. (1999, 2006), Keller et al. (2009a, b), Sharma and Khosla (2009), Khosla (2015) and this study
	Parasubbotina pseudobulloides (Plummer 1926)	Jhilmili	Brackish-marine waters of low to high latitudes across the globe, lived in muddy water of shallow marine environments	Olsson et al. (1999, 2006), Keller et al. (2009a, b), Sharma and Khosla (2009), Khosla (2015) and this study
	Globigerinelloides aspera (Ehrenberg 1854)	Jhilmili	Brackish-marine waters of low to high latitude marine environments	Gawor-Biedowa (1992), Keller et al. (2009a), Sharma and Khosla (2009), Khosla (2015) and this study
	Globigerina (*Eoglobigerina*) *pentagona* (Morozova 1961)	Jhilmili	Brackish-marine waters of low to high latitude marine environments	Olsson et al. (1999, 2006), Sharma and Khosla (2009), Khosla (2015) and this study

Woodringina hornerstownensis (Olsson 1960), *W. claytonensis* (Loeblich and Tappan 1957b), *Hedbergella holmdelensis* (Olsson 1964), *Guembelitria cretacea* (Cushman 1933), *Parasubbotina pseudobulloides* (Plummer 1926), *Globigerinelloides aspera* (Ehrenberg, 1854), *Globigerina* (*Eoglobigerina*) *pentagona* (Morozova 1961) and by several unidentified tests (Table 5.3).

The species *Subbotina triloculinoides* Plummer (1926) is a planktic foraminiferan and interpreted to represent a deeper dwelling form of brackish-marine water of low to high latitudes (Leckie 1987; Olsson et al. 1999, 2006). It shows the presence of planktic foraminiferal P1b to P4 zones in Jhilmili and shallow marine palaeoenvironments (Keller et al. 2008, 2009a, b; Khosla 2015). Another species, *Globanomalina compressa* (Plummer 1926), has been reported from brackish-marine water of low to high latitudes across the globe, occupied the same ecological niche as *Subbotina triloculinoides* and is restricted to P1c to P3 planktic foraminiferal zones.

Two species of the genus *Woodringina* – *W. hornerstownensis* (Olsson 1960) and *W. claytonensis* (Loeblich and Tappan 1957b) – were recorded from Jhilmili. Palaeoecologically, *Woodringina* was found to thrive in brackish-marine water of low to high latitudes worldwide. However, *W. hornerstownensis* is a good indicator of intense and stressful palaeoecological conditions, and *W. claytonensis* rarely indicates high latitude open marine environments (Olsson et al. 1999 2006; Khosla 2015). Furthermore, *W. hornerstownensis* and *W. claytonensis* represent the presence of the Pa to P3b and basal Pa to Plb planktic foraminiferal zones, respectively, of the Palaeocene Epoch at Jhilmili.

Hedbergella holmdelensis (Olsson 1964) was a habitant of brackish-marine water of low to high latitudes across the globe. It has been documented that the globular tests of *Hedbergella* are considered to be a good indicator of marine transgression, and they may have disappeared during the process of regression (Leckie 1987). It also reflects marginal and epeiric sea palaeosettings in shallow marine warm water environments.

Guembelitria cretacea (Cushman 1933) has generally been recovered in the near-shore marine environments and only rarely was found in the open oceans (Smith and Pessagno 1973; Olsson et al. 1999). Palaeoecologically, it is regarded as a disaster opportunist and was found capable to survive in very stressful environmental conditions where other species failed to cope (Keller et al. 2009a, b; Sharma and Khosla 2009; Malarkodi et al. 2010; Khosla 2015). Broadly, the genus *Guembelitria* is associated with regions of high rainfall, low salinity, low oxygen content and warm humid together with rich organic matter supply from nearby land sources (Malarkodi et al. 2010). *Guembelitria cretacea* rapidly diversified in open oceans after the Cretaceous-Palaeogene boundary and became abundant at low and middle latitudes (Olsson et al. 1999, 2006).

Parasubbotina pseudobulloides (Plummer 1926) was recovered from sediments of planktic foraminiferal zones Pa to P3a of the Palaeocene Epoch in low to high paleolatitudes of a marine environment. In fact, some species of *Parasubbotina* have been reported from the upwelling environmental regions of the coastal zones (Pearson et al. 2006). *Parasubbotina pseudobulloides* also lived in muddy water of shallow marine environments (Khosla 2015). *Globigerinelloides aspera* (Ehrenberg 1854) is characteristic of an open marine environment and mainly occurs at the Cretaceous-Palaeogene boundary. Its remains are broadly documented from brackish-marine waters of low to high latitude marine environments across the globe (Olsson et al. 1999, 2006). It is considered as indicative of the presence of marginal and epeiric seas of shallow depth (less than 100 m). *Globigerina* (*Eoglobigerina*) *pentagona* (Morozova 1961) occurs in sediments of the Pa to P2 planktic foraminiferal zones deposited in low and high latitude marine environments across the globe. It also indicates the presence of a brackish-marine water environment. The presence of the above-mentioned species of planktic foraminiferans in unit 3 is clearly indicating the advancing and retreating of the sea during the Cretaceous-Palaeogene transition in peninsular India. Their presence also shows the existence of shallow marine and brackish water depositional environmental conditions for unit 3.

5.5 Fishes

Fishes live in almost all aquatic environments such as marine and non-marine (Dudgeon et al. 2006). Thus, their fossil remains provide invaluable information about ecology and environment of the geological past (Grande 2010). Fragmentary fish remains assigned to *Igdabatis indicus* (Prasad and Cappetta 1993), *Lepisosteus indicus* (Woodward 1908) and Osteoglossidae indeterminate were recovered from the Shiraj (= Shriwas) well, Government well and the Ghat Parasia and Jhilmili intertrappean beds (Table 5.4).

The species *Igdabatis indicus* (Prasad and Cappetta 1993) belongs to the family Myliobatidae, which are pelagic forms that like to reside in tropical and warm waters. In addition, members of this family also undertake seasonal migrations in temperate waters (Nelson 2006; Claeson et al. 2010; Verma 2015, 2018; Verma et al. 2016, 2017). The majority of *Igdabatis* occurrences are from Upper Cretaceous

Table 5.4 Palaeoecological and palaeoenvironmental inferences of fishes

Group	Taxon	Sections	Palaeoecology and palaeoenvironments	References
Fishes	*Igdabatis indicus* (Prasad and Cappetta 1993)	Ghat Parasia and Shiraj (=Shriwas) well	Non-marine, coastal, fluvio-lacustrine environments	Prasad and Cappetta (1993), Verma et al. (2017) and this study
	Lepisosteus indicus (Woodward 1908)	Jhilmili, Ghat Parasia, Shiraj (=Shriwas) well and Government well	Non-marine, fluvio-lacustrine environments	*Grande* (2010) and this study
	Osteoglossidae Genus et Species indeterminate	Ghat Parasia and Government well	Non-marine, fluvio-lacustrine environments	Kumar et al. (2005) and this study

sediments of shallow marine to brackish/estuarine origin. However, some representatives are known from a deltaic environment, where the mouth of the river opens into the continental shelf or an epeiric seaway (Wroblewski 2004).

The occurrence of a lepisosteid (*Lepisosteus indicus* Woodward 1908) and an osteoglossid (Osteoglossidae genus et species indeterminate) together points towards the presence of a freshwater environment. The living members of the lepisosteid and osteoglossid fishes are mainly known from freshwater, but a few can tolerate saline conditions to some extent (Grande 2010). Based on the presence of fish remains of myliobats, lepisosteids and osteoglossids, it has been inferred that a freshwater environment prevailed during the deposition of fish-bearing units of the Shiraj (=Shriwas) well, Government well, Ghat Parasia and Jhilmili intertrappean beds. Further, their occurrence also indicates the nearby presence of a near-shore or coastal-plain environment.

At Jhilmili, the fresh water microbiota is also associated with two brackish water ostracod genera (JH 17–21, *Neocyprideis* and *Buntonia*), planktic foraminiferans (*Eoglobigerina eobulloides*, *E. edita*, *Globigerina* (*Eoglobigerina*) *pentagona*, *Globanomalina compressa*, *Globigerinelloides aspera*, *Globoconusa daubjergensis*, *Guembelitria cretacea*, *Hedbergella* cf. *holmdelensis*, *Parvularugoglobigerina eugubina, Parasubbotina pseudobulloides*, *Praemurica taurica*, *Subbotina triloculinoides* and *Woodringina hornerstownensis*) and marine benthic calcareous algae, which are suggestive of the incursion of marine waters of a shallow marine to brackish environment (Keller et al. 2009a, b; Sharma and Khosla 2009; Khosla 2015; Khosla and Verma 2015; Kundal et al. 2018). Furthermore, apparently brief or momentary marine invasions into freshwater environments brought about shallow marine to brackish estuarine conditions and may have moved and deposited planktic foraminiferans, brackish water ostracods and calcareous algae (Khosla 2015; Khosla and Verma 2015; Kundal et al. 2018).

At Jhilmili, the marine water came for a short period of time and transported planktic foraminiferans (Keller et al. 2009a, b; Khosla 2015; Kania et al. 2022; Khosla et al. 2022) and enabled the growth of a benthic marine dasycladalean and codiacean calcareous algal assemblage (*Boueina* sp., *Neomeris*? sp., *Dissocladella* sp., *Terquemella* sp. and *Trinocladus* sp.). The occurrence of these benthic marine dasycladaleans, udoteaceans and sea urchin spines in unit 3 is indicative of a short-term incursion of marine waters (Kundal et al. 2018).

The presence of non-marine ostracods, charophytes and fishes, as well as the absence of foraminiferans, in the various intertrappean horizons of the Ghat Parasia, Shiraj (=Shriwas) well and Government well, indicate the dominance of non-marine aquatic conditions in the above-mentioned localities because ostracods and charophytes have been considered as inhabitants of freshwater lakes/ponds of a low energy environment (Whatley et al. 2012; Khosla 2015). In addition, the gar fishes, *Lepisosteus indicus*, are also considered species of freshwater environments ranging from fluvial to lacustrine (Grande 2010; Verma et al. 2017). As a result, the presence of the above biotic elements indicates that the fossil-bearing units were deposited in a fluvial to lacustrine environment. Further, the presence of the dental remains of Late Cretaceous myliobats, *Igdabatis indicus,* at the Ghat Parasia and Shiraj well localities indicate a brackish to shallow marine coastal environment (Prasad and Cappetta 1993; Verma et al. 2017). As a result, if more focused palaeontological investigations are conducted in the near future, it is suggested that marine biotic signatures may also be found in the Ghat Parasia section and at Shiraj (=Shriwas) well.

5.6 Age of the Microbiota-Bearing Intertrappean Beds

Previous biostratigraphic studies based on planktic foraminiferans, ostracods and marine benthic calcareous algae were carried out to deduce the age of the Jhilmili intertrappean beds (Sharma and Khosla 2009; Keller et al. 2009a, b; Khosla 2015). These studies indicate that this intertrappean locality is of Late Cretaceous–Early Palaeocene in age (Khosla, 2015). The charophyte flora (JH 17, 19), consisting of *Platychara perlata*, *P. raoi*, *P. sahnii*, *Peckichara* cf. *varians* (occurs also in European Danian deposits), *Nemegtichara* cf. *grambasti*, *Microchara* sp. and *Chara* sp. (Table 5.1, Fig. 5.2), indicates a Late Cretaceous–Early Palaeocene age. It is necessary to mention here that in Europe, *P. varians* is rarely found in Upper Cretaceous deposits, and only with dwarf morphotypes, and, on the contrary, it is very common in the Palaeocene. In China, it is apparently only Palaeocene in age. The results obtained here agree with the previous biostratigraphic studies based on chemostratigraphic and magnetostratigraphic analysis that suggests a Late Cretaceous–Early Palaeocene age for the Jhilmili intertrappean beds. Planktic foraminiferans were originally recorded in washed residues in the Jhilmili intertrappean sediments by ostracod workers (Sharma and Khosla 2009; Keller et al. 2009a, b; Khosla 2015). The majority of the specimens found are larger than 150 μm. Following sample processing for smaller species (36–63 μm, 63–100 μm, 100–150 μm), only extremely rare and poorly preserved specimens were found (Keller et al. 2009a, b). This revealed that during laboratory processing, the small, delicate, and dissolution-prone early Danian species were likely destroyed. The discovery of small early Danian species in thin sections backed up this theory (Keller et al. 2009a, b).

Very small planktic foraminiferans (*Globoconusa daubjergensis, Eoglobigerina eobulloides, E. edita, Woodringina hornerstownensis, Parvularugoglobigerina extensa* and *P. eugubina*) ranging in size from 40 to 100 μm were recovered from clayey limestone and a nodular limestone band (JH 16, 17, 19 and JH 25) of middle unit 3. Some larger planktic foraminiferans, up to 150 μm in diameter, were recovered, too, using sieving methods. This assemblage of larger planktic foraminiferans is composed of *Praemurica taurica, Globigerina* (*Eoglobigerina*) *pentagona, Subbotina triloculinoides* and *Parasubbotina pseudobulloides* (Khosla et al. 2022, Fig. 5.2). The presence of the above-listed large and small planktic foraminiferans found in samples JH 16, 17, 19 and JH 25 indicates an early Danian age. In the interval 6.05–6.08 m and 6.15 m (samples JH17–18, JH 20), well-rounded clay clasts with *Guembelitria cretacea* and other small Danian species are present, representing corrosion and redeposition. *E. edita*, *P. eugubina*, *P. extensa* and *W.*

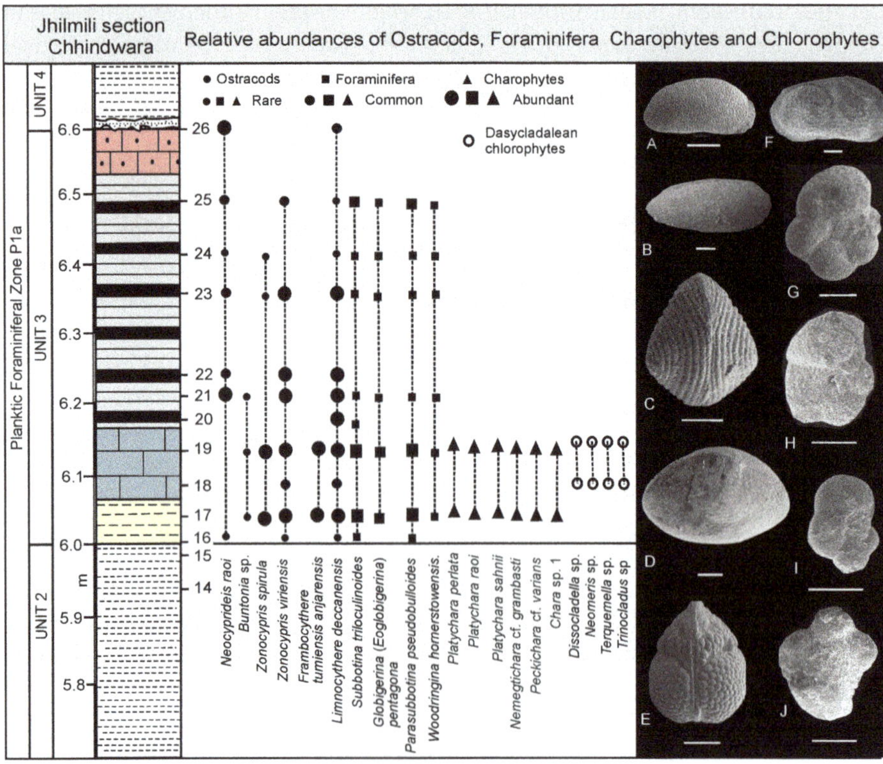

Fig. 5.2 Biostratigraphic occurrences of ostracods, foraminiferans charophytes and chlorophytes in unit 3 of the Jhilmili intertrappean beds. The occurrences of ostracods and foraminiferans are modified after Keller et al. (2009a). (**A**–**F**), Ostracods, (**A**) *Neocyprideis raoi* (MPL/SK/JML/2001) from JH 17, (**B**) *Buntonia whittakerensis* sp. nov. (MPL/SK/JML/2009) from JH 17, (**C**) *Zonocypris spirula* (MPL/SK/JML/2010) from JH 19, (**D**) *Zonocypris viriensis* (MPL/SK/JML/2011) from JH 19, (**E**) *Frambocythere tumiensis anjarensis* (MPL/SK/JML/2012) from JH 19 and (**F**) *Limnocythere deccanensis* (MPL/SK/JML/2013) from JH 17. (**G**–**J**) Foraminifera, (**G**) *Subbotina triloculinoides* (MPL/SK/JML/2014) from JH 19, (**H**) *Globigerina* (*Eoglobigerina*) *pentagona* (MPL/SK/JML/2015) from JH 19, (**I**) *Parasubbotina pseudobulloides* (MPL/SK/JML/2016) from JH 17, and (**J**) *Woodringina hornerstowensis* (MPL/SK/JML/2017) from JH 19. Scale bars = 100 μm. Note rare represents equal to or less than 10 specimens, common means equal to 10 to 20 specimens, and abundant is equal to or more than 20 specimens of ostracods and charophytes, and larger rectangles show samples with common foraminiferans. (Reproduced and modified from Khosla et al. 2022 with permission from Wiley)

hornerstownensis are among the early Danian zone P1a species found in the clasts and adjacent matrix. Between 6.40 and 6.55 m, foraminiferal assemblages are rather abundant (samples JH 24–25; Keller et al. 2009a, b).

Apart from the early Danian assemblage, planktic foraminiferans, mostly of Cretaceous age, have also been recorded from Jhilmili, for instance, *Guembelitria cretacea* (disaster opportunist), *Hedbergella* cf. *holmdelensis* and *Globigerinelloides aspera* (Keller et al. 2009a, b; Khosla 2015). The stratigraphic ranges of these characteristically Cretaceous species (*Guembelitria cretacea*) extend into the Danian, so they are commonly known as Cretaceous–Palaeogene survivors (Keller et al. 2002; Keller et al. 2009a, b; Khosla 2015). Sharma and Khosla (2009) and Keller et al. (2009a, b) assigned an early Danian P0–P1a (1) age to the lower part of the 6 m thick palaeosols (red clayey siltstone), while the upper part of the intertrappeans is 6 m thick and also consists lithologically of red shales (palaeosols); it has been assigned to the upper part of zone P1a (2). The presence of *Parvularugoglobigerina eugubina* indicates that the deposition of the foraminiferan–bearing horizon took place around 100–150 ka later than the Cretaceous-Palaeogene boundary (Keller et al. 2009a, b).

The intertrappean beds of Jhilmili also contain ostracod assemblages of Maastrichtian to early Palaeocene age. The freshwater ostracods of Late Cretaceous (Maastrichtian) age are in the basal part of unit 3 (JH17–21): *Limnocythere deccanensis, Cypria cyrtonidion, Paracypretta verruculosa, Paracypretta jonesi, Zonocypris viriensis, Zonocypris* sp., *Darwinula torpedo* and *Cypridopsis hyperectyphos*. The upper part of unit 3 (JH 25–26) also contains some freshwater ostracods such as *Zonocypris spirula, Eucypris pelasgicos, Strandesia jhilmiliensis, Frambocythere tumiensis anjarensis, Stenocypris cylindrica, Gomphocythere strangulata, Gomphocythere paucisulcatus, Centrocypris megalopos* and *Cyclocypris amphibolos* (Table 5.2; Sharma and Khosla 2009; Khosla et al. 2011a, b; Khosla, 2015). Aside from the freshwater assemblage, two brackish water ostracods (JH 17–19 and 21), *Neocypreides raoi* and *Buntonia whittakerensis* sp. nov., are abundant in unit 3 and can be used to identify the start of the Palaeocene and, more specifically, an Early Danian (Palaeocene) age (Khosla et al., 2022).

The Ghat Parasia, Government well and Shriwas well sections have produced a diverse micro-fauna and flora consisting of invertebrates, vertebrates and plants, including charophytes. The charophyte remains described from the Ghat Parasia section came from the topmost unit, a hard clayey limestone. Aside from charophytes, the unit has produced ostracods (*Frambocythere tumiensis anjarensis* and *Periosocypris megistus*) and gar fragments (*Lepisosteus indicus*). The dental remains of Late Cretaceous myliobatis, *Igdabatis indicus*, had been discovered in the fossiliferous greenish limestone unit. The presence of the above fauna in the intertrappean beds of Ghat Parasia supports a Late Cretaceous (Maastrichtian) age.

Vicente et al. (2019) investigated the evolution of European charophytes across the Cretaceous-Palaeogene boundary. Many charophyte species thought to exist only in the Maastrichtian were later discovered in the basal Danian, demonstrating how resilient charophytes were to the global Cretaceous-Palaeogene biotic crisis. They also found a step-wise extinction pattern within the Characeae family, with only a few members were to going extinct at the Cretaceous-Palaeogene boundary. Furthermore, their European data show that the charophytes experienced low extinction at the Cretaceous-Palaeogene boundary and that some of their forms, such as *Platychara* and *Peckichara* (also known from India), were well adapted to the Late Cretaceous environmental stress and survived the Cretaceous-Palaeogene crisis, successfully transitioning into the Early Danian (Khosla et al. 2022).

Although the genera *Platychara* and *Peckichara* are well known from India's infra- and intertrappean beds, a lack of reliable biostratigraphic analysis from these beds makes determining whether the Jhilmili and Ghat Parasia charophytes, particularly *Platychara* and *Peckichara*, are Late Cretaceous or Early Danian representatives difficult. To understand the floral dynamics associated with the Cretaceous-Palaeogene boundary mass extinction, more comprehensive biostratigraphic studies of all charophyte-bearing infra- and intertrappean beds are required. The Jhilmili section has been assigned a relative age ranging from the latest Cretaceous (Maastrichtian) to the Palaeocene (Early Danian) based on the recovery of charophytes along with some ostracods and planktic foraminiferans from the Cretaceous-Palaeogene boundary interval (Keller et al. 2002; Keller et al. 2009a; Khosla 2015; Khosla et al. 2022). On the basis of the presence of *Platychara* in Ghat Parasia section, an Upper Cretaceous to ?Lower Palaeocene age has been assigned to it.

5.7 Conclusions

1. The majority of recovered ostracod taxa are known in freshwater environments such as rivers, streams, lakes and ponds, and their presence in Jhilmili, Ghat Parasia, Government well and Shriwas (=Shiraj) well suggests that ostracod-bearing infra- and intertrappean sites may have been inter-linked by a palaeodrainage network.
2. Palaeoenvironmentally, the Jhilmili and Ghat Parasia intertrappean beds of the Chhindwara District, Madhya Pradesh, have yielded a prolific microfossil fossil assemblage of both non-marine aquatic organisms and those indicative of the temporary advances of a freshwater/lake environment over a shallow marine environment during low tide intervals in Jhilmili. The presence of planktic foraminiferans such as *Subbotina triloculinoides*, *Globanomalina compressa*, *Woodringina hornerstownensis*, *Woodringina claytonensis*, *Hedbergella holmdelensis*, *Guembelitria cretacea*, *Parasubbotina pseudobulloides*, *Globigerinelloides aspera*, *Globigerina* (*Eoglobigerina*) *pentagona* and benthic chlorophytes in the charophyte-bearing horizon of unit 3 of the Jhilmili intertrappean beds provides strong evidence of episodic advances and retreats of marine as well as freshwater conditions with a shallow depth of water.
3. The non-marine aquatic charophyte *Platychara closasi* sp. nov. is recovered from the hard clayey limestone of the Ghat Parasia intertrappean beds. This intertrappean site also yielded ostracods belonging to *Limnocythere deccanensis*, *Frambocythere tumiensis anjarensis*, *Gomphocythere paucisulcatus*, *Periosocypris megistus*, *Candona* sp., *Eucypris* sp. 1, *Cyclocypris amphibolos*, *Cyprois rostellum*, *Cyprois* sp. and *Darwinula* sp. of freshwater lakes/ponds with a low energy environment. The presence of ostracod genera such as *Limnocythere*, *Gomphocythere*, *Pericocypris*, *Candona*, *Eucypris*, *Cyclocypris*, *Cyprois* and *Darwinula* in the Ghat Parasia section favours fluvio-lacustrine low energy dep-

ositional environmental settings. The occurrences of ostracod genera such as *Frambocythere*, *Gomphocythere*, *Zonocypris*, *Cypridopsis*, *Eucypris*, *Cypris* and *Cyprois* in the Shiraj and Government wells sections suggest a fluvio-lacustrine low energy depositional environment. It is noted that no brackish and shallow marine ostracod taxa were recovered from Ghat Parasia, Shiraj (=Shriwas) well and Government well during the present study.

4. At Ghat Parasia, there are fragmentary remains of a fluvial to lacustrine gar, *Lepisosteus indicus,* and *Platychara closasi,* a non-marine aquatic charophyte. The presence of these biotic elements indicates the deposition of Ghat Parasia intertrappean beds under fluvial to lacustrine environments. Further, the Ghat Parasia intertrappean beds have also yielded the dental remains of myliobats (*Igdabatis indicus*). These myliobats commonly preferred to live in a brackish to shallow marine coastal environment. Therefore, it is expected that marine biotic signatures can be found in the Ghat Parasia site.
5. The presence of *Buntonia* and *Neocyprideis* at Jhilmili suggests that during the Cretaceous-Palaeogene interval, nearby shallow marine waters either from the Bay of Bengal via the Pranhita-Godavari rift or from the Arabian Sea via the Narmada-Tapti rift systems, infiltrated to the continental interior region of Central India.
6. It was found that intertrappean sediments of the Shriwas and Government wells yielded a smaller number of ostracod forms compared to the Jhilmili and Ghat Parasia assemblages.
7. On the basis of the presence of fish remains consisting of myliobats, lepisosteids and osteoglossids, it has been inferred that a freshwater environment prevailed during the deposition of fish-bearing units of the Shriwas (=Shiraj) well, Government well, Ghat Parasia and Jhilmili intertrappean beds. Further, their occurrences also indicate the nearby presence of a near-shore or coastal-plain environment.
8. Biostratigraphically, the recovered taxa suggest a Late Cretaceous to Early Palaeocene age of the studied Jhilmili intertrappean beds. A Late Cretaceous, especially Maastrichtian age, can be assigned to the intertrappean beds at Ghat Parasia, Government well and Shiraj (=Shriwas) well. These intertrappean beds extended up to an Early Palaeocene age.

References

Barbosa M, Lefler F, Berthold DE, Laughing house HD (2021) The ecology of charophyte algae (Charales). UF/IFAS Extension, SS-AGR-448:1–4. https://doi.org/10.32473/edis-ag448-2021

Bhandari A, Colin JP (1999) Limnic ostracods of the intertrappean sediments (terminal Maastrichtian-basal Paleocene) of the Anjar region (Kachchh, Gujarat state), India: Systematics, paleoecology, and paleobiogeographic affinities. J Micropaleontol 42(1):3–20

Bhandari N, Gupta M, Pandey J, Shukla PN (1994) Chemical profiles in K/T boundary section of Meghalaya, India: Cometary, asteroidal or volcanic. Chem Geol 113:45–60

Bhatia SB, Mannikeri MS (1976) Some Charophyta from the Deccan intertrappean beds near Nagpur, central India. Geophytology 6(1):75–81
Bhatia SB, Rana RS (1984) Palaeogeographic implications of the Charophyta and Ostracoda of the intertrappean beds of peninsular India. Mém Soc Geol Fr 147:29–35
Bhatia SB, Prasad GVR, Rana RS (1990a) Deccan volcanism, a Late Cretaceous event: conclusive evidence of ostracodes. In: Sahni A, Jolly A (eds) Cretaceous event stratigraphy and the correlation of the Indian nonmarine strata. A Seminar cum Workshop IGCP 216 and 245, Chandigarh, pp 47–49
Bhatia SB, Riveline J, Rana RS (1990b) Charophyta from the Deccan intertrappean beds near Rangapur, Andhra Pradesh, India. The Palaeobotanist 37:316–323
Bottjer DJ (2016) Paleoecology: Past, Present, and Future. Wiley, West Sussex
Charrière A, Haddoumi H, Mojon PO, Ferrière J, Cuche D, Zili L (2009) Miseen évidence par charophytes et ostracodes de l'âge Paléocène des depots discordants sur les rides anticlinales de la régiond' Imilchil (Haut Atlas, Maroc): consequences paléogéographiques et structurales. C R Palevol 8(1):9–19
Claeson KM, O'Leary MA, Roberts EM, Sissoko F, Bouaré M, Tapanila L, Goodwin D, Gottfried MD (2010) First Mesozoic record of the stingray *Myliobatis wurnoensis* from Mali and a phylogenetic analysis of Myliobatidae incorporating dental characters. Acta Palaeontol Pol 55:655–674
Cohuo S, Hernández M del C, Pérez L, Alcocer J (2017) *Candona alchichica* (Podocopida: Candonidae), a new ostracod species from saline, tropical Lake Alchichica, Mexico. J Limnol 76(1). https://doi.org/10.4081/jlimnol.2016.1480
Conkin JE, Conkin BM (1992) Late Silurian (Ludlovian) charophyte *Moellerina laufeldi* n. sp., from the Hamra beds of the Isle of Gotland, Sweden. Univ Louisville Notes Paleonotol Stratigr, pp 1–15
Cushman JA (1933) Some new foraminiferal genera. Contrib Cushman Lab Foraminifer Res 8:4–7
Cusminsky GC, Pérez PA, Schwalb A, Whatley R (2005) Recent lacustrine ostracods from Patagonia, Argentina. Rev Esp de Micropaleontol 37:431–450
de Sowerby JC (1840) Explanations of the plates and wood-cuts. In: Malcolmson JC (ed), On the fossils of the eastern portion of the Great Basaltic District of India. Trans Geol Soc London 2(5):511–567
Dodd JR, Stanton RJ (1990) Palaeoecology: concepts and applications. Wiley India Pvt Ltd, Delh, India
Dudgeon D, Arthington AH, Gessner MO, Kawabata ZII, Knowler DJ, Léveque C, Naiman RJ (2006) Freshwater biodiversity: importance, threats, status and conservation challenges. Biol Rev 81(2):163–182
Ehrenberg CG (1854) Mikrogeologie. Das Erden und Felsen schaffende Wirken des unsichtbar kleinen selbstständigen Lebens auf der Erde, Leopold Voss, Leipzig, p xxviii + 374
El Hajj L, Horne DJ, Baudin F, Gèze R, Azar D (2021) The first non-marine ostracod fauna from the Lower Barremian dysodiles of Lebanon. Lethaia 54:127–139
Feist M (1979) Charophytes at the Cretaceous-Tertiary boundary. New data and present state of knowledge. In: Cretaceous-Tertiary boundary events symposium. II Proceedings Copenhagen, pp 88–94
Feist M, Grambast-Fessard N (1991) The genus concept in Charophyta: evidence from Palaeozoic to Recent, vol 11. In: Riding R (ed) Calcareous algae and stromatolites, pp 189–203
Feist M, Grambast-Fessard N, Guesquerlin M, Karol K, Lu H, McCourt RM, Wang Q, Shenzen Z (2005a) Treatise on invertebrate paleontology, Part B Protoctista 1, vol 1. Charophyta. Geol Soc Am, Boulder, Colorado, pp 1–170
Feist M, Liu J, Tafforeau P (2005b) New insights into Paleozoic charophyte morphology and phylogeny. Am J Bot 92(7):1152–1160
Feist-Castel M (1975) Répartitions des Charophytes dans le Paléocène et l'Éocène du basind' Aix-en-Provence. Bull Soc Géol Fr 17(1):88–97
Feist-Castel M (1977) Etude floristique et biostratigraphique des Charophytes dans les series Du Paleogene de Provence. Géol Méditerr 4:109–138

Feist PMR, Carreño AL, Ortega-Ramírez JR, Alvarado-Valdez G (2002) A paleoenvironmental reconstruction of Laguna Babicora, Chihuahua, Mexico based on ostracode paleoecology and trace element shell chemistry. J Paleolimnol 27:185–206

Galbrun B, Feist M, Colombo F, Rocchia R, Tambareau Y (1993) Magnetostratigraphy and biostratigraphy of Cretaceous-Tertiary continental deposits, Ager basin, province of Lerida, Spain. Palaeogeog Palaeoclimat Palaeoecol 102(1):41–52

Garcia A (1994) Charophyta: their use in paleolimnology. J Paleolimnol 10:43–52

Gawor-Biedowa E (1992) New foraminifera of the Campanian and Maastrichtian in the Lublin Region (Eastern Poland)). Kwartalnik Geologiczny 36(1):75–96

Grambast L (1957) Ornamentation de la gyrogonite et systematique chez les charophytes fossiles. Rev Gén Bot 64:339–362

Grande L (2010) An empirical synthetic pattern study of gars (Lepisosteiformes) and closely related species, based mostly on skeletal anatomy: The resurrection of Holostei. Copeia 10(2A):1–871

Jaillard E, Cappetta H, Ellenberger P, Feist M, Grambast-Fessard N, Lefranc JP, Sige B (1993) Sedimentology, palaeontology, biostratigraphy and correlation of the Late Cretaceous Vilquechico Group of southern Peru. Cretac Res 14:623–661

Jain SP (1978) Ostracodes des "inter-trappean beds" (Eocene Infeerieur) de Kateru, Rajahmundry, Côte est de l' Inde. Rev de Microplaeontol 21(2):51–58

Jones T (1860) Notes on the fossil Cypride. Quart J Geol Soc London 16:154–189

Jones RW (2011) Applications of Palaeontology. Cambridge University Press, Cambridge, UK

Kania S, Khosla A, Verma O, Prashanth M (2022) Charophyte assemblage in the Cretaceous–Palaeogene boundary of the Deccan intertrappeans from of Chhindwara (Madhya Pradesh), Central India. Him Geol 43(2):383–396

Kapur VV, Khosla A, Tiwari N (2019) Paleoenvironmental and paleobiogeographical implications of the microfossil assemblage from the Late Cretaceous intertrappean beds of the Manawar area, District Dhar, Madhya Pradesh, Central India. Hist Biol 31(9):1145–1160

Karczewska J, Ziembińska-Tworzydlo M (1983) Age of the Upper Cretaceous Nemegt Formation (Mongolia) on charophyte evidence. Acta Palaeontol Pol 28:1317–1146

Keen MC (2004) The origin of the modern tropical West African marine Ostracoda fauna with a description of Ruggieriini n. tribe. Bull Ital Paleontol Soc 43(1–2):201–216

Keller G, Adatte T, Burns SJ, Tantaway A (2002) High stress paleoenvironment during the late Maastrichtian to early Paleocene in Central Egypt. Palaeogeogr Palaeoclimatol Palaeoecol 187:35–60

Keller G, Adatte T, Gardin S, Bartolini A, Bajpai S (2008) Main Deccan volcanism phase ends near the K-T boundary: evidence from the Krishna-Godavari Basin, SE India. Earth Planet Sci Lett 268:293–311

Keller G, Khosla SC, Sharma R, Khosla A, Bajpai S, Adatte T (2009a) Early Danian planktic foraminifera from Cretaceous-Tertiary intertrappean beds at Jhilmili, Chhindwara District, Madhya Pradesh, India. J Foram Res 39(1):40–55

Keller G, Adatte T, Bajpai S, Mohabey DM, Widdowson M, Khosla A, Sharma R, Khosla SC, Gertsch B, Fleitmann D, Sahni A (2009b) K-T transition in Deccan Traps of central India marks major marine seaway across India. Earth Planet Sci Lett 282:10–23

Khosla A (2014) Upper Cretaceous (Maastrichtian) charophyte gyrogonites from the Lameta Formation of Jabalpur. Central India: palaeobiogeographic and palaeoecological implications. Acta Geol Pol 64(3):311–323

Khosla A (2015) Palaeoenvironmental, palaeoecological and palaeobiogeographical implications of mixed fresh water and brackish marine assemblages from the Cretaceous–Palaeogene Deccan intertrappean beds at Jhilmili, Chhindwara District, Central India. Rev Mex de Cien Geol 32(2):344–357

Khosla A, Lucas SG (2020) Introduction of Indian Late Cretaceous dinosaur eggs and eggshells of Peninsular India. In: Khosla A, Lucas SG (eds) Late Cretaceous dinosaur eggs and eggshells of Peninsular India: Oospecies diversity and taphonomical, palaeoenvironmental, biostratigraphical and palaeobiogeographical inferences, Topics in Geobiology, vol 51. Springer, Cham, pp 1–30

Khosla A, Lucas, SG (2021) End-Cretaceous Extinctions. In: Elias S, Alderton, D (eds) Encyclopedia of Geology, 2nd edition. Earth Systems and Environmental Sciences (Academic Press, Elsevier) , pp. 665–678. https://doi.org/10.1016/B978-0-12-409548-9.124
Khosla SC, Nagori ML (2005) A restudy of ostracode fauna from the intertrappean beds of the Anjar, Kachchh District, Gujarat. J Geol Soc Ind 66(5):573–580
Khosla A, Sahni A (2003) Biodiversity during the Deccan volcanic eruptive episode. J Asi Earth Sci 21(8):895–908
Khosla A, Verma O (2015) Paleobiota from the Deccan volcano-sedimentary sequences of India: paleoenvironments, age and paleobiogeographic implications. Hist Biol 27(7):898–914
Khosla SC, Nagori ML, Mohabey DM (2005) Effect of Deccan volcanism on non-marine Late Cretaceous Ostracode fauna: a case study from Lameta Formation of Dongargaon area (Nand-Dongargaon basin), Chandrapur District, Maharashtra. Gond Geol Magaz 8:133–146
Khosla SC, Nagori ML, Jakhar SR, Rathore AS (2011a) Early Danian lacustrine–brackish water Ostracoda from the Deccan Intertrappean beds near Jhilmili, Chhindwara District, Madhya Pradesh, India. Micropaleontology 57(3):223–245
Khosla SC, Rathore AS, Nagori ML, Jakhar SR (2011b) Non-marine Ostracoda from the Lameta Formation (Maastrichtian) of Jabalpur, Madhya Pradesh and Nand-Dongargaon Basin, Maharashtra, India: their correlation, age and taxonomy. Rev Esp de Micropaleontol 43:209–260
Khosla A, Chin K, Verma O, Alimohammadin H, Dutta D (2016) Paleobiogeographical and paleoenvironmental implications of the freshwater Late Cretaceous ostracods, charophytes and distinctive residues from coprolites of Lameta Formation at Pisdura, Chandrapur District (Maharashtra), Central India. In: Khosla A, Lucas SG (eds) Cretaceous period: Biotic diversity and biogeography, vol 71. New Mexico Museum of Natural History & Science, Albuquerque, pp 173–184
Khosla A, Kania S, Lucas SG, Verma O (2022) Charophytes from the Cretaceous–Palaeogene transition in the Jhilmili intertrappean beds of Central India. Geol J 57(11):4412–4438. https://doi.org/10.1002/gj.4528
Khosla SC, Rathore AS, Nagori ML, Jakahr SL (2013) Palaeoecology and Affinity of Ostracod fauna from the classic localities of Lameta Formation of Jabalpur, Madhya Pradesh, India. J Geol Soc India 1:401–415
Krishnan MS (1949) Geology of India and Burma. The Madras Law Journal Office, Madras
Külköylüoğlu O (2003) Ecology of freshwater Ostracoda (Crustacea) from lakes and reservoirs in Bolu, Turkey. J Freshw Eco 18(3):343–347
Külköylüoğlu O, Yavuzatmaca M, Cabral MC, Colin JP (2015) *Gomphocythere besnin.* sp. (Crustacea, Ostracoda) from a man-made pool (Adıyaman, Turkey). Zootaxa 3937(3):456–470
Kumar R (1996) Fundamentals of Historical Geology and Stratigraphy of India, Fourth Reprint. New Age International Publishers, New Delhi
Kumar K, Rana RS, Paliwal BS (2005) Osteoglossid and lepisosteid fish remains from the Paleocene Palana Formation, Rajasthan, India. Palaeontology 48(6):1187–1209
Kundal P, Humane S, Humane SK, Petkar SP (2018) Discovery of marine benthic chlorophycean algae in Early Danian Deccan intertrappean at Jhilmili, Central India: new insights into existence of marine seaway close to Cretaceous-Paleogene boundary. J Palaeontol Soc India 63(2):203–211
Leckie RM (1987) Paleoecology of Mid-Cretaceous planktonic foraminifera: A comparison of open ocean and epicontinental sea assemblages. Micropaleontology 33(2):164–176
Li S, Wang Q, Zhang H, Lu H, Martín-Closas C (2016) Charophytes from the Cretaceous–Paleogene transition in the Pingyi Basin (Eastern China) and their Eurasian correlation. Cretac Res 59:179–200
Li S, Wang QF, Zhang HC, Wan XQ, Martín-Closas C (2019) Charophytes from the Cretaceous–Paleocene boundary in the Songliao Basin (north-eastern China): a Chinese biozonation and its calibration to the Geomagnetic Polarity Time Scale. Pap Palaeontol 5:47–81
Li S, Fang YN, Wang H, Wang QF, Zhang HC (2020a) Maastrichtian charophyte flora from the Jiankou Section of the Jiaolai Basin. Acta Geol Sin 94:556–557

Li S, Wang QF, Zhang HC, Wang H, Wan XQ (2020b) Latest Campanian to Maastrichtian charophytes in the Jiaolai Basin (Eastern China). Cretac Res 106:104–266

Loeblich AR, Tappan H (1957b) *Woodringina,* a new foraminiferal genus (Heterohelicidae) from the Paleocene of Alabama. J Wash Acad Sci 47:39–40

Lucas SG (2018) Permian-Triassic charophytes. Distribution, biostratigraphy and bioevents. J Earth Sci 29(4):778–793

Mahalakshmi P, Hussain SM (2015) Distribution of ostracods from Pazhaverkadu (Pulicat lagoon), Tamil Nadu, India. In: Vasudevan S (ed) Lakes and Wetlands. Partridge, India, pp 268–287

Malarkodi N, Keller G, Fayazudeen PJ, Mallikarjuna UB (2010) Foraminifera from the Early Danian intertrappean beds in Rajahmundry Quarries, Andhra Pradesh. J Geol Soc Ind 75:851–863

Martens K, Davies BR, Baxter AJ, Meadows ME (1996) A contribution to the taxonomy and ecology of the Ostracoda (Crustacea) from Verlorenvlei (Western Cape, South Africa). S Afr J Zool 31(1):23–36

Masriera A, Ullastre J (1988) Nuevos datos sobre las capas maestrichtiens es con *Septorella*: su presencia al norte del Montsec (Pirineocatalan). Acta Geol Hisp 23:71–77

Mazzini I (2011) The genus *Zonocypris* Müller, 1898 (Crustacea, Ostracoda, Cyprididae) from continental Miocene deposits of Central Anatolia (Turkey): palaeoecological and palaeogeographical implications. J Geol Paläontol 11:124–125

Mccourt RM, Karol KG, Guerlesquin M, Feist M (1996) Phylogeny of extant genera in the family Characeae (Charales, Charophyceae) based on rbcl, sequences and morphology. Am J Bot 83(1):125–131

McKenzie KG (1971) Entomostraca of Aldabra, with special reference to the genus *Heterocypris* (Crustacea, Ostracoda). Philos Trans R Soc London B 260:257–297

McKenzie KG (1981) Palaebiogeography of some salt lake faunas. Hydrobiologia 81:407–418

Meisch C (2000) Freshwater Ostracoda of Western and Central Europe. Spektrum Akademischer Verlag GmbH, Heidelberg, Berlin, pp 1–522

Morozova VG (1961) Datsko-Montskie planktonnye foraminifery yuga SSSR (Danian-Montian Planktonic Foraminifera of the Southern USSR). Paleontologicheskiy Zhurnal 2:8–19

Naqvi SM (2005) Geology and Evolution of the Indian Plate (From Hadean to Holocene – 4 Ga to 4 Ka). Capital Publishing Company, New Delhi

Nelson JS (2006) Fishes of the World. Wiley, New Jersey, pp 1–601

Oldham T (1856) Memoir of the Geological Survey of India, vol 1. Bengal Military Orphan Press, Calcutta, pp 1–309

Olsson RK (1960) Foraminifera of latest Cretaceous and earliest Tertiary age in the New Jersey coastal plain. J Paleontol 34:1–58

Olsson RK (1964) Late Cretaceous planktonic foraminifera from New Jersey and Delaware. Micropaleontology 10:157–188

Olsson RK, Berggren WA, Hemleben C, Huber BT (1999) Atlas of Paleocene Planktonic Foraminifera. Smithsonian Contrib Paleobiol 85:1–252

Olsson RK, Hemleben C, Huber BT, Berggren WA (2006) Taxonomy, biostratigraphy, and phylogeny of Eocene *Globigerina*, *Globoturborotalita*, *Subbotina*, and *Turborotalita*. In: Pearson PN, Olsson RK, Huber BT, Hemleben C, Berggren WA (eds), Atlas of Eocene Planktonic Foraminifera. Cushman Found Spec Publ 41:111–168

Ozawa H, Tanaka G (2019) Paleoenvironmental analysis from fossil ostracod assemblages of the Middle Pleistocene Naganuma Formation in the Sagami Group, central Japan. Bull Geol Surv Japan 70(1/2):5–16

Pearson PN, Olsson RK, Huber BT, Hemleben C, Berggren WA, Coxall HK (2006) Overview of Eocene planktonic foraminiferal taxonomy, paleoecology, phylogeny, and biostratigraphy. Cushman Found Spec Publ 41:11–28

Peck RE, Forester RM (1979) The genus *Platychara* from the Western Hemisphere. Rev Palaeobot Palynol 28:223–236

Peck RE, Reker CC (1947) Cretaceous and Lower Cenozoic charophyta from Peru. Novit Am Mus Nat Hist 1369:1–6

Peck RE, Reker CC (1948) Eocene Charophyta from North America. J Paleontol 22(1):85–90
Plummer HJ (1926) Foraminifera of the Midway Formation in Texas. Univ Tex Bull 2644:1–206
Prasad GVR, Cappetta H (1993) Late Cretaceous selachians from India and the age of the Deccan Traps. Palaeontology 36(1):231–248
Ramón MJ, Laprida C, Massaferro J, Rogora M, Tartari G, Maidana NI (2012) Patagonian ostracods as indicators of climate-related hydrological variables: implications for paleoenvironmental reconstructions in Southern South America. Hydrobiologia 694(1):235–251
Ramos L, Cusminsky G, Schwalb A, Alperin M (2017) Morphotypes of lacustrine ostracoda *Limnocythere rionegroensis* Cusminsky and Whatley from Patagonia, Argentina, shaped by aquatic environments. Hydobiologica 786(1):137–148
Raup DM, Stanely SM (1985) Principles of paleontology. CBS Publishers and Distributors, Delhi
Riveline J, Berger JP, Bilan W, Feist M, Martin-Closas C, Schudack M, Soulie-Marsche I (1996) European Mesozoic-Cenozoic charophyte biozonation. Bull Soc Géol Fr 167:453–468
Saber AA, Gontcharov AA, Nikulin AY, Nikulin VY, Rayan WA, Cantonati M (2021) Integrative taxonomic, ecological and genotyping study of charophyte populations from the Egyptian Western-Desert Oases and Sinai Peninsula. Plan Theory 10:1157. https://doi.org/10.3390/plants10061157
Sanjuan J, Alqudah M (2018) Charophyte flora from the Miocene of Zahle (Beeka Valley, Lebanon). Biostratigraphic, palaeoenvironmental and palaeobiogeographical implications. Geodiversitas 40(10):195–209
Sanjuan J, Martín-Closas C, Barberà X, Costa E, Garcés M (2015) Els caròfits de les comarques centrals de Catalunya. Nova proposta de biozonació Misc Aqualata 16:73–120
Sanjuan J, Vicente A, Eaton JG (2020) New charophyte flora from the Pine Hollow and Claron formations (southwestern Utah). Taxonomic, biostratigraphic, and paleobiogeographic implications. Rev Palaeobot Palynol 282:04289. https://doi.org/10.1016/j.revpalbo.2020.104289
Schubert H, Marquardt R, Schories D, Blindow I (2015) Biogeography of Chilean charophytes Aquat Bot 120:129–141
Sharma R, Khosla A (2009) Early Palaeocene ostracoda from the Cretaceous-Tertiary (K-T) Deccan intertrappean sequence at Jhilmili, District Chhindwara, Central India. J Palaeontol Soc Ind 54(2):197–208
Smith CC, Pessagno EA (1973) Planktonic foraminifera and stratigraphy of the Corsicana Formation (Maestrichtian) North-central Texas. Spec Publ Cushman Found Foram Res 12:1–68
Soulié-Märsche I (1989) Étude comparée de gyrogonites de charophyte sactuelleset fossils et phylogénie des genres actuels. PhD thesis. Imprimerie des Tilleuls, Millau, France
Soulie-Marsche I (1994) The paleoecological implications of the charophyte flora of the Trinity Division, Junction. Texas J Paleontol 68(6):1145–1157
Soulié-Märsche I (1999) Chirality in charophytes: Stability and evolution from 400 million years to present. In: Pály G, Zucchi C, Caglioti L (eds) Advances in Biochirality. Elsevier, Amsterdam, pp 191–207
Srinivasan S, Sahni A, Bajpai S (1992) Fossil Charophyta from the Deccan intertrappean beds of Gurmatkal, Gulbarga District, Karnataka. Curr Sci 63:396–398
Srinivasan S, Bajpai S, Sahni A (1994) Charophytes from Deccan intertrappean beds of peninsular India: implications for age and correlation of Deccan volcanic. Geobios 27:559–571
Stoliczka F (1873) Cretaceous fauna of southern India. Mem Geol Surv India Palaeontol Indica 4:66–69
Tian JX, Chen J, Lu NH, Chen C, Wang H, Li T, Li S, Wang QF, Zhang HC, Wan XQ (2021) Discovery of charophyte flora across the Cretaceous-Paleocene transition in the Jiaolai basin. Palaeoworld 30(3):538–550
Vaidyanadhan R, Ramakrishnan M (2010) Geology of India, vol II. Geological Society of India, Bangalore
Van Itterbeeck J, Horne DJ, Bultynck P, Vandenberghe N (2005) Stratigraphy and palaeoenvironment of the dinosaur-bearing Upper Cretaceous Iren Dabau Formation, Inner Mongolia, People's Republic of China. Cretac Res 26:699–725

Van Itterbeeck J, Missien P, Folie A, Markevich VS, Damme DV, Yong GD, Smith T (2007) Woodland in a fluvio-lacustrine environment on the dry Mongolian Plateau during the late Paleocene: Evidence from the mammal bearing Subeng section (Inner Mongolia, P.R. China). Palaeogeogr Palaeoclimatol Palaeoecol 243:55–78
Verma O (2015) Cretaceous vertebrate fauna of the Cauvery Basin, southern India: palaeodiversity and palaeobiogeographic implications. Palaeogeog Palaeoclimat Palaeoecol 431:53–67
Verma O (2018) Palaeoecology: Past, Present, and Future. J Geol Soc India 91:115–116
Verma O, Khosla A (2018) Application of internet technology in assembling literature for palaeontological research. Iran J Sci Technol Trans A Sci 4:1715–1723
Verma O, Khosla A (2019) Developments in the stratigraphy of the Deccan Volcanic Province,peninsular India. Compt Rend Geosci 351:461–476
Verma O, Prasad GVR, Khosla A, Parmar V (2012) Late Cretaceous Gondwanatherian mammals of India: Distribution, interrelationships and biogeographic implications. J Paleontol Soc India 57:95–104
Verma O, Khosla A, Goin FJ, Kaur J (2016) Historical biogeography of the Late Cretaceous vertebrates of India: Comparison of geophysical and paleontological data. In: Khosla A, Lucas SG (eds) Cretaceous period: Biotic diversity and biogeography, vol 71. New Mexico Museum of Natural History & Science, Albuquerque, pp 317–330
Verma O, Khosla A, Kaur J, Prasanth M (2017) Myliobatid and pycnodont fish from the Late Cretaceous of central India and their paleobiogeographic implications. Hist Biol 29(2):253–265
Vicente A, Martín-Closas C, Arz JA, Oms O (2015) Maastrichtian-basal Paleocene charophyte biozonation and its calibration to the Global Polarity Time Scale in the southern Pyrenees (Catalonia, Spain). Cretac Res 52:268–285
Vicente A, Expósito M, Sanjuan J, Martín-Closas C (2016a) Small sized charophyte gyrogonites in the Maastrichtian of Coll de Nargó, Eastern Pyrenees: An adaptation to temporary floodplain ponds. Cretac Res 57:443–456
Vicente A, Villalba-Breva S, Ferràndez-Cañadell C, Martín-Closas C (2016b) Charophyte biostratigraphy in the Fontllonga type section. Geol Acta 14:349–362
Vicente A, Csiki-Sava Z, Martín-Closas C (2019) European charophyte evolution across the Cretaceous–Paleogene boundary. Palaeogeog Palaeoclimat Palaeoecol 533:109–244
Villalba-Breva S, Martín-Closas C (2011) A characean thallus with attached gyrogonites and associated fossil charophytes from the Maastrichtian of the Eastern Pyrenees (Catalonia, Spain). J Phycol 47:131–143
Villalba-Breva S, Martín-Closas C (2013) Upper Cretaceous paleogeography of the Central Southern Pyrenean Basins (Catalonia, Spain) from microfacies analysis and charophyte biostratigraphy. Facies 59(2):319–345
Villalba-Breva S, Martín-Closas C, Marmi J, Gomez B, Fernandez-Marron MT (2012) Peat-forming plants in the Maastrichtian coals of the Eastern Pyrenees. Geol Acta 10:189–207
Wadia DN (1966) Geology of India. Macmillan Press, London
Wang Z, Lu HN, Zhao CB (1985) In: Wang TQ (ed) Cretaceous charophytes from Songliao Basin and adjacent areas. Heilongjiang Scientific and Technical Publishing House, Harbin, pp 1–80
Wang C, Wang H, Kuang X, Guo G (2021) Life stages and morphological variations of *Limnocythere inopinata* (Crustacea, Ostracoda) from Lake Jiang-Co (northern Tibet): a bioculture experiment. ZooKeys 1011:25–40
Weller JM (1960) Stratigraphic principles and practice. Harpers and Brothers, Publishers, New York
Whatley RC (2012) The 'Out of India' hypothesis: further supporting evidence from the extensive endemism of Maastrichtian non-marine Ostracoda from the Deccan volcanic region of peninsular India. Rev de Paleobiol 11:229–248
Whatley RC, Bajpai S (2000a) A new fauna of Late Cretaceous non-marine Ostracoda from the Deccan intertrappean beds of Lakshmipur, Kachchh (Kutch District), Gujarat, western India. Rev Esp de Micropaleontol 32(3):385–409
Whatley RC, Bajpai S (2000b) Further nonmarine Ostracoda from the Late Cretaceous intertrappean deposits of the Anjar region, Kachchh, Gujarat, India. Rev de Micropaleontol 43(1):173–178

Whatley RC, Bajpai S, Srinivasan S (2002a) Upper Cretaceous nonmarine Ostracoda from intertrappean horizons in Gulbarga district, Karnataka state, South India. Rev Esp de Micropaleontol 34(2):163–186

Whatley RC, Bajpai S, Srinivasan S (2002b) Upper Cretaceous intertrappean nonmarine Ostracoda from Mohgaonkala (Mohgaon-Kalan), Chhindwara District, Madhya Pradesh state, Central India. J Micropaleontol 21:105–114

Whatley RC, Khosla SC, Rathore AS (2012) *Periosocypris megistus* n. gen. and n. sp.: A new gigantic non-marine cyprid ostracod from the Maastrichtian Lameta Formation of India. J Palaeontol Soc Ind 57(2):113–117

Woodward AS (1908) On some fish remains from the Lameta Beds at Dongargaon, Central Province. Mem Geol Sur India, Palaeontol Indica NS 3:1–6

Wroblewski AFJ (2004) New selachian paleofaunas from the "Fluvial" deposits of the Ferris and Lower Hanna formations (Maastrichtian-Selandian: 66–58 Ma) Southern Wyoming. PALAIOS 19:249–258

Yavuzatmaca M, Külköylüoğlu O (2019) Fossil and recent distribution and ecology of ancient asexual ostracod *Darwinula stevensoni* (Ostracoda, Crustacea) in Turkey. J Limnol Freshw Fish Res 5(1):47–59

Yin Y, Geiger W, Martens K (1999) Effects of genotype and environment on phenotypic variability in *Limnocythere inopinata (Crustacea: Ostracoda). Hydrobiologia 400, 85–114*

Zhamangara AK, Lucas SG (1999) Revision of some Eocene charophytes from the Zaysan Basin, Eastern Kazakhstan. Aust J Bot 47(3):297–304

Chapter 6
Palaeobiogeographical Implications of Late Cretaceous-Early Palaeocene Microbiota from the Deccan Intertrappean Beds of the Chhindwara District, Madhya Pradesh, India

6.1 Introduction

Geography, biogeography, palaeogeography and palaeobiogeography are considered to be interrelated scientific domains and are useful for understanding faunal dynamics associated with moving lithospheric plates (e.g. Fosberg 1976; Jablonski et al. 1985; Lieberman 2000; Parenti and Ebach 2009; Sanmartín 2012; Servais et al. 2013; Verma et al. 2016; Markwick 2019). Geography involves the study of earth's surface features and their patterns, distribution and interaction together with processes that give rise to them and their influence on earth's biological system (Fosberg 1976; Oxford Dictionary 2005). Biogeography deals with spatial (geographical) and temporal (time) distribution of organisms on the earth's surface (Lieberman 2000; Servais et al. 2013; Verma et al. 2016). It is a comparative scientific discipline and serves as a powerful tool to understand geology and geography of the planet earth and life on it, which are useful to unravel geological, biological and climatological evolutionary history of the planet earth (Parenti and Ebach 2009). However, it should be noted that biogeography largely deals with faunas and floras of the modern world (Servais et al. 2013). The science that deals with the reconstruction of the past earth's surface through geological time is known as palaeogeography (Willis 2010). It represents spatial re-organisation of the past landscapes of the earth and helps scientists to understand how earth evolves through geological time under the influence of its own internal dynamics (Markwick 2019). The main concerns of palaeogeography are threefold: firstly, it focuses on the mapping of the past positions of the oceanic and continental basins, secondly, to show how earth's evolved spatial features through time and thirdly, to reconstruct the drift of continents during earth history (Lieberman 2000; Servais et al. 2013). Therefore, it uses data derived from numerous geological sub-disciplines, notably structural geology, geotectonics, sedimentology, palaeomagnetism, geochronology,

A. Khosla et al., *Microbiota from the Late Cretaceous-Early Palaeocene Boundary Transition in the Deccan Intertrappean Beds of Central India*, Topics in Geobiology 54, https://doi.org/10.1007/978-3-031-28855-5_6

stratigraphy, palaeontology, petrology, geophysics and geochemistry (Servais et al. 2013). The study of the geographical distribution of faunas and floras of the earth's past is referred to as palaeobiogeography, which is a combination of palaeogeography and biogeography (Lieberman 2000; Servais et al. 2013; Verma 2018). Thus, palaeobiogeography is regarded as an interdisciplinary science that employs data from biology, geology, palaeontology, palaeoecology and palaeobiology.

Many biogeographers, palaeobiogeographers, palaeontologists and molecular biologists are interested in the Indian subcontinent's Late Mesozoic and Early Cenozoic palaeobiogeography because, during this time period, the Indian plate was fragmented from the Southern Gondwanan continents and later docked with the Northern Laurasian supercontinents, specifically Asia, which has many palaeobiogeographical implications (Ali and Aitchison 2008; Chatterjee et al. 2013; Verma 2016; Verma and Khosla 2018, 2019; Khosla 2021). As a result, numerous researchers have investigated the palaeobiogeographic relationships of the Indian plate (Khosla and Lucas 2020e and references therein).

6.2 Geotectonic Evolution of the Indian Plate

The Indian plate's geotectonic journey began with the rifting of the Gondwana supercontinent and ended with its ultimate docking to the Asia mainland (Barron and Harrison 1980; Ali and Aitchison 2008; Chatterjee and Scotese 2010; Chatterjee et al. 2013; Kapur and Khosla 2016, 2019; Khosla and Verma 2015; Verma et al. 2016; Khosla 2021; Khosla and Lucas 2021). Numerous studies have shown that the Indian plate was a part of the former Gondwanan continents and was located in the Southern Hemisphere as part of East Gondwana for much of the Mesozoic Era (Khosla 2021). Apart from India, the East Gondwana continental block included Antarctica, Australia, Madagascar and the Seychelles. During the Mesozoic, South America and Africa joined to form West Gondwana. Africa, Madagascar, and the Seychelles were adjacent landmasses to the west of the Indian plate; Australia to the southeast; Antarctica and Sri Lanka to the south; and the Neo-Tethys Sea covered the entire northern margin of the plate (Scotese 1991; Smith et al. 1994; Storey 1996; Ali and Aitchison 2008; Chatterjee et al. 2013; Verma et al. 2016; Khosla and Lucas 2020e, Fig. 6.1). Numerous workers have summarised the rifting, drifting and docking history of the Indian plate (Rabinowitz et al. 1983; Powell et al. 1988; Storey 1996; Hay et al. 1999; Ali and Aitchison 2008; Chatterjee et al. 2013; Gibbons et al. 2013; Verma et al. 2016; Khosla and Lucas 2020e).

In the Middle Jurassic, around 165–150 million years ago (Ma), at the time of the initial rift between East and West Gondwana, India first began to separate from Africa (Rabinowitz et al. 1983; Storey 1996; Chatterjee et al. 2013). Following this, around 130–120 Ma (mid-Early Cretaceous), the Indo-Madagascar block began to break away from Antarctica-Australia (Powell et al. 1988; Chand et al. 2001). The Rajmahal-Kerguelen mantle plumes caused this separation, which resulted in the formation of the Rajmahal-Sylhet, Bunbury and Kerguelen Island basalts in the

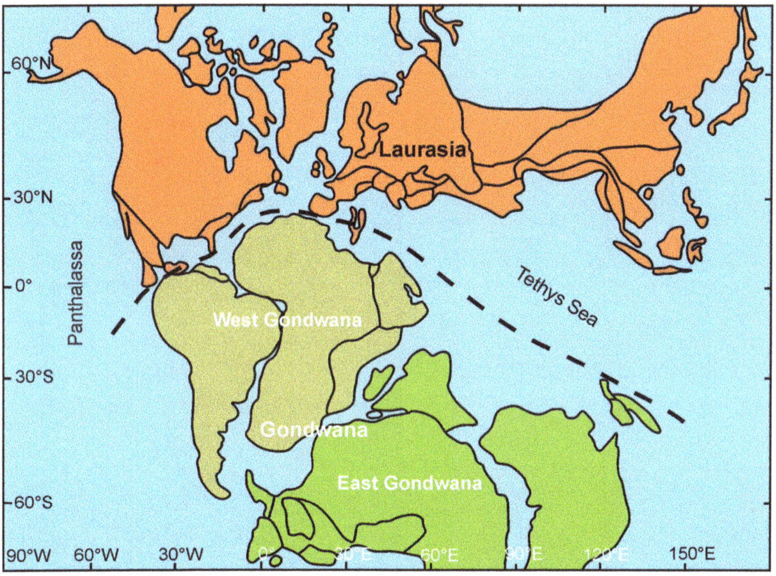

Fig. 6.1 Palaeogeographic map showing position of the Indian plate in the Southern Hemisphere. (Modified after Smith 1992)

Indian Ocean to the south of the Indian plate. Among these plateau basalts, the Kerguelen Plateau and Gunnerus Ridge are thought to have provided stepping stones between Indo-Madagascar and Antarctica for dispersal of Gondwanan biotas from South America to Indo-Madagascar via Antarctica during the Late Cretaceous (Krause et al. 1997; Hay et al. 1999; Case 2002; Verma et al. 2012). While reconstructing the Cretaceous palaeogeography, Hay et al. (1999) stated that Indo-Madagascar was connected to Antarctica as late as 80 Ma by a large, sub-aerially exposed portion of the Kerguelen Plateau. Further, Case (2002) proposed that Antarctica and Madagascar were connected together by the Gunnerus Ridge in the mid-Late Cretaceous. However, palaeoceanographical, palaeobathymetrical, geophysical, sedimentalogical and palaeontological studies have revealed that the Kerguelen Plateau and Gunnerus Ridge margins were covered by deep-sea waters, so they were not available as land connections between Antarctica and Indo-Madagascar during the Late Cretaceous (Ali and Aitchison 2009; Ali and Krause 2011; Khosla 2021).

The Indo-Madagascar blocks remained joined until about 88 Ma, when India physically separated from the Madagascan Island in the Late Cretaceous about 90–88 Ma and began moving rapidly northeastward within the Indian Ocean (Storey et al. 1995; Torsvik et al. 2000; Reeves 2014). Finally, the Indian plate collided with the Asian mainland around 52 Ma, but the collisional process may have begun in the middle of the Palaeocene around 59 Ma (Ali and Aitchison 2009; Chatterjee et al. 2013; Hu et al. 2016; Fig. 6.2).

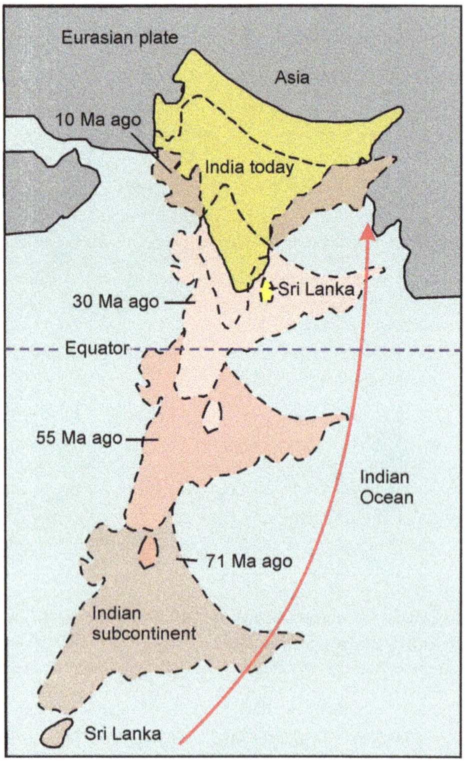

Fig. 6.2 Map showing the northward drift of the Indian plate. (Modified after Dèzes 1999)

During the Late Cretaceous, the Indian plate attained its maximum geographic isolation relative to its other neighbouring landmasses (Briggs 2003; Chatterjee et al. 2013; Verma et al. 2016; Krause et al. 2019). During the same time interval, while crossing over the Reunion mantle plume, the Indian plate was the site of voluminous and numerous volcanic eruptions that gave rise to the Deccan Volcanic Province (also known as Deccan traps) in peninsular India (Courtillot et al. 1986; Verma and Khosla 2018, 2019).

Recent studies contend that the eruptions of this volcanic province occurred between 69 and 64 Ma, and that nearly 80% of the total basaltic lava erupted rapidly in a time span of less than one million years at or near geomagnetic chron 29R, straddling the Cretaceous-Palaeogene (K-Pg) boundary, and contributed to the K-Pg mass extinction (Courtillot et al. 1986; Keller et al. 2009a, b; Courtillot and Fluteau 2014; Khosla and Verma 2015; Verma and Khosla 2019; Khosla 2021; Verma et al. 2022). Based on the Indian plate‘s long period of geographic isolation, changing climatic and latitude zones and severe environmental stress caused by the volcanic eruptions of the Deccan traps, it has been predicted that the Indian plate had sufficient time to develop unusual biotic elements (Rana and Wilson 2003; Verma et al. 2016; Khosla 2021). In addition, some marine incursions have been recorded in peninsular India in the Narmada-Tapti and Pranhita-Godavari rift valleys during the K-Pg transition (Keller et al. 2008, 2009a, b; Kania et al. 2022; Khosla et al. 2022).

Thus, the Late Cretaceous-Early Palaeocene appears to be the most critical time interval for understanding the Indian plate's geographical isolation, and spatial connectedness and for evaluating the effects of environmental changes on the onboard biotic elements, as well as establishing the position of the K-Pg boundary.

As a result, it is necessary to describe the various biogeographic models pertinent to the drifting Indian plate. Endemism, cosmopolitanism, dispersal, vagility, land bridge, landspan, vicariance, corridor, filter, sweepstake, extinction, Noah's arks, docked Noah's arks and beached Viking funeral ships are examples of these models (Verma et al. 2016). Endemism is defined as a taxonomic group that is restricted to a specific geographic area, and the organisms that live there are known as endemic organisms because they are not found anywhere else on the planet (Humphries 2001). A cosmopolitan organism has a wide or global spatial distribution, and no endemism occurs in this case. Dispersal involves the spread of organisms from one place to a new place (Humphries 2001; Verma et al. 2016). Vagility refers to the dispersal ability of the organisms. A land bridge refers to a temporary land connection, which connects two separated landmasses and allows biota of one landmass to disperse and colonise another landmass. Landspan is a sub-aerial connection between continents, and island arcs, islands and micro-continents are land areas between continents but separated from them by short seaways (Iturralde-Vinent and MacPhee 1999). Vicariance is defined as a process of splitting of a population into two or more geographical subgroups by the development of a physical barrier (Humphries 2001). A corridor is a path through which almost all organisms can pass. A filter route involves the highly selective dispersal of organisms. A sweepstake is a path that involves biotic exchanges between isolated regions that are difficult to reach. Extinction is the complete elimination of a biotic group. A piece of continental crust that broke away from its continent of origin, along with its biota is referred to as a Noah's Ark. The beached Viking funeral ship represents that portion of continental crust that rifted from one continent and may later dock and meet a new landmass, introducing onboard fossil biota to the new landmass.

The Indian plate's palaeobiogeographic relationships in the Late Mesozoic and Early Cenozoic are complex, with biotic elements from three domains: Gondwanan, Laurasian and endemic. Many researchers have advanced models to explain the biogeographic history of the Indian plate, including Khosla and Sahni (2003), Whatley and Bajpai (2006), Whatley et al. (2012), Khosla (2015), Khosla and Verma (2015), Verma (2015) Kapur and Khosla (2016), Verma et al. (2016), Kapur and Khosla (2016, 2019) and Khosla and Lucas (2020e). The palaeobiogeographic implications of fossil biota, such as charophytes, ostracods, planktic foraminiferans and fishes, are discussed below.

6.3 Charophytes

Charophytes of the family Characeae largely dominated the Late Cretaceous charophyte flora globally, and the extinction of the clavatoraceans and porocharaceans took place close to the K-Pg boundary (Grambast 1974; Feist et al. 2005a, b; Vicente et al. 2019). The charophytes *Platychara perlata* (Peck and Reker 1947), *P. raoi* (Bhatia and Mannikeri 1976), *P. sahnii* (Bhatia and Mannikeri 1976), *P. compressa* (Peck and Reker 1948), *P.* sp., *P. closasi* sp. nov., *Peckichara* cf. *varians* (Grambast 1957), *Nemegtichara* cf. *grambasti* (Bhatia et al. 1990b),?*Grambastichara* sp., *Microchara shivarudrappai* sp. nov. and *Chara chhindwaraensis* sp. nov. have been recovered from the intertrappean beds of the Jhilmili and Ghat Parasia localities, Chhindwara District of Madhya Pradesh, together with endemic ostracods.

Charophytes represent a poorly studied group from the infra- and intertrappean sediments compared to contemporary biotic groups such as other microfossils (ostracods, foraminiferans, pollens and spores), vertebrates (fishes, amphibians, reptiles and mammals) and plant macrofossils (Khosla 2014, 2015; Khosla and Verma 2015; Kapur and Khosla 2016, 2019; Kania et al. 2022; Khosla et al. 2022). Numerous sediments associated with the Deccan Volcanic Province, such as infratrappeans (Jabalpur, Papro, and Nand-Dongargaon) and intertrappeans (Kachchh, Manawar, Mamoni, Jhilmili, Ghat Parasia, Nagpur, Rangapur, Gurmatkal and Rajahmundry) in peninsular India, have yielded some genera of charophytes such as *Platychara*, *Peckichara*, *Nemegtichara*, *Stephanochara*, *Harrisichara*, *Pseudoharrisichara*, *Microchara*, *Chara*, and ?*Grambastichara* sp. (Bhatia and Mannikeri 1976; Bhatia and Rana 1984; Bhatia et al. 1990a, b; Bhatia 1992; Mohabey et al. 1993; Srinivasan et al. 1992, 1994; Udhoji and Mohabey 1996; Khosla 2014; Khosla et al. 2016, 2022; Kapur et al. 2019; Kania et al. 2022; Fig. 6.3). A Late Cretaceous, particularly a Maastrichtian age, has been assigned to the majority of these infra- and intertrappean sites (Khosla and Sahni 2003; Khosla and Verma 2015), excluding Jhilmili and Rajahmundry, which are dated Late Cretaceous (Maastrichtian) to Early Palaeocene (Danian) (Keller et al. 2008, 2009a, b; Sharma and Khosla 2009; Khosla 2015). In India, the K-Pg transition has been recorded at four sites, Jhilmili, Rajahmundry (at four quarries, Duddukuru lake, Government quarry, Balaji quarry and Church quarry), the Um Sohryngkew River section (Meghalaya, northeast India) and Anjar (Kachchh, Gujarat) (Bhandari et al. 1987, 1994, 1995, 1996; Mukhopadhyay 2008; Keller et al. 2008, 2009a, b; Kania et al. 2022; Fig. 6.3). The Danian planktic foraminiferal zone P1a has been recorded from the stratigraphic successions exposed at Jhilmili, Rajahmundry and the Um Sohryngkew River section, which led to the demarcation of the K-Pg boundary transition in these areas (Mukhopadhyay 2008; Keller et al. 2008, 2009a, b). The foraminiferan, ostracod and charophyte association has only been observed in the Jhilmili and Rajahmundry intertrappean sections (Fig. 6.3). Furthermore, it has been observed that the charophyte genera *Platychara*, *Microchara* and *Chara* were widely distributed in infra- and intertrappean beds of peninsular India (Fig. 6.3), and the genus *Peckichara* was mainly reported from the intertrappean beds of southern India (Fig. 6.3). Hence, it is proposed that charophyte-bearing sites may have been interconnected to each other by means of a palaeodrainage network of streams or rivers (Kania et al. 2022).

The Jhilmili assemblage is dominated by six species of *Platychara namely, P. perlata*, *P. raoi*, *P. sahnii*, *P. compressa*, *P.* sp. and *P. closasi* sp. nov. Several workers, notably Bhatia and Rana (1984), Bhatia et al. (1990a, b), Srinivasan et al.

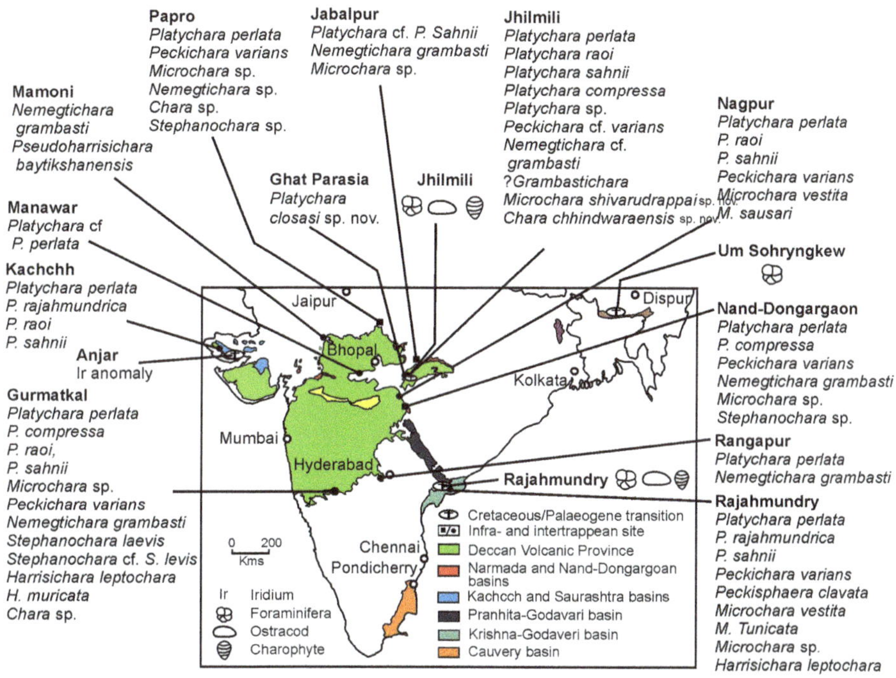

Fig. 6.3 Map depicting the spatial distribution of charophytes and recorded K-Pg transition beds in peninsular India (compiled after various sources: Bhatia and Mannikeri 1976; Singh 1980; Bhatia and Rana 1984; Bhatia et al. 1990a, 1990b; Bhatia 1992; Srinivasan et al. 1992, 1994; Mohabey et al. 1993; Udhoji and Mohabey 1996; Shome and Chandel 2013; Khosla 2014, 2015; Khosla et al. 2016; Kapur et al. 2019; Kania et al. 2022, and the present study). (Modified after Kania et al. 2022)

(1992, 1994) and Khosla (2014), considered the Indian species of the genus *Platychara* to be endemic to the Indian peninsular region. In India, *Platychara* was firstly recorded from the Upper Cretaceous infratrappean beds (~ Lameta Formation) of Nand in the Nand-Dongargaon basin, Maharashtra (Mohabey et al. 1993) and Jabalpur in Madhya Pradesh (Khosla 2014). This genus was subsequently recorded from the Maastrichtian intertrappean deposits, where it has an extensive distribution in the Nagpur (Maharashtra), Naskal, Asifabad and Rangapur (Telangana) and Kachchh (Gujarat) areas (Bhatia and Rana 1984; Bhatia et al. 1990a, b; Srinivasan et al. 1994; Khosla 2014; Khosla et al. 2022).

The species *Platychara perlata, P. raoi* and *P. sahnii* have been widely reported from the intertrappean beds of Gurmatkal, Chandarki and Yanagudi (South India), Kora (Western Gujarat) and more recently, from Manawar in District Dhar, Madhya Pradesh, Central India (Srinivasan et al. 1992, 1994; Khosla 2014; Kapur et al. 2019).

Globally, the genus *Platychara* ranges from mid-Campanian to Early Danian (Vicente et al. 2019) and has been well documented from the Late Cretaceous of Europe (Villalba-Breva and Martín-Closas 2011), especially from the Maastrichtian of Spain (Villalba-Breva and Martín-Closas 2013) and France (Villalba-Breva and

Martín-Closas 2013), Late Cretaceous of North and South America (Peck and Forester 1979; Villalba and Martín-Closas 2013; Vicente et al. 2019) and the Late Cretaceous in Tendra, Morocco (Chassagne-Manoukian et al. 2013). *Platychara* is also well documented from a number of localities in China (Van Itterbeeck et al. 2005).

Peckichara is not a very common taxon in the Jhilmili assemblage. Previously, the genus *Peckichara* has been reported from the intertrappean beds of Gurmatkal, Yanagundi and Chandarki (South India), and Kora in Gujarat (Srinivasan et al. 1992, 1994). *Peckichara* has been widely recorded globally and ranges in age from mid-Campanian to mid-Danian, and the species *P. varians* is a cosmopolitan species and inhabited freshwater lakes of the Eurasian Palaeocene ecosystems (Li et al. 2016; Khosla et al. 2022).

Nemegtichara cf. *grambasti* is a common species in the Jhilmili assemblage. In India, *N. grambasti* has been recorded from the infratrappean beds of the Nand-Dongargaon basin in the Chandrapur District, Maharashtra (Mohabey et al. 1993), and from the Lameta Formation of the Jabalpur region, Madhya Pradesh (Khosla 2014), and the Upper Cretaceous intertrappean beds of Mamoni (Rajasthan), Rangapur and Naskal in Telangana (Bhatia et al. 1990a, b) and Gurmatkal in Karnataka (Srinivasan et al. 1994). *Microchara* has been recorded in the Jhilmili assemblage. In India, *Microchara* was first recorded from the Upper Cretaceous intertrappean beds of Gurmatkal in Karnataka (Srinivasan et al. 1992, 1994).

Chara is also abundant in the Jhilmili assemblage. *Chara* is well documented from different intertrappean beds of the Indian peninsular region, such as Nagpur (Maharashtra), and Papro, and Lalitpur in Uttar Pradesh (Sahni and Rao 1934; Rao and Rao 1939; Mahadevan and Sarma 1948; Rao 1955; Bhatia and Mannikeri 1976; Shivarudrappa 1972a, b; Singh and Mathur 1979; Singh 1980). The genus has also been widely recorded from the west European localities of the Palaeogene and Neogene basins of Spain, the Swiss Molasse of Switzerland, the Rhine Graben basin of Germany and the Paris and Aquitaine basins of France (Riveline 1986; Sanjuan and Martín-Closas 2014). More recently, *Chara* has been reported from the Neogene of the Zahle basin in Lebanon (Sanjuan and Alqudah 2018). In Asia, Wang (1978) first reported the genus *Chara* and the species *Chara changzhouensis* in the Yangqi, Zhaizishan and Fangjiahe formations of?Late Palaeocene to Early Eocene age of the Jianghan Basin. Later, this genus was widely recorded from the Upper Cretaceous to Palaeocene deposits in the Mingshui and Taizhou formations in the North Jiangsu Basin (Wang et al. 1982), the Songliao Basin (Wang et al. 1985; Li et al. 2019), the Erli and Abusu formations of Inner Mongolia (Liu 1987), the Nanxiong and Shanghu formations in the Nanxiong Basin (Huang 1988), the Ziniquanzi Formation in the Junggar Basin (Yang et al. 2005) and the Pingyi Basin of eastern China (Li et al. 2016). Li et al. (2019) indicated that the age of these Chinese formations is approximate and needs refinement/confirmation in future work.

The charophyte flora of the infra- and intertrappean beds of east-west and central peninsular India helps to fill a gap in the palaeobiogeographic affinities (Fig. 6.4) and constrains the position of the isolated Indian plate in the context of its

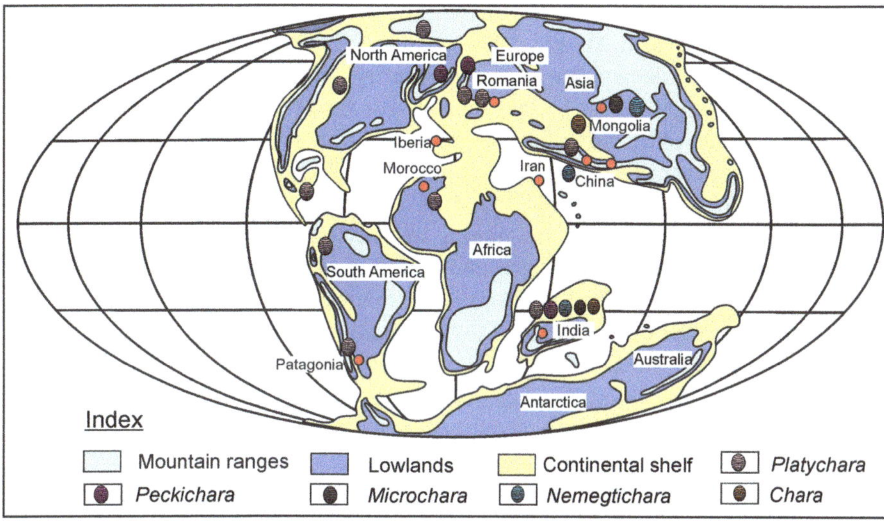

Fig. 6.4 Map displaying global palaeobiogeographic dispersal of recovered charophyte genera during the Late Cretaceous-Early Palaeogene. (Base map modified after Wing and Sues 1992; Khosla et al. 2022 reproduced from Khosla et al. 2022 with permission from Wiley)

northward drift during the Late Cretaceous-Palaeogene (Khosla and Sahni 2003; Khosla 2014; Khosla and Verma 2015; Kapur and Khosla 2016, 2019; Verma and Khosla 2019; Kapur et al. 2019; Khosla and Lucas 2020a, b; Khosla et al. 2022). The charophyte assemblages of the Jhilmili area in Central India exemplify a strong relationship or affinities with the K_-Pg transition charophyte assemblages known from different continents such as Asia, America, Europe and Africa, both at the generic and specific level and support some palaeobiogeographic inferences (Fig. 6.4).

The charophyte assemblage from the Upper Cretaceous of the Jhilmili section reflects the fact that during the K-Pg time interval the Indian subcontinent experienced exchanges of species from different corners of the world, chiefly Laurasian ones. Thus, similarities are noted between the charophytes of Jhilmili and those from North–South Americas, Europe and North Africa, which point to dispersals of charophytes among these continental masses across some migration routes. It is worth noting that during the Cretaceous, India experienced some continental palaeogeographic events, notably two supercontinent break-ups and subsequent isolation (Khosla et al. 2022).

The majority of the charophyte genera recovered from peninsular India (*Nemegtichara, Microchara, Harrisichara and Pseudoharrisichara, Platychara*, *Peckichara* and *Chara* among others) have Laurasian affinities (Bhatia and Rana 1984; Bhatia et al. 1990a, b; Mohabey et al. 1993; Srinivasan et al. 1992, 1994; Bhatia et al. 1996; Khosla and Sahni 2000, 2003; Khosla 2014; Khosla and Verma 2015; Kapur et al. 2019; Khosla et al. 2022; Fig. 6.4). Even at the species level, all charophyte species from Jabalpur, Kachchh, Rangapur, Manawar, Pisdura and

Gurmatkal share striking similarities with those from Europe, Asia (Mongolia and China), Africa and North America. The presence of Laurasian elements in India suggests that biotic dispersal occurred primarily through sweepstake and filter modes, either through India's initial contact with Asia (Jaeger et al. 1989; Khosla 2014) or through another route, for example, intermittent islands bordered by India and Asia (Bhatia et al. 1990a, b, 1996; Khosla 2014; Khosla 2015, 2019; Khosla and Verma 2015). Thewissen and McKenna (1992) and McKenna (1995) argued against an early India-Asia collision model, noting that due to a scarcity of fossils, achieving a higher-quality systematic resolution of the dispersing taxa than eutherian warm-blooded animals and pelobatid frogs was unrealistic.

Several researchers (Goswami et al. 2011; Khosla 2014, 2019; Khosla and Verma 2015; Kapur and Khosla 2019) argued that the presence of Laurasiatic components in India supports a northern sweepstakes dispersal route bordered by India and Eurasia over the Kohistan-Dras volcanic island-arc framework that may have provided an important land entry (dispersal route) to India for the movement of pelobatid and discoglossid frogs, anguimorph reptiles and eutherian mammals, of Laurasiatic origin during the Late Cretaceous (Khosla et al. 2004; Ali and Aitchison 2008; Prasad et al. 2010; Goswami et al. 2011; Khosla 2014; Kapur and Khosla 2016, 2019; Kapur et al. 2019; Khosla and Verma 2015; Verma et al. 2016; Khosla 2019; Khosla and Lucas 2020a, b, c, d, e; Khosla et al. 2022). According to the Jhilmili charophytes, the genera *Platychara, Peckichara, Nemegtichara, Microchara* and *Chara dispersed from Laurasia to India via a sweepstakes dispersal* route through the Kohistan-Dras volcanic island-arc system (Fig. 6.5). Furthermore, based on the current fossil record of the southern continents, it is possible that these forms had a vicariant distribution associated with the fragmentation of the Gondwana continents.

6.4 Ostracods

A total of 34 ostracod taxa have been recovered from the investigated intertrappean sites in the Chhindwara region, Madhya Pradesh – *Buntonia whittakerensis* sp. nov., *Neocyprideis raoi* (Jain 1978), *Limnocythere deccanensis* (Khosla et al. 2005), *Limnocythere whatleyi* sp. nov., *Frambocythere tumiensis anjarensis* (Bhandari and Colin 1999), *Frambocythere tumiensis lakshmiae* (Whatley and Bajpai 2000a), *Gomphocythere strangulata* (Jones 1860), *Gomphocythere paucisulcatus* (Whatley et al. 2002b), *Gomphocythere dasyderma* (Whatley et al. 2002a), *Gomphocythere* sp. 1, *Paracypretta subglobosa* (Sowerby 1840), *Paracypretta jonesi* (Bhatia and Rana 1984), *Paracypretta verruculosa* (Whatley et al. 2002a), *Strandesia jhilmiliensis* (Khosla et al. 2011a), *Stenocypris cylindrica* (Sowerby in Malcolmson 1840), *Periosocypris megistus* (Whatley et al. 2012), *Zonocypris spirula* (Whatley and Bajpai 2000), *Zonocypris viriensis* (Khosla and Nagori 2005), *Zonocypris labyrinthicos* (Whatley et al. 2002b), *Zonocypris gujaratensis* (Bhandari and Colin 1999), *Zonocypris penchi* sp. nov., *Cypridopsis astralos* (Whatley et al. 2002a), *Cypridopsis hyperectyphos* (Whatley and Bajpai 2000), *Cypridopsis elachistos* (Whatley et al.

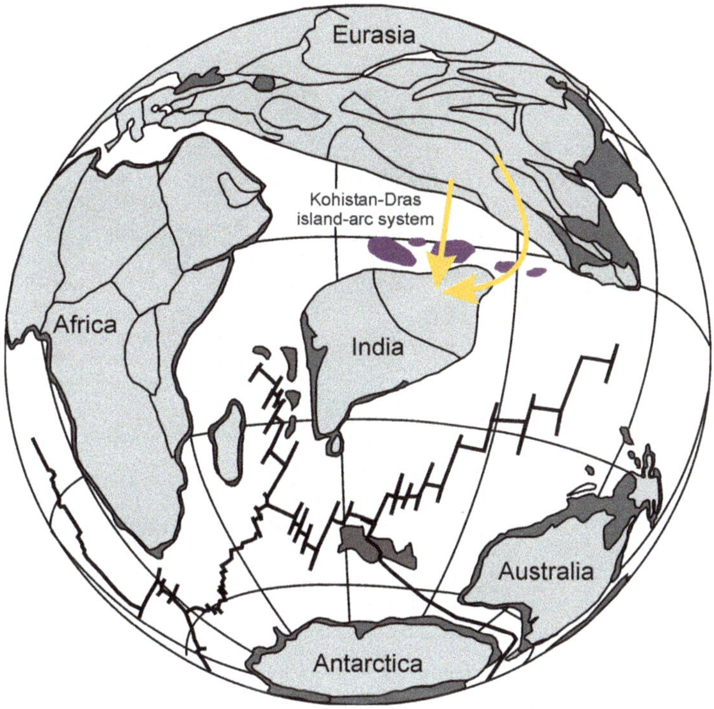

Fig. 6.5 Map showing palaeobiogeographic reconstruction portraying presumed Laurasian connections with the Indian Subcontinent. (Reproduced from Khosla et al. 2022 with permission from Wiley)

2002b), *Candona* sp., *Eucypris pelasgicos* (Whatley and Bajpai 2000), *Eucypris* sp. 1, ?*Eucypris verruculosa* (Whatley et al. 2002a), *Cyclocypris amphibolos* (Whatley et al. 2002a), *Cypria cyrtonidion* (Whatley and Bajpai 2000), *Talicypridea pavnaensis* (Khosla et al. 2005), *Cyprois rostellum* (Whatley and Bajpai 2000), *Cyprois* sp. and *Darwinula* sp. Among these species, 34 taxa were recovered from the two surface sites exposed at Jhilmili (25 taxa) and Ghat Parasia (09 taxa). And 16 species were recovered from two subsurface intertrappeans sites, which are in the Government well (08 species) and Shriwas (=Shiraj) well (08 species). These taxa have a broad spatial distribution in the infra- and intertrappean beds of peninsular India (Figs. 6.6, 6.7, 6.8; Table 6.1).

The species *Buntonia whittakerensis* sp. nov., *Limnocythere whatleyi* sp. nov., *Zonocypris penchi* sp. nov. and *Neocyprideis raoi* are only known from K-Pg transition yielding Jhilmili intertrappean beds (Fig. 6.5). Except for brackish water genera, *Buntonia* and *Neocyprideis*, most recovered ostracod taxa were found in freshwater environments such as rivers, streams, lakes, and ponds, and their presence in Jhilmili, Ghat Parasia, Government well and Shriwas (=Shiraj) well suggests that ostracod-bearing infra- and intertrappean sites may have been inter-linked by a palaeodrainage network. The presence of *Buntonia* and *Neocyprideis* in Jhilmili suggests that during the K-Pg interval, nearby shallow marine waters, either from

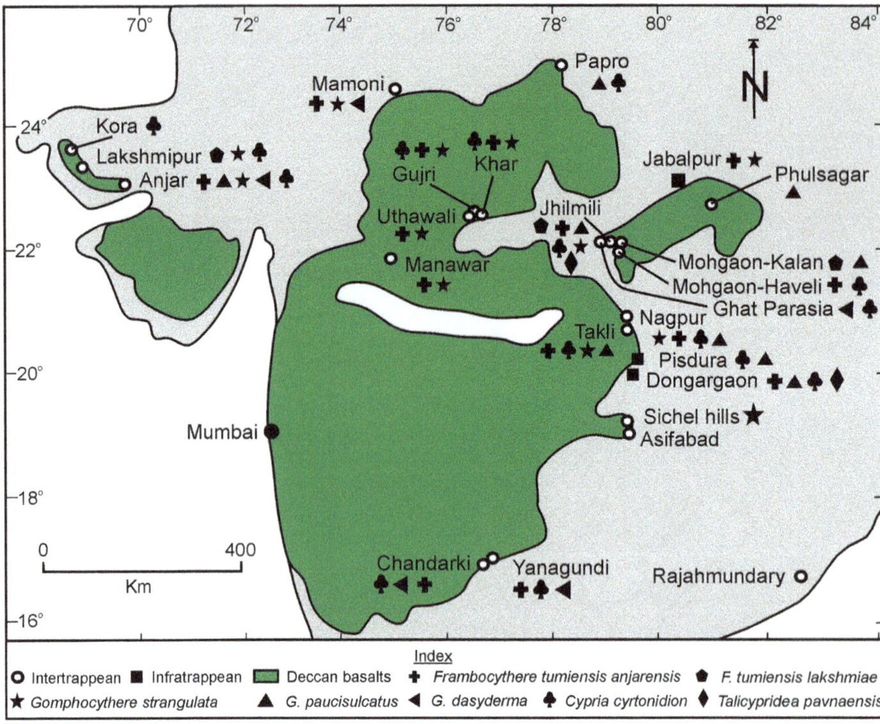

Fig. 6.6 Map showing the spatial distribution of ostracods in the Upper Cretaceous–? Lower Palaeocene infra- and intertrappean beds, peninsular India

the Bay of Bengal via the Pranhita-Godavari rift or from the Arabian Sea via the Narmada-Tapti rift systems, infiltrated the continental interior region of Central India. Planktic foraminiferans from the K-Pg transition-bearing intertrappeans of Rajahmundry Traps, Andhra Pradesh, and Jhilmili support marine water incursions from south-eastern India (Keller et al. 2008, 2009a, b).

The vast majority of fossil ostracod species from India are endemic to the Indian subcontinent (Khosla 2015). Several Indian taxa, such as *Cyclocypris, Eucypris, Limnocythere* and *Mongolianella*, appear to have spread from India to various parts of the Maastrichtian world during the latest Cretaceous (Whatley and Bajpai 2006; Khosla 2015). *Gomphocythere* and *Paracypretta* are endemic to infra- and intertrappean localities in the east, west, central and southern parts of peninsular India (Whatley and Bajpai 2006; Khosla 2015). Although the species *Frambocythere tumiensis* was first discovered in the Late Cretaceous of Spain (Helmdach 1978), the genus *Frambocythere* has also been widely reported from Indian Upper Cretaceous intertrappeans. Furthermore, it is derived from the Montian of Belgium (Tambareau 1984) and Albian of Africa (Colin 1993). *Frambocythere* has also been found in the middle Eocene deposits of China, Iran and Europe (Hou et al. 1978; Tambareau 1984; Tambareau et al. 1991; Colin 2011; Kapur et al. 2019).

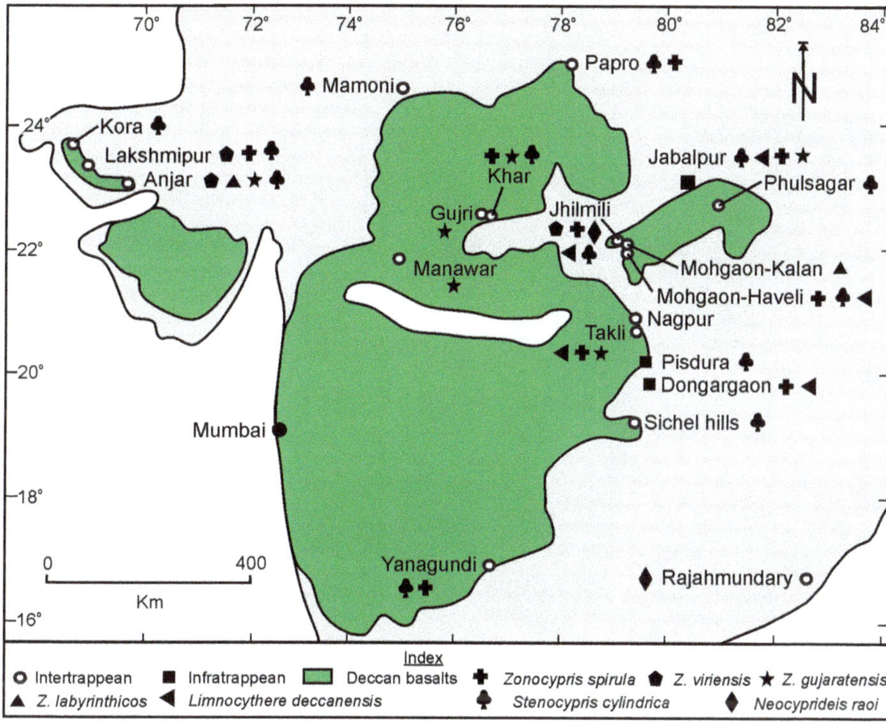

Fig. 6.7 Map showing the spatial distribution of ostracods in the Upper Cretaceous–Lower Palaeocene infra- and intertrappean beds, peninsular India

Different genera, for example, the limnocytherids *Limnocythere* and *Gomphocythere*, and cyprids like *Cypria* and *Paracandona* are found to have originated in and advanced out from India (Whatley and Bajpai 2006; Khosla 2015). *Neocyprideis* (brackish water) has been extensively recorded from the Rajahmundry intertrappean beds of the south-eastern region of India (Jain 1978). Outside India, Keen (1977) has also recorded the same species from the upper Eocene beds of the Hampshire Basin, England. A level of 98% endemism has been recorded in the Indian ostracod fauna, and one genus, *Limnocypridea*, which has been broadly recorded from the Upper Cretaceous deposits of Mongolia and China, colonised India (Whatley and Bajpai 2006; Khosla 2015).

Three of the ostracod genera – *Gomphocythere, Eucypris* and *Cypridopsis* – have been considered as Indian dominants. *Gomphocythere*, with nine species previously known from India, is supposed to have travelled "Out-of-India" to North China, Alaska and Africa. Eleven species of *Eucypris* are known from Indian Upper Cretaceous deposits, which possibly originated in India and then evolved out from India to China, Mongolia and Europe. *Cypridopsis*, with 17 species known from India, is accepted to have relocated "Out-of-India" to colonise Alaska and Mongolia (Kapur et al. 2019). Furthermore, the diversity of the Indian *Eucypris* is much higher

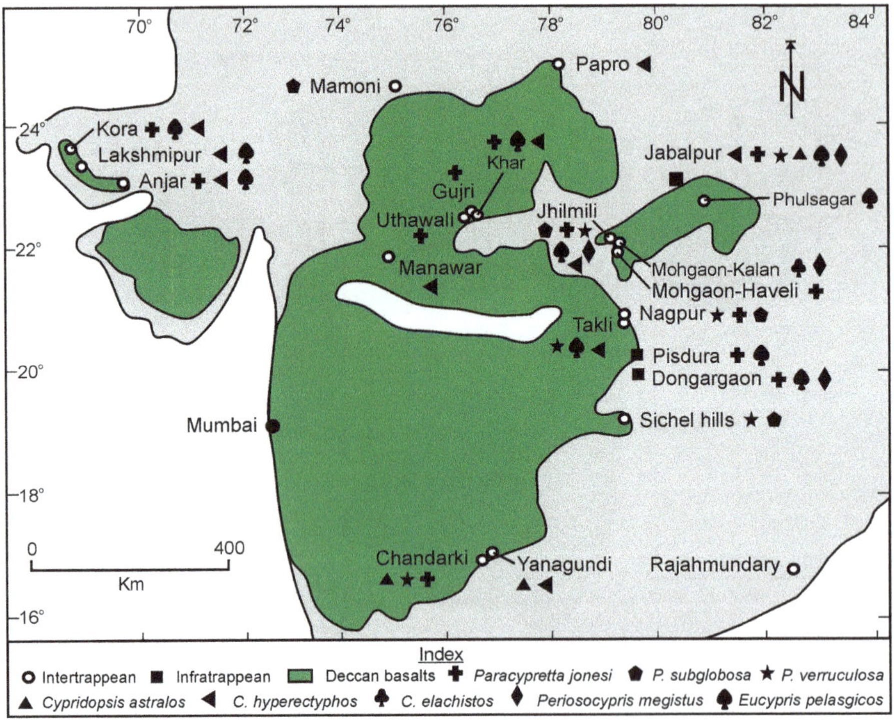

Fig. 6.8 Map showing the spatial distribution of ostracods in the Upper Cretaceous of peninsular India

when contrasted with that in Mongolia during the Maastrichtian (Whatley 2012). Considering the greater diversity of *Gomphocythere, Eucypris* and *Cypridopsis* in the Indian Late Cretaceous (Maastrichtian), Whatley et al. (2012) and Kapur et al. (2019) supported an Indian origin and an ensuing "Out-of-India" dispersal of these genera.

Buntonia (Howe 1935) is a unique genus for reconstructing palaeobiogeography. It is found on all continents except Australia and Antarctica and ranges from the "middle" Cretaceous to the Recent (Puckett et al. 2012; Elewa and Abdelhady 2020). It is widely distributed on the African continent and is thought to be a common genus of ostracods in the continent's Maastrichtian to Eocene deposits (Elewa and Abdelhady 2020). *Buntonia* is known from the Late Cretaceous to Palaeocene of Libya (El Sogher 1996) and the Senonian to Palaeogene of Egypt (Sarr 2015). During the Early Cenozoic, Africa produced a rich fossil record of *Buntonia*, primarily from the Palaeocene of Libya, Dahomey, Ivory Coast, Mali, Egypt and Nigeria, and the Eocene of Algeria, Togo, Nigeria, Senegal, Mali and Egypt (Shahin 2005; Elewa and Abdelhady 2020; Fig. 6.9). Apart from Africa, it has been reported from Maastrichtian deposits in Jamaica and Madagascar (Puckett et al. 2012; Benmansour et al. 2016; Fig. 6.9), as well as Eocene sediments of the Shubuta Clay

Table 6.1 Distribution of recovered ostracod taxa in the infra- and intertrappean beds of peninsular India

Localities	Madhya Pradesh										MR				KN		GR			RS	TG	AP	UP
Ostracod taxa	Jhilmili	Mohgaon-Kalan	Mohgaon-Haveli	Ghat Parasia	Phul-sagar	Jabalpur	Manawar	Khar	Uthawali	Gujri	Takli	Nand	Pinjaurni	Don-gargaon	Chandarki	Yana-gundi	Anjar	Lakshmi-pur	Kora	Mamoni	Asifabad	Dud-dukuru	Papro
Buntonia whitta-kerensis sp. nov.	√	×	×	×	×	×	×	×	×	×	×	×	×	×	×	×	×	×	×	×	×	×	×
Neocyprideis raoi	√	×	×	×	×	×	×	×	×	×	×	×	×	×	×	×	×	×	×	×	×	√	×
Limno-cythere deccanensis	√	×	√	√	√	√	×	×	×	×	√	×	√	√	×	×	×	×	×	×	×	×	×
Limno-cythere martensi sp. nov.	√	×	×	×	×	×	×	×	×	×	×	×	×	×	×	×	×	×	×	×	×	×	×
Frambo-cythere tumiensis anjarensis	√	×	√	√	×	√	√	√	√	√	√	×	√	×	√	√	√	×	×	√	×	×	×
Frambo-cythere tumiensis lakshmiae	×	√	×	×	×	×	×	×	×	×	×	×	×	×	×	×	×	√	×	×	×	×	×
Gompho-cythere strangulata	√	√	×	×	×	√	√	√	√	√	√	×	√	√	×	×	√	√	×	√	√	×	×
Gompho-cythere paucisulcatus	√	√	×	√	×	√	×	×	×	×	√	×	√	×	×	×	√	×	×	×	×	×	√

(continued)

Table 6.1 (continued)

Localities	Madhya Pradesh										MR				KN		GR			RS	TG	AP	UP
Ostracod taxa	Jhilmili	Mohgaon-Kalan	Mohgaon-Haveli	Ghat Parasia	Phul-sagar	Jabalpur	Manawar	Khar	Uthawali	Gujri	Takli	Nand	Pinjaurni	Don-gargaon	Chandarki	Yana-gundi	Anjar	Lakshmi-pur	Kora	Mamoni	Asifabad	Dud-dukuru	Papro
Gompho-cythere dasyderma	×	×	×	√	×	×	×	×	×	×	×	×	×	×	√	√	×	×	×	√	×	×	×
Gompho-cythere sp. 1	×	√	×	×	×	×	×	×	×	×	×	×	×	×	×	×	×	×	×	×	×	×	×
Paracypretta subglobosa	√	×	×	√	×	×	×	×	×	×	×	×	×	×	×	×	×	×	×	√	×	×	×
Paracypretta jonesi	√	×	√	×	×	√	×	√	√	√	×	×	×	√	√	×	√	×	√	×	×	×	×
Paracypretta verruculosa	√	×	×	×	×	√	×	×	×	×	√	×	×	×	√	×	×	×	×	×	×	×	×
Strandesia jhilmiliensis	√	×	×	×	×	×	×	×	×	×	×	×	×	×	×	×	×	×	×	×	×	×	×
Stenocypris cylindrica	√	×	×	×	√	√	×	√	×	×	√	×	×	×	√	√	√	√	√	√	×	×	×
Periosocypris megistus	√	×	×	√	×	√	×	×	×	×	×	×	×	×	×	×	×	×	×	×	×	×	×
Zonocypris spirula	√	×	√	×	×	×	×	√	×	×	√	×	√	√	×	√	×	√	×	×	×	×	√
Zonocypris viriensis	√	×	×	×	×	×	×	×	×	×	×	×	×	×	×	×	×	√	×	×	×	×	×
Zonocypris labyrinthicos	×	√	√	×	×	×	×	×	×	×	×	×	×	×	×	×	√	√	×	×	×	×	×
Zonocypris gujaratensis	×	√	√	×	×	√	√	√	×	√	√	×	×	×	×	×	√	√	×	×	×	×	×

Zonocypris penchi sp. nov.	√	×	×	×	×	×	×	×	×	×	×	×	×	×	×	×	×	×	×	×	×	×	×
Cypridopsis astralos	√	×	×	×	×	√	×	×	×	×	×	×	×	×	√	√	×	×	×	×	×	×	×
Cypridopsis hyperectyphos	√	×	×	√	×	√	√	√	×	×	√	×	√	×	×	√	√	√	√	√	×	×	√
Cypridopsis elachistos	×	×	×	×	×	×	×	×	×	×	×	×	×	×	×	×	×	×	×	×	×	×	×
Candona sp.	×	×	×	√	×	×	×	×	×	×	×	×	×	×	×	×	×	×	×	×	×	×	×
Eucypris-pelasgicos	√	√	×	×	×	√	×	√	×	×	√	×	×	√	×	×	√	√	√	×	×	×	×
Eucypris sp. 1	×	×	×	√	×	×	×	×	×	×	×	×	×	×	×	×	×	×	×	×	×	×	×
? *Eucypris verruculosa*	×	√	×	×	×	×	×	×	×	×	×	×	×	×	×	×	×	×	×	×	×	×	×
Cyclocypris amphibolos	√	×	√	√	×	√	×	√	×	×	√	×	×	√	×	√	×	×	√	√	×	×	×
Talicypridea pavnaensis	√	×	×	×	×	×	×	×	×	×	×	×	×	√	×	×	×	×	×	×	×	×	×
Cyprois rostellum	×	√	×	×	×	√	×	×	×	×	×	×	×	×	×	×	×	√	√	×	×	×	√
Cyprois sp.	×	×	×	√	×	×	×	×	×	×	×	×	×	×	×	×	×	×	×	×	×	×	×
Darwinula sp.	×	×	×	√	×	×	×	×	×	×	×	×	×	×	×	×	×	×	×	×	×	×	×

Compiled after Bhatia and Rana (1984), Bhandari and Colin (1999), Whatley and Bajpai (2000a, b), Whatley et al. (2002a, b, 2003a, b, 2012), Khosla et al. (2005, 2011a, b, 2015); Kshetrimayum et al. 2021 and present study. *MR* stands for Maharashtra, *KN* for Karnataka, *GR* for Gujarat, *RS* for Rajasthan, *TG* for Telangana, *AP* for Andhra Pradesh and *UP* for Uttar Pradesh

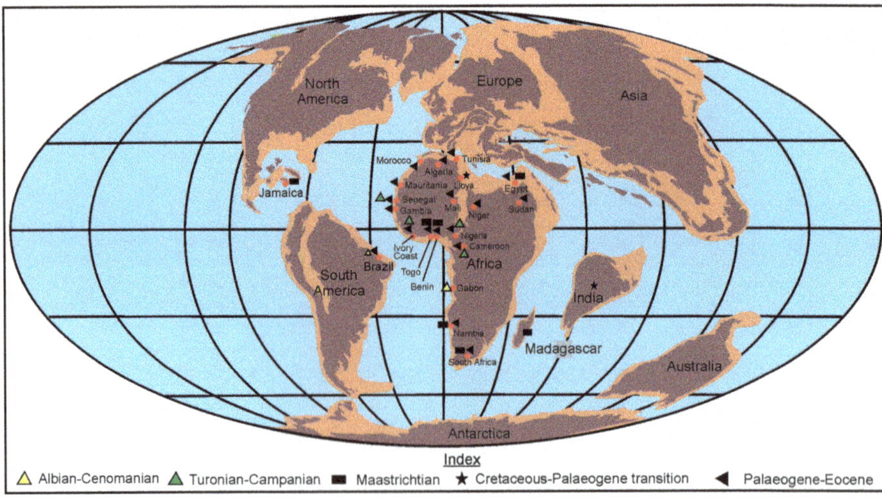

Fig. 6.9 Map showing global palaeobiogeographic distribution of the genus *Buntonia* during the Late Cretaceous to Eocene. (Base map modified after Wing and Sues 1992)

in eastern Mississippi, USA (Howe and Chambers 1935; Puckett et al. 2012; Benmansour et al. 2016).

Elewa and Abdelhady (2020) undertook a detailed investigation of various species of *Buntonia*, known from Maastrichtian to Late Eocene deposits of Africa. Based on the hierarchical clustering of characteristic *Buntonia* species (*B. bopaensis*, *B. punctata* and *B. tatteuliensis*), these workers identified three bioregions for the African *Buntonia* during the Palaeocene-Eocene. These bioregions are: (i) North Africa comprising Algeria, Tunisia, Libya, Sudan, and Morocco; (ii) West Africa consisting of Mali, Benin, Cameron, Côte d'Ivoire, Ghana, Gambia, Mauritania, Nigeria, Senegal, and Togo; and (iii) South Africa including Namibia and South Africa. Thus, two migration trends for *Buntonia* on the African continent have been reported from the Late Palaeocene to Late Eocene, the first in West Africa from south to north and the second from West Africa to North Africa (Elewa and Abdelhady 2020). Furthermore, two dispersal routes for *Buntonia* in Africa were visualised in the Late Cretaceous, one from South to North Africa via the Trans Sahara route and the other from West to North Africa via the Atlantic Tethyan route (Elewa and Abdelhady 2020).

The presence of *Buntonia* in the Maastrichtian deposits of Madagascar and K-Pg boundary transition sediments of the Jhilmili section (India) indicates that *Buntonia* dispersed in shallow marine waters from Africa to India via Madagascar during the latest Cretaceous by following the Seychelles block, Amirante Ridge and Providence bank. The presence of *Buntonia* in India also suggests the possibility of another dispersal route from Africa to India via the Oman-Kohistan-Dras Island Arc near or at the K-Pg transition (Chatterjee et al. 2013; Goswami et al. 2011, 2013; Verma et al. 2017).

Limnocythere is known from the Bajocian of Tunisia (Africa), Early Cretaceous of Brazil, Late Cretaceous of Mongolia and Late Cretaceous-Late Palaeogene of China (Zhang 1984; Moura 1988; Khand et al. 2000; Tiss et al. 2019). *Zonocypris* has been reported from the Cretaceous of Europe, India, China and South America. Therefore, these genera had a cosmopolitan distribution during the Late Cretaceous-Early Palaeocene.

According to a few workers such as Story et al. (1995) and Karanth (2006), the Indian Maastrichtian freshwater ostracod fauna demonstrates that around 88 Ma India was isolated from Madagascar and remained detached during its northward drift through the Indian Ocean for 40 Ma prior to slamming into the Asian landmass around 50 Ma, conveying with it Gondwanan forms to Asia (Karanth 2006). In this way, the ostracod fauna from the investigated sites and different areas of peninsular India upholds the "Out-of-India" dispersal hypothesis (Whatley and Bajpai 2006; Khosla 2015; Kapur et al. 2019).

6.5 Foraminiferans

Planktic foraminiferans are important in understanding ancient biogeography because they provide useful information about patterns of oceanic surface water palaeocirculation, palaeoclimate zones, palaeotemperature, palaeobiogeographic realms, sea level fluctuations, the nature of the upper water column in ancient oceans and biotic episodes of extinction/recovery (Malmgren 1991; Olsson et al. 2006). The biogeographic provincialisation in planktic foraminiferans has been observed since the "middle" Cretaceous (Malmgren 1991). As a consequence, four biogeographic realms have been established for Cretaceous planktic foraminiferans: tropical-subtropical Tethyan realm, cool temperature boreal realm in the Northern Hemisphere, austral realm in the Southern Hemisphere and warm temperature transitional realm (located between the boreal/austral and Tethyan realms) (Scheibnerova 1971; Malmgren 1991).

The planktic foraminiferal assemblage consisting of *Subbotina triloculinoides* (Plummer 1926), *Globanomalina compressa* (Plummer 1926), *Woodringina hornerstownensis* (Olsson 1960), *W. claytonensis* (Loeblich and Tappan 1957), *Hedbergella holmdelensis* (Olsson 1964), *Guembelitria cretacea* (Cushman 1933), *Parasubbotina pseudobulloides* (Plummer 1926), *Globigerinelloides aspera* (Ehrenberg 1854) and *Globigerina* (*Eoglobigerina*) *pentagona* (Morozova 1961), and foraminiferam genus and species indeterminate were recovered from the Jhilmili section, India. Among these species, *Hedbergella holmdelensis*, and *Guembelitria cretacea* range from Late Cretaceous (Maastrichtian) to Palaeocene and *Globigerinelloides aspera* is restricted to the Late Cretaceous. The remaining species, *Subbotina triloculinoides* (planktic foraminiferal zones P1b to P4), *Globanomalina compressa* (zones P1c to P3), *W. hornerstownensis* (zones Pa to P3b), *W. claytonensis* (zones Pa to Plb), *Parasubbotina pseudobulloides* (zones Pa to P3a) and *Globigerina* (*Eoglobigerina*) *pentagona* (zones Pa to P2), are dominantly of Palaeocene age.

The presence of the above-mentioned mixed Late Cretaceous to Palaeocene planktic foraminiferal assemblage at Jhilmili clearly represents the presence of the K-Pg boundary interval during the time of the Deccan volcanic eruptions in peninsular India. This assemblage is largely known from low to high latitudes and has a global distribution during the Late Cretaceous-Palaeocene (Olsson et al. 1999, 2006; Fig. 6.10).

The presence of this assemblage at Jhilmili indicates that the Jhilmili area of Central India was episodically well-connected with marine environments, and marine incursion/s occurred inland during eustatic sea-level changes or high tides. Furthermore, regional tectonic activity associated with the Pranhita-Godavari and the Narmada-Tapti rift systems may have facilitated marine incursions at or near the K-Pg boundary (Kumari et al. 2020; Khosla et al. 2021, 2022). The discovery of a similar planktic foraminiferal assemblage from coeval intertrappeans of the Rajmundary Traps in the Pranhita-Godavari valley supports the possibility of marine incursion/s from the Bay of Bengal, southeastern India (Keller et al. 2008). The

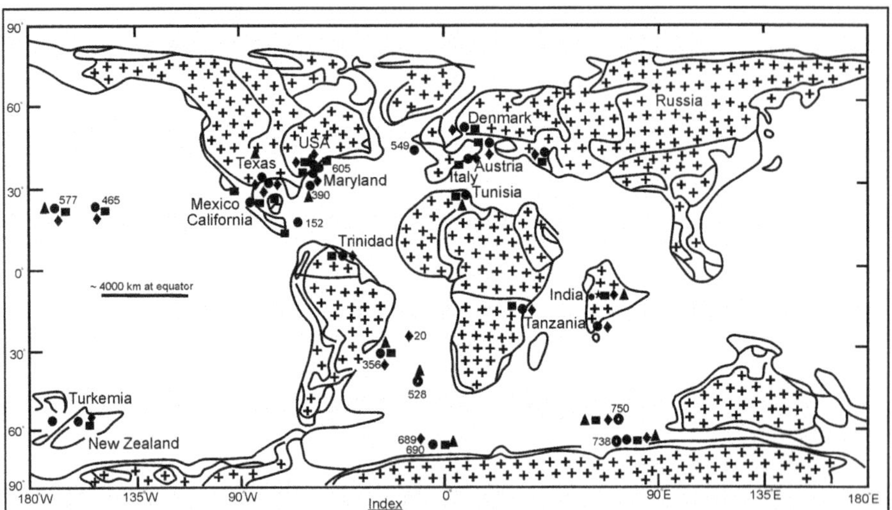

Fig. 6.10 Map showing global distribution Cretaceous to Palaeocene of planktic foraminiferal taxa recovered from Jhilmili (modified after Olsson et al. 1999, 2006). Whereas 20: DSDP hole, Brazil Basin, South Atlantic Ocean; 152: DSDP site, Nicaragua Rise, Caribbean Sea; 356: DSDP site, Sao Paulo Plateau, South Atlantic Ocean; 390: DSDP hole, Blake-Bahama Basin, North Atlantic Ocean; 465: DSDP site, Hess Rise, Central North Pacific Ocean; 528: DSDP site, Walvis Ridge, South Atlantic Ocean; 549: DSDP site, Goban Spur, eastern North Atlantic Ocean; 577: DSDP site, Shatsky Rise, western Pacific Ocean; 605: DSDP site, New Jersey margin; 689: ODP hole, Maud Rise, Weddell Sea, Southern Ocean; 690: ODP hole 690, Maud Rise, Wedell Sea, Southern Ocean and 750: ODP hole, Kerguelen Plateau, Southern Indian Ocean. DSDP stands for Deep Sea Drilling Project, and ODP for Ocean Drilling Program

absence of a planktic foraminiferal assemblage from the intertrappean beds in the westernmost part of the Narmada-Tapti rift reduces the likelihood of marine incursions from the Arabian Sea in western India. As a result, more effort will be required to identify K-Pg transition-bearing foraminiferal assemblages from the western part of the Narmada-Tapti rift systems.

6.6 Fishes

Fishes live in a variety of aquatic environments, including the deep sea, shallow sea, coastal, freshwater and lacustrine. As a result, their fossil remains provide numerous clues for reconstructing geological and biogeographic patterns. Because freshwater fishes are unable to cross marine barriers and have limited dispersal abilities, they are an important fauna for reconstructing terrestrial palaeobiogeography, whereas marine fishes are of the opposite utility (Lieberman 2000; Nelson 2006; Claeson et al. 2010). In the current study, three fish taxa were recovered from the intertrappean sediments at Jhilmili, Ghat Parasia, the Shiraj (= Shriwas) well and Government well, Chhindwara District, Madhya Pradesh, Central India. They are *Igdabatis indicus* (Prasad and Cappetta 1993 Myliobatidae), *Lepisosteus indicus* (Woodward 1908, Lepisosteidae) and Osteoglossidae indeterminate.

The living representatives of Myliobatidae are inhabitants of pelagic environments and commonly live in tropical and warm waters and undertake seasonal migration in temperate waters (Nelson 2006; Claeson et al. 2010; Verma et al. 2017). The genus *Igdabatis* has been documented from the Upper Cretaceous of Africa, Spain (Europe) and India (Fig. 6.11). *Igdabatis sigmodon* is an African

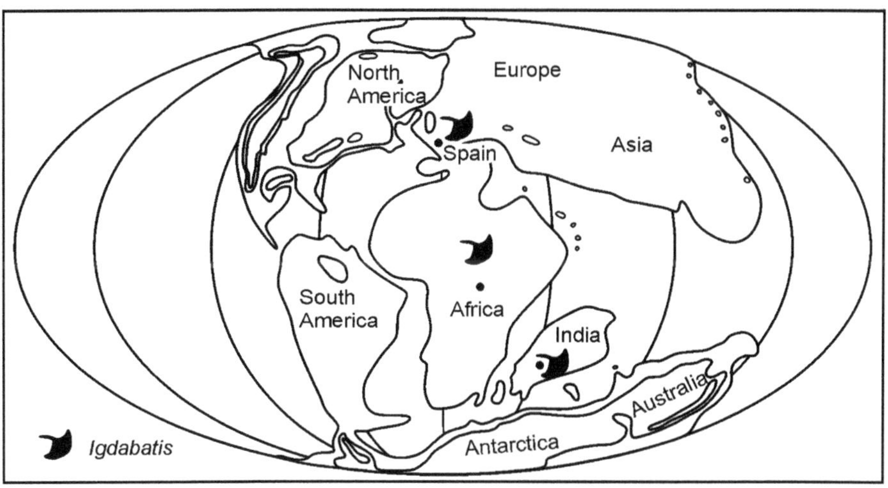

Fig. 6.11 Map showing the spatial distribution of the genus *Igdabatis* during the Late Cretaceous. (Base map modified after Wing and Sues 1992)

species and is reported from the Maastrichtian of Niger (Cappetta 1972). The Indian form, *Igdabatis indicus,* is known from many Maastrichtian sites in peninsular India associated with the Deccan Traps such as Pisdura, Jabalpur, Asifabad, Nagpur, Naskal, Timsanpali, Marepalli, Kisalpuri, Lothkheri and Piplanarayanwar (Khosla et al. 2004; Verma et al. 2016; Lourembamm et al. 2017). In Europe, the genus *Igdabatis* is represented by two forms, *Igdabatis marmii* and *Igdabatis* cf. *indicus*, which are documented from the Campanian-Maastrichtian deposits in the Tremp Formation, Spain (Gijon and Martinez 1998; Kriwet et al. 2007; Blanco 2019). This distribution of the genus *Igdabatis* in Spain, Niger and India reflects shallow marine water dispersal events between northern Africa and southern Europe and between eastern Africa and western India. The presence of the Oman-Kohistan-Dras Island Arc between India and Africa near or at the K-Pg transition would have facilitated the shallow marine dispersal of *Igdabatis* from Africa to India or vice versa (Chatterjee et al. 2013; Verma et al. 2017).

In peninsular India, the fossil record of the genus *Lepisosteus* is known from several infratrappean sites Dongargaon, Pisdura, Nand-Dongargaon and Marepalli – as well as intertrappean localities such as Nagpur, Asifabad, Naskal, Kisalpuri and Piplanarayanwar (Woodward 1908; Gayet et al. 1984; Prasad and Khajuria 1990; Mohabey and Udhoji 1996; Khosla 2004; Verma et al. 2016; Lourembam et al. 2017). In addition, it is also known from the Palana and Fatehgarh formations of Palaeocene age, Rajasthan (Kumar et al. 2005; Rana et al. 2006). Apart from India, the genus *Lepisosteus* has a global distribution (Fig. 6.12) as it is known from the Upper Cretaceous (? Campanian) Maevarano Formation of the Mahajanga Basin, Madagascar (Gotterfried and Krause 1998), Campanian deposits of the Champ-Garimond (Gard), France (Sige et al. 1997), Maastrichtian deposits of Oarda de Jos, Romania (Codrea et al. 2010), Upper Maastrichtian sediments of Serrat del Pelleu

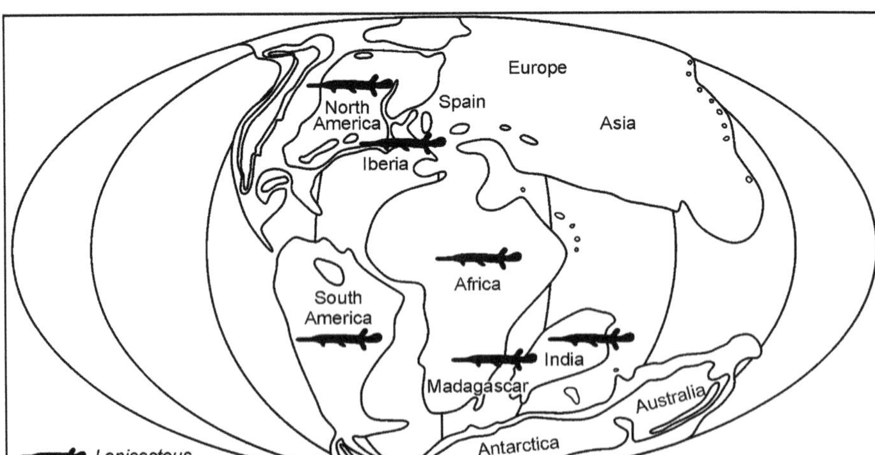

Fig. 6.12 Map showing the spatial distribution of the genus *Lepisosetus* in the Late Cretaceous. (Base map modified after Wing and Sues 1992)

(southern Pyrenees), Spain (Astibia et al. 1990; Blanco and Bolet 2014) and Uppermost Maastrichtian deposits of Budurone (Hateg Basin), Western Romania (Csiki et al. 2008). It has also been found in the USA in the Palaeocene Cannonball Formation of North and South Dakota (Cvancara and Hoganson 1993), the Palaeocene of South Carolina (Weems 1998) and the Eocene of Virginia (Westgate 1989). Thus, it is proposed that this genus may have achieved a global distribution, and its presence in India may have resulted from a dispersal event from the south or north during the Late Cretaceous.

Except for Antarctica, the living and fossil forms of osteoglossids are known from North America, South America, Australia, Africa, Europe and Asia (Hilton and Carpenter 2020). This group has more fossil genera than extant ones, and its fossil record is diverse, extending all the way back to the Late Cretaceous and known from both marine and freshwater deposits around the world (Hilton and Carpenter 2020). As a result, assessing the palaeobiogeographic significance of the current find of osteoglossids would be difficult to ascertain because the recovered material is fragmentary and poor and cannot be identified at the species level.

In conclusion, the investigated biota, which included charophytes, ostracods, foraminiferans and fishes from the intertrappean beds of the Chhindwara District, Madhya Pradesh, reveal a complex biogeographic relationship with neighbouring landmasses during the Late Cretaceous, or at the K-Pg boundary. This is due to their Gondwanan and Laurasian affinities. Similar conclusions were reached based on vertebrate fossil records from the Late Cretaceous of India. Consequently, the palaeobiogeographical affinities of the charophyte flora, ostracods, vertebrates and invertebrates from peninsular India are intricate and display a mixed and varied pattern that resulted from the accretion of Laurasian elements (e.g. charophytes, palynomorph assemblages, vertebrates like discoglossid frogs, anguid lizards and alligatorid crocodiles) and endemic forms (ostracods) to Gondwanan taxa, for example, nigerophiid and madtosiid snakes, baurusuchid and notosuchian crocodiles, bothremydid and pelomedusid turtles, ranoid, hylid and leptodactylid frogs, haramyid and gondwanatherian mammals and sauropod eggs and abelisaurid dinosaur skeletal material (Khosla 2014, 2021; Khosla and Verma 2015; Kapur and Khosla 2016, 2019).

6.7 Conclusions

1. Palaeobiogeographically, the Jhilmili assemblage is dominated by five species of *Platychara* (*Platychara perlata*, *P. raoi*, *P. sahnii*, *P. compressa*, and *P.* sp.). The Indian species of the genus *Platychara* are endemic to the Indian peninsular region. The charophyte assemblages of the Jhilmili area in Central India exemplify strong relationship or affinities with the Late Cretaceous charophyte assemblages globally known from different continents such as Asia, the Americas, Europe and Africa, both at the generic and specific level, and support some palaeobiogeographic inferences. *Chara* is a rare charophyte genus of Asian affini-

ties, and its presence in India indicates dispersal events from Asia to India near the Cretaceous-Palaeogene boundary.

2. The charophyte assemblage from the Upper Cretaceous of the Jhilmili section reflects the fact that during the Cretaceous-Palaeogene time interval, the Indian subcontinent experienced exchanges of species from different corners of the world, chiefly Laurasian ones. Thus, similarities are noted between the charophytes of Jhilmili and those from North–South America, Europe and North Africa, which point to dispersals of charophytes among these continental masses across some migration routes.
3. On the basis of the Jhilmili charophytes, it is proposed that the genera *Platychara, Peckichara*, *Nemegtichara*, *Microchara* and *Chara* dispersed from Laurasia to India by following a sweepstakes dispersal route via the Kohistan–Dras volcanic island-arc system. Further, there is also a possibility that these forms may have a vicariant distribution associated with fragmentation of the Gondwana continents, but it remains inconclusive based on the current fossil record of the southern continents.
4. Three of the ostracod genera *Gomphocythere, Eucypris* and *Cypridopsis* have been considered as Indian dominants. *Gomphocythere*, with nine species with their earliest records in India, is supposed to have travelled "Out-of-India" to North China, Alaska and Africa.
5. Eleven species of *Eucypris* are known from Indian Upper Cretaceous deposits, which possibly originated in India and then evolved out from India to China, Mongolia and Europe.
6. The presence of *Buntonia* in Maastrichtian deposits of Madagascar and K-Pg boundary-sediments of the Jhilmili section (India) indicates that *Buntonia* dispersed in shallow marine waters from Africa to India via Madagascar during the latest Cretaceous by following the Seychelles block, Amirante Ridge and Providence bank. The presence of *Buntonia* in India also suggests the possibility of another dispersal route from Africa to India via the Oman-Kohistan-Dras Island Arc near or at the K-Pg transition.
7. The presence of a planktic foraminiferal assemblage at Jhilmili indicates that the Jhilmili area of Central India was episodically well-connected with marine environments, and marine incursions occurred inland during eustatic sea-level changes or high tides.
8. The presence of the Oman-Kohistan-Dras Island Arc near or at the K-Pg transition between India and Africa would have been facilitated the shallow marine dispersal of *Igdabatis* from Africa to India or vice versa. The genus *Lepisosteus* may have achieved a global distribution during the Cretaceous, and its presence in India may have resulted from a dispersal event from the south or north during the Late Cretaceous.
9. More discoveries of microbiota will be needed from the intertrappean beds situated in the adjoining areas of Chhindwara to evaluate the role of the environmental impacts of the Deccan Volcanic Province on the contemporary biota across the K-Pg boundary.

References

Ali JR, Aitchison JC (2008) Gondwana to Asia: plate tectonics, paleogeography and the biological connectivity of the Indian subcontinent from the middle Jurassic through latest Eocene (166–35 Ma). Earth-Sci Rev 88:145–166

Ali JR, Aitchison JC (2009) Kerguelen Plateau and the Late Cretaceous southern-continent bioconnection hypothesis: Tales from a topographical ocean. J Biogeogr 36:1778–1784

Ali JR, Krause DW (2011) Late Cretaceous bioconnections between Indo-Madagascar and Antarctica: refutation of the Gunnerus ridge causeway hypothesis. J Biogeogr 38:1855–1872

Astibia H, Buffetaut E, Buscalioni AD, Cappetta H, Corral C, Estes R, Garciagarmilla F, Jaeger JJ, Jimenez-Fuentes E, Leloeuff J, Mazin JM, Orue-Extebarria X, Perereda-Suberblola J, Powell JE, Rage JC, Rodriguez-Lazaro J, Sanz JL, Tong H (1990) The fossil vertebrates from Lano (Basque Country, Spain): new evidence on the composition and affinities of the Late Cretaceous continental faunas of Europe. Terra Nova 2:460–466

Barron EJ, Harrison CGA (1980) An analysis of past plate motions: the South Atlantic and Indian oceans. In: Davies PA, Runcorn SK, editors. Mechanisms of continental drift and plate tectonics. London: Academic Press; p. 89–109

Benmansour S, Andreu BA, Yahiaoui A (2016) The Campanian–Maastrichtian of the Aures Basin, Algeria: paleobiogeographical distribution of ostracods. Cretac Res 58:86–107

Bhandari A, Colin JP (1999) Limnic ostracods of the intertrappean sediments (terminal Maastrichtian-basal Paleocene) of the Anjar region (Kachchh, Gujarat state), India: systematics, paleoecology, and paleobiogeographic affinities. J Micropaleontol 42(1):3–20

Bhandari M, Shukla PM, Pandey J (1987) Iridium enrichment at Cretaceous–Tertiary boundary in Meghalaya. Curr Sci 56:1003–1005

Bhandari N, Gupta M, Pandey J, Shukla PN (1994) Chemical profiles in K/T boundary section of Meghalaya, India: cometary, asteroidal or volcanic. Chem Geol 113:45–60

Bhandari N, Shukla PN, Ghevariya Z, Sundaram SM (1995) Impact did not trigger Deccan volcanism: evidence from Anjar K/T boundary intertrappean sediments. Geophys Res Lett 22:433–436

Bhandari N, Shukla PN, Ghevariya ZG, Sundaram SM (1996) K/T boundary layer in Deccan intertrappeans at Anjar, Kutch. In: Ryder G, Fastovsky DS, Gartner S (eds), The Cretaceous-Tertiary Event and other catastrophes in earth history. Geol Soc Am Spec Pap 307:417–424

Bhatia SB (1992) Recent advances in studies on fossil Charophyta. Palaeobotanist 40:413–419

Bhatia SB, Mannikeri MS (1976) Some charophyta from the Deccan intertrappean beds near Nagpur, Central India. Geophytology 6(1):75–81

Bhatia SB, Rana RS (1984) Palaeogeographic implications of the charophyta and ostracoda of the intertrappean beds of peninsular India. Mém Soc Geol Fr 147:29–35

Bhatia SB, Prasad GVR, Rana RS (1990a) Deccan volcanism, a Late Cretaceous event: conclusive evidence of ostracodes. In: Sahni A, Jolly A (eds) Cretaceous event stratigraphy and the correlation of the Indian nonmarine strata. A Seminar cum Workshop IGCP 216 and 245, Chandigarh, pp 47–49

Bhatia SB, Riveline J, Rana RS (1990b) Charophyta from the Deccan intertrappean beds near Rangapur, Andhra Pradesh, India. The Palaeobotanist 37:316–323

Bhatia SB, Prasad GVR, Rana RS (1996) Maastrichtian non-marine ostracodes from peninsular India: Palaeobiogeographic and age implications. Mem Geol Soc Ind 37:297–311

Blanco A (2019) *Igdabatis marmii* sp. nov. (Myliobatiformes) from the lower Maastrichtian (upper cretaceous) of North-Eastern Spain: An Ibero-Armorican origin for a Gondwanan batoid. J Syst Palaeontol 17(10):865–879

Blanco A, Bolet A (2014) Updating the knowledge on faunal assemblages from the Tremp formation (Late Cretaceous, upper Maastrichtian): microvertebrate remains from four new sites. In: Marmi J, Oms O, Vila B, Galobart A, Estrada R, Dinares-Turell J (eds) Reconstructing the terrestrial end-Cretaceous paleoenvironments in Europe, vol 7. Paleontologia i Evolucio Special Issue, p 60

Briggs JC (2003) The biogeographic and tectonic history of India. J Biogeogr 30:381–388
Cappetta H (1972) Les poisons cretaces et tertiaires du bassin des Iullemmeden (Republique du Niger). Palaeovertebrata 5:179–251
Case JA (2002) A new biogeographic model for dispersal of Late Cretaceous vertebrates into Madagascar and India. J Vertebr Paleontol 22(3):42A
Chand S, Radhakrishna M, Subrahmanyan C (2001) India-East Antarctica conjugate margins: rift-shear tectonic setting inferred from gravity and bathymetric data. Earth Planet Sci Lett 185:225–236
Chassagne-Manoukian M, Haddoumi H, Cappetta H, Charriere A, Feist M, Tabuce R, Vianey-Liaud M (2013) Dating the "red beds" of Eastern Moroccan high plateaus, evidence from Late Cretaceous charophytes and dinosaur eggshells. Geobios 46(5):371–379
Chatterjee S, Scotese CR (2010) The wandering Indian plate and its changing biogeography during the Late Cretaceous–early Tertiary period. Lect Not Ear Sci, New Asp Mes Biod 132:105–126
Chatterjee S, Goswami A, Scotese CR (2013) The longest voyage: tectonic, magmatic, and paleoclimatic evolution of the Indian plate during its northward flight from Gondwana to Asia. Gondwana Res 23(1):238–267
Claeson KM, O'Leary MA, Roberts EM, Sissoko F, Bouaré M, Tapanila L, Goodwin D, Gottfried MD (2010) First Mesozoic record of the stingray *Myliobatis wurnoensis* from Mali and a phylogenetic analysis of Myliobatidae incorporating dental characters. Acta Palaeontol Pol 55:655–674
Codrea V, Barbu O, Jipa-Murzea C (2010) Upper Cretaceous (Maastrichtian) land vertebrate diversity in Alba District (Romania). In BGSG, XLIII 2:594–601
Colin JP (1993) An early representative of the genus *Frambocythere Colin, 1980: Frambocythere pustulosa (Grekoff, 1957) from the Albian of Zaire. J Micropal 12:170*
Colin JP (2011) From light to darkness: *Frambocythere* Colin 1980 to *Kovalevskiella* Klein, 1963 (Limnocytheridae: Timiriasevinae). J Geol Paläontol 11:44–47
Courtillot V, Fluteau F (2014) A review of the embedded time scales of flood basalt volcanism with special emphasis on dramatically short magmatic pulses. Geol Soc Am Spec Pap 505:301–317
Courtillot V, Besse J, Vandamme D, Montigny R, Jaeger JJ, Cappetta H (1986) Deccan flood basalts at the Cretaceous/Tertiary boundary? Earth Planet Sci Lett 80:361–374
Csiki Z, Ionescu A, Grigorescu D (2008) The Budurone microvertebrate site from the Maastrichtian of the Haţeg Basin-flora, fauna, taphonomy and paleoenvironment. Acta Palaeontol Roman 6(4):49–66
Cushman JA (1933) Some new foraminiferal genera. Contrib Cushman Lab Foraminifer Res 8:4–7
Cvancara AM, Hoganson JW (1993) Vertebrates of the cannonball formation (Paleocene) in North and South Dakota. J Vertebr Paleontol 13(1):1–23
Sowerby J d C (1840) Explanations of the plates and wood-cuts. In: Malcolmson JC (ed), On the fossils of the eastern portion of the Great Basaltic District of India. Trans Geol Soc London 2(5):511–567
Dèzes P (1999) Tectonic and metamorphic evolution of the central Himalayan domain in southeast Zanskar (Kashmir, India). Mém Géol (Lausanne) 32:1–150
Ehrenberg CG (1854) Mikrogeologie. Das Erden und Felsen schaffende Wirken des unsichtbar kleinen selbstständigen Lebens auf der Erde. Leopold Voss, Leipzig, pp xxviii + 374
El Sogher A (1996) Late Cretaceous and Paleocene ostracodes from the Waha limestone and Hagfa shale formations of the Sirt Basin, Libya. In: Salim MJ, Mouzughi AJ, Hammuda OS (eds) The geology of Sirt Basin, Libya, vol 1. Elsevier, Amsterdam, pp 287–382
Elewa AMT, Abdelhady AA (2020) Migration routes of the genus *Buntonia* Howe, 1935 (Ostracoda, Crustacea) during the Late Cretaceous–Paleogene in Africa. Arab J Geosci 13:1182. https://doi.org/10.1007/s12517-020-06203-y
Feist M, Grambast-Fessard N, Guesquerlin M, Karol K, Lu H, McCourt RM, Wang Q, Shenzen Z (2005a) Treatise on invertebrate paleontology, part B Protoctista 1, vol 1. Charophyta. Geol Soc Am, Boulder, Colorado, pp 1–170
Feist M, Liu J, Tafforeau P (2005b) New insights into Paleozoic charophyte morphology and phylogeny. Am J Bot 92(7):1152–1160

Fosberg FR (1976) Geography, ecology, and biogeography. Ann Assoc Am Geogr 66(1):117–128
Gayet M, Rage JC, Rana RS (1984) Nouvelles ichthyofaune et herpétofaune de Gitti Khadan, le plus ancien gisement connu du Deccan (Crétacé/Paléocène) à microvertébrés. Implications paléogéographiques. Mém Soc Géol Fr 147:55–65
Gibbons AD, Whittaker JM, Müller RD (2013) The breakup of East Gondwana: assimilating constraints from Cretaceous Ocean basins around India into a best-fit tectonic model. J Geophys Res Solid Earth 118:1–15. https://doi.org/10.1002/jgrb.50079
Gijón RS, Martínez NL (1998) Sharks and rays (chondrichthyes) from the Upper Cretaceous red beds of the south-Central Pyrenees (Lleida, Spain): indices of an India-Eurasia connection. Palaeogeogr Palaeoclimatol Palaeoecol 141:1–12
Goswami A, Prasad GVR, Upchurch P, Boyer DM, Seiffert ER, Verma O, Gheerbrant E, Flynn JJ (2011) A radiation of arboreal basal eutherian mammals beginning in the Late Cretaceous of India. Proc Natl Acad Sci U S A 108:16333–16338
Goswami A, Prasad GVR, Verma O, Flynn JJ, Benson RBJ (2013) A troodontid dinosaur from the latest Cretaceous of India. Nat Commun 4:1–5. https://doi.org/10.1038/ncomms2716
Gottfried MD, Krause DW (1998) First record of gars (Lepisosteidae, Actinopterygii) on Madagascar: late cretaceous remains from the Mahajanga Basin. J Vertebr Paleontol 18(2):275–279
Grambast L (1957) Ornamentation de la gyrogonite et systematique chez les charophytes fossiles. Rev Gén Bot 64:339–362
Grambast L (1974) Phylogeny of the Charophyta. Taxon 23:463–481
Hay WW, DeConto RM, Wold CN, Wilson KM, Voigt S, Schulz M, Wold AR, Dullo WC, Ronov AB, Balukhovsky AN, Söding E (1999) Alternative global Cretaceous paleogeography. Geol Soc Am Spec Pap 332:1–47
Helmdach FF (1978) Nicht marine ostracodes aus der Spanischer Oberkreide. Berl Geowiss Abh 3:71–78
Hilton EJ, Carpenter J (2020) Bony-tongue fishes (Teleostei: Osteoglossomorpha) from the Eocene Nanjemoy Formation, Virginia. Northeast Nat 27(1):25–34
Hou Y, Ho JD, Ye CH (1978) The Cretaceous–Tertiary ostracodes from the marginal region of the Yangtze-Han river plain in Central Hubei. Mem Nanj Inst Geol and Pal Acad Sin 9:121–206. in Chinese
Howe HV (1935) In: Howe HV, Chambers J 1935. Louisiana Jacksone Eocene Ostracoda. Louisiana Geol Surv Geol Bull 5:1–65
Howe HV, Chambers J (1935) Louisiana Jacksone Eocene Ostracoda. Louisiana Geol Surv Geol Bull 5:1–65
Hu X, Garzanti E, Wang J, Huang W, An W, Webb A (2016) The timing of India-Asia collision onset – facts, theories, controversies. Earth-Sci Rev 160:264–299
Huang RJ (1988) Charophytes of Nanxiong Basin, Guangdong and its Cretaceous–Tertiary Boundary. Acta Micropaleontol Sin 27(4):457–474
Humphries CJ (2001) Vicariance biogeography. In: Levin SA (ed) Encyclopedia of biodiversity, vol 5. Elsevier Academic Press, pp 767–779
Iturralde-Vinent MA, MacPhee RDE (1999) Paleogeography of the Caribbean region: implications for Cenozoic biogeography. Bull Am Mus Nat Hist 238:1–95
Jablonski D, Flessa KW, Valentine JW (1985) Biogeography and paleobiology. Paleobiology 11(1):75–90
Jaeger JJ, Courtillot V, Tapponier P (1989) Palaeontological view of the ages of the Deccan Traps, the Cretaceous–Tertiary boundary, and the India–Asia collision. Geology 17:316–319
Jain SP (1978) Ostracodes des "inter-trappean beds" (Eocene Infeerieur) de Kateru, Rajahmundry, Côte est de l' Inde. Rev de Microplaeontol 21(2):51–58
Jones T (1860) Notes on the fossil Cypride. Quart J Geol Soc London 16:154–189
Kania S, Khosla A, Verma O, Prashanth M (2022) Charophyte assemblage in the Cretaceous–Palaeogene boundary of the Deccan intertrappeans from Chhindwara District (Madhya Pradesh). Central India Him Geol 43(2):383–396

Kapur VV, Khosla A (2016) Late Cretaceous terrestrial biota from India with special references to vertebrates and their implications for biogeographic connections. In: Khosla A, Lucas SG (eds) Cretaceous period: biotic diversity and biogeography, vol 71. New Mexico Museum of Natural History & Science, Albuquerque, pp 161–172

Kapur VV, Khosla A (2019) Faunal elements from the Deccan volcano-sedimentary sequences of India: a reappraisal of biostratigraphic, palaeoecologic, and palaeobiogeographic aspects. Geol J 54(5):2797–2828

Kapur VV, Khosla A, Tiwari N (2019) Paleoenvironmental and paleobiogeographical implications of the microfossil assemblage from the late Cretaceous intertrappean beds of the Manawar area, district Dhar, Madhya Pradesh. Central India. Hist Biol 31(9):1145–1160

Karanth KP (2006) Out-of-India Gondwanan origin of some tropical Asian biota. Curr Sci 90(6):789–792

Keen MC (1977) Ostracode assemblages and the depositional environments of the Headon, Osborne and Bembridge beds (upper Eocene) of the Hampshire Basin. Palaeontology 20(2):405–445

Keller G, Adatte T, Gardin S, Bartolini A, Bajpai S (2008) Main Deccan volcanism phase ends near the K-T boundary: evidence from the Krishna-Godavari Basin, SE India. Earth Planet Sci Lett 268:293–311

Keller G, Khosla SC, Sharma R, Khosla A, Bajpai S, Adatte T (2009a) Early Danian planktic foraminifera from Cretaceous-Tertiary intertrappean beds at Jhilmili, Chhindwara District, Madhya Pradesh. India. J Foram Res 39(1):40–55

Keller G, Adatte T, Bajpai S, Mohabey DM, Widdowson M, Khosla A, Sharma R, Khosla SC, Gertsch B, Fleitmann D, Sahni A (2009b) K-T transition in Deccan traps of Central India marks major marine seaway across India. Earth Planet Sci Lett 282:10–23

Khand Y, Badamgarav D, Ariunchimeg Y, Barsbold R (2000) Cretaceous system in Mongolia and depositional environments. In: Okada H, Mateer NJ (eds) Cretaceous environments of Asia. Elsevier, Amsterdam, pp 49–79

Khosla A (2014) Upper Cretaceous (Maastrichtian) charophyte gyrogonites from the Lameta formation of Jabalpur. Central India: palaeobiogeographic and palaeoecological implications. Acta Geol Pol 64(3):311–323

Khosla A (2015) Palaeoenvironmental, palaeoecological and palaeobiogeographical implications of mixed fresh water and brackish marine assemblages from the Cretaceous–Palaeogene Deccan intertrappean beds at Jhilmili, Chhindwara District, Central India. Rev Mex de Cien Geol 32(2):344–357

Khosla A (2021) Paleobiogeographical inferences of Indian late cretaceous vertebrates with special reference to dinosaurs. Hist Biol 33:1431–1442

Khosla A, Chin K, Alimohammadin H, Dutta D (2015) Ostracods, plant tissues, and other inclusions in coprolites from the Late Cretaceous Lameta Formation at Pisdura, India: Taphonomical and palaeoecological implications. Palaeogeog Paleoclimat Palaeoecol 418:90–100

Khosla A, Lucas SG (2020a) Introduction of Indian Late Cretaceous dinosaur eggs and eggshells of Peninsular India. In: Khosla A, Lucas SG (eds) Late Cretaceous dinosaur eggs and eggshells of Peninsular India: oospecies diversity and taphonomical, palaeoenvironmental, biostratigraphical and palaeobiogeographical inferences, Topics in Geobiology, vol 51. Springer Nature International Publishing, Cham, pp 1–30

Khosla A, Lucas SG (2020b) Historical background of Late Cretaceous dinosaur studies and associated Biota in India. In: Khosla A, Lucas SG (eds) Late Cretaceous dinosaur eggs and eggshells of Peninsular India: oospecies diversity and taphonomical, palaeoenvironmental, biostratigraphical and palaeobiogeographical inferences, Topics in Geobiology, vol 51. Springer Nature International Publishing, Cham, pp 31–56

Khosla A, Lucas SG (2020c) Geology and stratigraphy of dinosaur eggs and eggshell-bearing infra- and intertrappean beds of peninsular India. In: Khosla A, Lucas SG (eds) Late Cretaceous dinosaur eggs and eggshells of Peninsular India: oospecies diversity and taphonomical, palaeoenvironmental, biostratigraphical and palaeobiogeographical inferences, Topics in Geobiology, vol 51. Springer Nature International Publishing, Cham, pp 57–115

Khosla A, Lucas SG (2020d) Indian Late Cretaceous dinosaur nesting sites and their systematic studies. In: Khosla A, Lucas SG (eds) Late Cretaceous dinosaur eggs and eggshells of Peninsular India: oospecies diversity and taphonomical, palaeoenvironmental, biostratigraphical and palaeobiogeographical inferences, Topics in Geobiology, vol 51. Springer Nature International Publishing, Cham, pp 117–205

Khosla A, Lucas SG (2020e) Discussion: oospecies diversity, biomineralisation aspects, taphonomical, biostratigraphical, palaeoenvironmental, palaeoecological and palaeobiogeographical inferences of the dinosaur-bearing Lameta formation of peninsular India. In: Khosla A, Lucas SG (eds) Late Cretaceous dinosaur eggs and eggshells of Peninsular India: oospecies diversity and taphonomical, palaeoenvironmental, biostratigraphical and palaeobiogeographical inferences, Topics in Geobiology, vol 51. Springer Nature International Publishing, Cham, pp 207–271

Khosla A, Lucas, SG (2021) End-Cretaceous Extinctions. In: Elias S, Alderton, D (eds) Encyclopedia of Geology, 2nd edition. Earth Systems and Environmental Sciences (Academic Press, Elsevier), pp. 665–678. https://doi.org/10.1016/B978-0-12-409548-9.124

Khosla SC, Nagori ML (2005) A restudy of ostracode fauna from the intertrappean beds of the Anjar, Kachchh, District Gujarat. J Geol Soc Ind 66(5):573–580

Khosla A, Sahni A (2000) Late Cretaceous (Maastrichtian) ostracodes from the Lameta Formation, Jabalpur cantonment area, Madhya Pradesh, India. J Palaeont Soc India 45:57–78

Khosla A, Sahni A (2003) Biodiversity during the Deccan volcanic eruptive episode. J Asi Earth Sci 21(8):895–908

Khosla A, Verma O (2015) Paleobiota from the Deccan volcano-sedimentary sequences of India: paleoenvironments, age and paleobiogeographic implications. Hist Biol 27(7):898–914

Khosla A, Prasad GVR, Verma O, Jain AK, Sahni A (2004) Discovery of a micromammal-yielding Deccan intertrappean site near Kisalpuri, Dindori District. Madhya Pradesh. Curr Sci 87(3):380–383

Khosla SC, Nagori ML, Mohabey DM (2005) Effect of Deccan volcanism on non-marine late cretaceous Ostracode fauna: a case study from Lameta formation of Dongargaon area (Nand-Dongargaon basin), Chandrapur District, Maharashtra. Gond Geol Magaz 8:133–146

Khosla SC, Nagori ML, Jakhar SR, Rathore AS (2011a) Early Danian lacustrine–brackish water Ostracoda from the Deccan Intertrappean beds near Jhilmili, Chhindwara District, Madhya Pradesh, India. Micropaleontology 57(3):223–245

Khosla SC, Rathore AS, Nagori ML, Jakhar SR (2011b) Non-marine Ostracoda from the Lameta Formation (Maastrichtian) of Jabalpur, Madhya Pradesh and Nand-Dongargaon Basin, Maharashtra, India: their correlation, age and taxonomy. Rev Esp de Micropaleontol 43:209–260

Khosla A, Chin K, Verma O, Alimohammadin H, Dutta D (2016) Paleobiogeographical and paleoenvironmental implications of the freshwater Late Cretaceous ostracods, charophytes and distinctive residues from coprolites of Lameta Formation at Pisdura, Chandrapur District (Maharashtra), Central India. In: Khosla A, Lucas SG (eds) Cretaceous period: biotic diversity and biogeography, vol 71. New Mexico Museum of Natural History & Science, Albuquerque, pp 173–184

Khosla A, Kania S, Lucas SG, Verma O (2022) Charophytes from the Cretaceous–Palaeogene transition in the Jhilmili intertrappean beds of Central India. Geol J 57(11):4412–4438. https://doi.org/10.1002/gj.4528

Krause DW, Prasad GVR, Koenigswald W, Sahni A, Grine FE (1997) Cosmopolitanism among Gondwanan Late Cretaceous mammals. Nature 390:504–507

Krause DW, Sertich JJW, O'Connor PM, Rogers KC, Rogers RR (2019) The Mesozoic biogeographic history of Gondwanan terrestrial vertebrates: insights from Madagascar's fossil record. Annu Rev Earth Planet Sci 47:519–553

Kriwet J, Soler-Gijón R, López-Martínez N (2007) Neoselachians from the upper Campanian and lower Maastrichtian (upper cretaceous) of the Southern Pyrenees, Northern Spain. Paléo 50:1051–1071

Kshetrimayum DS, Parmar V, Lourembam RS, Prasad GVR (2021) A diversified Ostracoda (Crustacea) assemblage from the Upper Cretaceous intertrappean beds of Gujri, Dhar District, Madhya Pradesh, India. Cretac Res 124:104784. https://doi.org/10.1016/j.cretres.2021.104784
Kumar K,Rana RS, Paliwal BS (2005) Osteoglossidand lepisosteid fish remains fromthe PalaeocenePalana Formation, Rajasthan,India. Palaeontology 48(6):1187–1209
Kumari V, Tandon S, Kumar N, Ghatak A (2020) Epicontinental Permian-Cretaceous seaways in Central India: the debate for the Narmada versus Godavari rifts for the Cretaceous-Tertiary incursion. Ear-Sci Rev 211(1):103284
Li S, Wang Q, Zhang H, Lu H, Martín-Closas C (2016) Charophytes from the Cretaceous–Paleogene transition in the Pingyi Basin (Eastern China) and their Eurasian correlation. Cretac Res 59:179–200
Li S, Wang QF, Zhang HC, Wan XQ, Martín-Closas C (2019) Charophytes from the Cretaceous–Paleocene boundary in the Songliao Basin (north-eastern China): a Chinese biozonation and its calibration to the Geomagnetic Polarity Time Scale. Pap Palaeontol 5:47–81
Lieberman BS (2000) Paleobiogeography. Kluwer Academic/Plenum Publishers, New York
Liu JY (1987) Late Cretaceous–Tertiary charophytes from the Shalamulun area of Inner Mongolia. Proc Stratigr Paleontol 19:129–148
Loeblich AR, Tappan H (1957) *Woodringina*, a new foraminiferal genus (Heterohelicidae) from the Paleocene of Alabama. J Wash Acad Sci 47:39–40
Lourembam RS, Prasad GVR, Grover P (2017) Ichthyofauna (Chondrichthyes, Osteichthyes) from the Upper Cretaceous intertrappean beds of Piplanarayanwar, Chhindwara District, Madhya Pradesh, India. Island Arc 26(1):e12180. https://doi.org/10.1111/iar.12180
Mahadevan C, Sarma SR (1948) Deccan intertrappean series (Vikarabad, Hyderabad State)-Charophyta from the Deccan-VI. J Indian Bot Soc 26(4):260
Malmgren BA (1991) Biogeographic patterns in terminal Cretaceous planktonic foraminifera from Tethyan and warm transitional waters. Mar Micropaleontol 18:73–99
Markwick PJ (2019) Palaeogeography in exploration. Geol Magaz 156(2):366–407
McKenna MC (1995) The mobile Indian raft: a reply to Rage and Jaeger. Syst Biol 44:265–271
Mohabey DM, Udhoji SG (1996) *Pycnodus lametae* (Pycnodontidae), a holostean fish from freshwater Upper Cretaceous Lameta Formation of Maharashtra. J Geol Soc India 47:593–598
Mohabey DM, Udhoji SG, Verma KK (1993) Palaeontological and sedimentological observations on non-marine Lameta Formation (Upper Cretaceous) of Maharashtra, India: their palaeontological and palaeoenvironmental significance. Palaeogeog Palaeoclimat Palaeoecol 105:83–94
Morozova VG (1961) Datsko-Montskie planktonnye foraminifery yuga SSSR (Danian-Montian planktonic foraminifera of the Southern USSR). Paleontol Z 2:8–19
Moura JA (1988) Ostracods from non-marine early Cretaceous sediments of Campos Basin, Brazil. In: Hanai T, Ikeya N, Ishizaki K (eds) Evolutionary biology of Ostracoda. Elsevier, Amsterdam, pp 1207–1216
Mukhopadhyay SK (2008) Planktonic foraminiferal succession in Late Cretaceous to Early Palaeocene strata in Meghalaya, India. Lethaia 41:71–84
Nelson JS (2006) Fishes of the world, 4th edn. Wiley, New Jersey, pp 1–601
Olsson RK (1960) Foraminifera of latest Cretaceous and earliest Tertiary age in the New Jersey coastal plain. J Paleontol 34:1–58
Olsson RK (1964) Late Cretaceous planktonic foraminifera from New Jersey and Delaware. Micropaleontology 10:157–188
Olsson RK, Berggren WA, Hemleben C, Huber BT (1999) Atlas of Paleocene planktonic foraminifera. Smithsonian Contrib Paleobiol 85:1–252
Olsson RK, Hemleben C, Huber BT, Berggren WA (2006) Taxonomy, biostratigraphy, and phylogeny of Eocene *Globigerina*, *Globoturborotalita*, *Subbotina*, and *Turborotalita*. In: Pearson PN, Olsson RK, Huber BT, Hemleben C, Berggren WA (eds), Atlas of Eocene planktonic foraminifera. Cushman Found Spec Publ 41:111–168
Oxford Dictionary (2005) A Dictionary of science. Oxford University Press, Oxford
Parenti LR, Ebach MC (2009) Comparative biogeography: discovering and classifying biogeographical patterns of a dynamic earth. University of California Press, California

Peck RE, Forester RM (1979) The genus *Platychara* from the Western hemisphere. Rev Palaeobot Palynol 28:223–236
Peck RE, Reker CC (1947) Cretaceous and lower Cenozoic charophyta from Peru. Novit Am Mus Nat Hist 1369:1–6
Peck RE, Reker CC (1948) Eocene Charophyta from North America. J Paleontol 22(1):85–90
Plummer HJ (1926) Foraminifera of the midway formation in Texas. Univ Tex Bull 2644:1–206
Powell CMCA, Roots SR, Veevers JJ (1988) Pre-breakup continental extension in east Gondwanaland and the early opening of the eastern Indian Ocean. Tectonophysics 155:261–283
Prasad GVR, Cappetta H (1993) Late Cretaceous selachians from India and the age of the Deccan Traps. Palaeontology 36(1):231–248
Prasad GVR, Khajuria CK (1990) A record of microvertebrate fauna from the inter-trappean beds of Naskal, Andhra Pradesh. J Palaeontol Soc Ind 35:151–161
Prasad GVR, Verma O, Gheerbrant E, Goswami A, Khosla A, Parmar V, Sahni A (2010) First mammal evidence from the Late Cretaceous of India for biotic dispersal between India and Africa at the K/T transition. C R Palevol 9:63–71
Puckett TM, Colin JP, Mitchell S (2012) New species and genera of Ostracoda from the Maastrichtian (Late Cretaceous) of Jamaica. Micropaleontology 58(5):397–455
Rabinowitz PD, Coffin MF, Falvey D (1983) The separation of Madagascar and Africa. Science 220:67–69
Rana RS, Wilson GP (2003) New Late Cretaceous mammals from the intertrappean beds of Rangapur, India and paleobiogeographic framework. Acta Palaeontol Pol 48:331–348
Rana RS, Kumar K, Singh H (2006) Paleocene vertebrate fauna from the Fatehgarh Formation of Barmer District, Rajasthan, Western India. In: Sinha DK (ed) Micropaleontology application in stratigraphy and paleoceanography. Narosa Publishing House, New Delhi, pp 113–130
Rao LR (1955) On charas from the Yellur intertrappean bed. J Poona Univ Sci Sect 6:108–109
Rao KS, Rao SRN (1939) Fossil charophyta of the Deccan intertrappeans near Rajahmundry, India. Mem Geol Surv India Palaeontol Indica 29:1–14
Reeves C (2014) The position of Madagascar within Gondwana and its movements during Gondwana dispersal. J Afr Earth Sci 94:45–57
Riveline J (1986) Les charophytes du Paleogeneet du Miocene Inferieur d'Europe occidentale, Biostratigraphie des formations continentales. Editions du Centre national de la recherche scientifique, Paris, pp 1–227
Sahni B, Rao SRN (1934) *Chara sausari* sp. nov. a *Chara* (sensu stricto) from the intertrappean cherts at Sausar in the Deccan. Proc Natl Acad Sci India 13(3):215–223
Sanjuan J, Alqudah M (2018) Charophyte flora from the Miocene of Zahle (Beeka Valley, Lebanon). Biostratigraphic, palaeoenvironmental and palaeobiogeographical implications. Geodiversitas 40(10):195–209
Sanjuan J, Martín-Closas C (2014) Biogeographic history of two Eurasian Cenozoic charophyte lineages. Aquat Bot 120(A):18–30
Sanmartín I (2012) Historical biogeography: evolution in time and space. Evol Educ Outreach 5:555–568
Sarr R (2015) Paleobiogeographic implications of Paleocene ostracods from Senegal and related West and North Africa basins. Rev Micropaleontol 58(4):333–349
Scheibnerova V (1971) Foraminifera and the Mesozoic biogeoprovinces. Geol Surv NSW 13:135–174
Scotese CR (1991) Jurassic and Cretaceous plate tectonic reconstructions. Palaeogeogr Palaeoclimatol Palaeoecol 87:493–501
Servais T, Cecca F, Harper DAT, Isozaki Y, Niocaill CM (2013) Palaeozoic palaeogeographical and palaeobiogeographical nomenclature. Mem Geol Soc London 38:25–33
Shahin A (2005) Maastrichtian to Middle Eocene ostracodes from Sinai, Egypt: systematics, biostratigraphy and paleobiogeography. Rev de Paléobiol 24(2):749–779
Sharma R, Khosla A (2009) Early Palaeocene ostracoda from the Cretaceous-Tertiary (K-T) Deccan intertrappean sequence at Jhilmili, district Chhindwara. Central India. J Palaeontol Soc India 54(2):197–208

Shivarudrappa TV (1972a) On the occurrence of charophytic remains from the intertrappeans of Gurmatkal, Gulbarga District. Mysore State Curr Sci 41(2):21–23

Shivarudrappa TV (1972b) On *Gyrogonites medicaginula* and *Chara wrighti* from the intertrappeans of Gurmatkal, Gulbarga District, Mysore state. Proc 2nd Indian Colloq Micropaleontol Stratigr, Lucknow, pp 115–119

Shome S, Chandel RS (2013) Palaeontological studies of Papro Formation (intertrappean) of Lalitpur District, Uttar Pradesh – its age, correlation and palaeoecology. Indian J Geosci 65:49–62

Sigé B, Buscalioni AD, Duffaud S, Gayet M, Orth B, Rage JC, Sanz JL (1997) Status of data on the continental Upper Cretaceous site of Champ-Garimond (Gard, South of France). Münchner Geowiss Abh 34:111–130

Singh MP (1980) Charophytes from the infratrappean beds of Papro, Lalitpur District, Uttar Pradesh. J Palaeontol Soc India 23:144–153

Singh MP, Mathur DK (1979) Discovery of some Charophyta from the fossiliferous infratrappean beds of Lalitpur. Uttar Pradesh Curr Sci 48(5):208–209

Smith AG (1992) Plate tectonics and continental drift. In: Brown GC, Hawkesworth CJ, Wilson RL (eds) Understanding the Earth: Cambridge University Press, Cambridge, pp 187–203

Smith AG, Smith DG, Funnell BM (1994) Atlas of Mesozoic and Cenozoic coastlines. Cambridge University Press, Cambridge

Srinivasan S, Sahni A, Bajpai S (1992) Fossil Charophyta from the Deccan intertrappean beds of Gurmatkal, Gulbarga District, Karnataka. Curr Sci 63:396–398

Srinivasan S, Bajpai S, Sahni A (1994) Charophytes from Deccan intertrappean beds of peninsular India: implications for age and correlation of Deccan volcanic. Geobios 27:559–571

Storey BC (1996) Microplates and mantle plumes in Antarctica. Terra Antarct 3:91–102

Storey M, Mahoney JJ, Saunders AD, Duncan RA, Kelley SP, Coffin MF (1995) Timing of hot spot-related volcanism and the breakup of Madagascar and India. Science 267:852–855

Tambareau YP (1984) Les ostracodes du "Montien Continental" de Hainin, Hainaut. Belgique Rev de Micropalaeontol 27(2):144–156

Tambareau Y, Gruas-Cavagnetto C, Feist M Villatte J (1991) Flores et faunes continentales ilerdiennes du versant sud de la Montagne Noire et de la Montagne d'Alaric. Rev de Micropal 34:69–89

Thewissen JGM, Mckenna MC (1992) Paleogeography of Indo-Pakistan: a response to Briggs, Patterson and Owen. Syst Biol 41:248–251

Tiss L, Trabelsi K, Kamoun F, Soussi M, Houla Y, Sames B, Martín-Closas C (2019) Middle Jurassic charophytes from southern Tunisia: implications on evolution and paleobiogeography. Rev Palaeobot Palynol 263:65–84

Torsvik TH, Tucker RD, Ashwal LD, Carter LM, Jamtveit B, Vidyadharan KT, Venkataramana P (2000) Late Cretaceous India-Madagascar fit and timing of break-up related magmatism. Terra Nova 1:220–224

Udhoji SG, Mohabey DM (1996) Ostracoda and charophyta from the Late Cretaceous Lameta formation of Maharashtra: paleobiogeographic and age implication. In: Azmi RJ, Bhandari A, Dave A (eds) Pandey J. Contr XV Indian Colloq Micropalaeontol Stratigr, Dehradun, pp 409–418

Van Itterbeeck J, Horne DJ, Bultynck P, Vandenberghe N (2005) Stratigraphy and palaeoenvironment of the dinosaur-bearing Upper Cretaceous Iren Dabau Formation, Inner Mongolia, People's Republic of China. Cretac Res 26:699–725

Villalba-Breva S, Martín-Closas C (2011) A characean thallus with attached gyrogonites and associated fossil charophytes from the Maastrichtian of the Eastern Pyrenees (Catalonia, Spain). J Phycol 47:131–143

Verma O (2015) Cretaceous vertebrate fauna of the Cauvery Basin, southern India: palaeodiversity and palaeobiogeographic implications. Palaeogeog Palaeoclimat Palaeoecol 431:53–67

Verma O (2018) Palaeoecology: Past, Present, and Future. J Geol Soc India 91:115–116

Verma O, Khosla A (2018) Application of internet technology in assembling literature for palaeontological research. Iran J Sci Technol Trans A Sci 4:1715–1723

Verma O, Khosla A (2019) Developments in the stratigraphy of the Deccan Volcanic Province, peninsular India. Compt Rendus Geosci 351:461–476

Verma O, Prasad GVR, Khosla A, Parmar V (2012) Late Cretaceous Gondwanatherian mammals of India: distribution, interrelationships and biogeographic implications. J Palaeontol Soc India 57:95–104

Verma O, Khosla A, Goin FJ, Kaur J (2016) Historical biogeography of the Late Cretaceous vertebrates of India: comparison of geophysical and paleontological data. In: Khosla A, Lucas SG (eds) Cretaceous period: biotic diversity and biogeography, vol 71. New Mexico Museum of Natural History & Science, Albuquerque, pp 317–330

Verma O, Prashanth M, Greco R, Khosla A, Singh K (2022) Geological education scenario in India and role of open educational resources in the light of COVID-19 pandemic. Earth Sci Res J 26(2):239–254

Verma O, Khosla A, Kaur J, Prasanth M (2017) Myliobatid and pycnodont fish from the Late Cretaceous of Central India and their paleobiogeographic implications. Hist Biol 29(2):253–265

Vicente A, Csiki-Sava Z, Martín-Closas C (2019) European charophyte evolution across the Cretaceous–Paleogene boundary. Palaeogeogr Palaeoclimat Palaeoecol 533:109–244

Villalba-Breva S, Martín-Closas C (2013) Upper Cretaceous paleogeography of the Central Southern Pyrenean Basins (Catalonia, Spain) from microfacies analysis and charophyte biostratigraphy. Facies 59(2):319–345

Wang Z (1978) Paleogene charophytes from the Yangtze-Han River basin. Characeae. Mem Nanjing Inst Geol Paleontol Acad Sin 9:101–120

Wang S, Wang Z, Lin XD, Zhang ZR, Xu XL (1982) Cretaceous and Cenozoic charophytes from Jiangsu. Geological Publishing House, Beijing, pp 1–94

Wang Z, Lu HN, Zhao CB (1985) In: Wang TQ (ed) Cretaceous charophytes from Songliao Basin and adjacent areas. Heilongjiang Scientific and Technical Publishing House, Harbin, pp 1–80

Weems R (1998) Actinopterygian fish remains from the Paleocene of South Carolina. In: Sanders AE (ed), Paleobiology of the Williamsburg formation (Black Mingo Group; Paleocene of South Carolina, USA). Trans Am Philos Soc 88(4):147–164

Westgate JW (1989) Lower vertebrates from estuarine facies of the middle Eocene Laredo Formation (Claiborne group), Webb County, Texas. J Vertebr Paleontol 9(3):282–294

Whatley RC (2012) The 'out-of-India' hypothesis: further supporting evidence from the extensive endemism of Maastrichtian non-marine Ostracoda from the Deccan volcanic region of peninsular India. Rev de Paléobiol 11:229–248

Whatley RC, Bajpai S (2000) A new fauna of Late Cretaceous non-marine Ostracoda from the Deccan intertrappean beds of Lakshmipur, Kachchh (Kutch District), Gujarat, western India. Rev Esp de Micropaleontol 32(3):385–409

Whatley RC, Bajpai S (2006) Extensive endemism among the Maastrichtian nonmarine Ostracoda of India with implications for palaeobiogeography and "out of India" dispersal. Rev Esp de Micropaleontol 38(2–3):229–244

Whatley RC, Bajpai S (2000a) A new fauna of Late Cretaceous non-marine ostracoda from the Deccan intertrappean beds of Lakshmipur, Kachchh (Kutch District), Gujarat, western India. Rev Esp de Micropaleontol 32(3):385–409

Whatley RC, Bajpai S (2000b) Further nonmarine Ostracoda from the Late Cretaceous intertrappean deposits of the Anjar region, Kachchh, Gujarat, India. Rev Micropaleontol 43(1):173–178

Whatley RC, Bajpai S, Srinivasan S (2002a) Upper Cretaceous nonmarine Ostracoda from intertrappean horizons in Gulbarga district, Karnataka state, South India. Rev Esp de Micropaleontol 34(2):163–186

Whatley RC, Bajpai S, Srinivasan S (2002b) Upper Cretaceous intertrappean nonmarine Ostracoda from Mohgaonkala (Mohgaon-Kalan), Chhindwara District, Madhya Pradesh state, Central India. J Micropaleontol 21:105–114

Whatley RC, Bajpai S, Whittaker J (2003a) Freshwater ostracoda from the Upper Cretaceous intertrappean beds at Mamoni (Kota district), southeastern Rajasthan, India. Rev Esp de Micropaleontol 35(1):75–86

Whatley RC, Khosla SC, Rathore AS (2012) *Periosocypris megistus* n. gen. and n. sp.: a new gigantic non-marine cyprid ostracod from the Maastrichtian Lameta Formation of India. J Palaeontol Soc Ind 57(2):113–117

Willis B (2010) Principles of paleogeography. Sci New Ser 31(790):241–260

Wing SL, Sues HD (1992) Mesozoic and early Cenozoic terrestrial ecosystems. In: Behrensmeyer AK, Damuth JD, DiMichele WA, Potts R, Sues HD, Wing SL (eds) Terrestrial ecosystems through time. University of Chicago Press, Chicago, pp 327–416

Woodward AS (1908) On some fish remains from the Lameta beds at Dongargaon, Central Province. Mem Geol Sur India Palaeontol Indica NS 3:1–6

Yang JL, Wang QF, Lu HN (2005) Discovery of Cretaceous and Paleocene charophyte floras from well Hu-2 in the southern edge of Junggar Basin. Acta Micropaleontol Sin 22(3):251–268

Zhang ZK (1984) Sino-Korean block and Yangtze block as part of the Pacifica continent in the late Palaeozoic. Bull Chin Acad Geol Sci 9:45–54

Index

A. Khosla et al., *Microbiota from the Late Cretaceous-Early Palaeocene Boundary Transition in the Deccan Intertrappean Beds of Central India*, Topics in Geobiology 54, https://doi.org/10.1007/978-3-031-28855-5

www.ingramcontent.com/pod-product-compliance
Ingram Content Group UK Ltd.
Pitfield, Milton Keynes, MK11 3LW, UK
UKHW022213051026
15624UKWH00001B/139
9783031288548